THE DEPTHS OF RUSSIA

THE DEPTHS OF RUSSIA

Oil, Power, and Culture after Socialism

Douglas Rogers

CORNELL UNIVERSITY PRESS   ITHACA AND LONDON

This book was published with the assistance of the Frederick W. Hilles Publication Fund of Yale University and a Director's Award from the Whitney and Beth MacMillan Center for International and Area Studies, Yale University.

First published 2015 by Cornell University Press
First printing, Cornell Paperbacks, 2015

Printed in the United States of America

Library of Congress Cataloging-in-Publication Data

Rogers, Douglas, 1972– author.
The depths of Russia : oil, power, and culture after socialism / Douglas Rogers.
pages cm
Includes bibliographical references and index.
ISBN 978-0-8014-5373-1 (cloth : alk. paper) —
ISBN 978-0-8014-5658-9 (pbk. : alk. paper)
1. Petroleum industry and trade—Social aspects—Russia (Federation)—Permskii krai. 2. Post-communism—Social aspects—Russia (Federation)—Permskii krai. 3. Social responsibility of business—Russia (Federation)—Permskii krai. 4. Lukoil (Firm)—History. 5. Permskii krai (Russia)—Social conditions. 6. Ethnology—Russia (Federation)—Permskii krai. I. Title.
HD9575.R83P477 2015
338.7'6655094747—dc23

2015009575

Cornell University Press strives to use environmentally responsible suppliers and materials to the fullest extent possible in the publishing of its books. Such materials include vegetable-based, low-VOC inks and acid-free papers that are recycled, totally chlorine-free, or partly composed of nonwood fibers. For further information, visit our website at www.cornellpress.cornell.edu.

Cloth printing 10 9 8 7 6 5 4 3 2 1
Paperback printing 10 9 8 7 6 5 4 3 2 1

For Melanie

Contents

Illustrations

Maps

Figures

Preface

The Soviet city of Perm was a defense and industrial hub. Its factories turned out rockets for the Soviet space program, airplane engines for military and civilian use, and artillery pieces for the Red Army. The surrounding Perm region, on the northern and western edge of the Ural Mountains, was home to dozens of metallurgy factories and oil production sites, one of the world's largest potash mines, and a high concentration of prison and labor camps in the Gulag system. Yet few in the Soviet Union—let alone the rest of the world—knew much about any of this, for Perm was also a closed city. Foreigners were not permitted to visit at all and national Soviet media outlets scarcely mentioned it. Indeed, several of my Russian acquaintances who had been transferred to Perm after working elsewhere in the Soviet Union told me that they were mystified by their assignments at first, having had no more than a vague sense that Perm was to be found somewhere in the Urals. Others insisted that Perm was not on Soviet-era maps at all, although I have yet to find an example. In one friend's memorable phrasing, Perm had "fallen out of the consciousness" of ordinary Soviet citizens.

Although the present-day Perm region occupies some 160,000 square kilometers, with a population of roughly two and a half million, nearly a million of those in Perm itself, it sometimes seemed as though the end of the Soviet Union had not changed geographical consciousness very much. In the early 2000s, a federal news channel memorably ran an entire news segment about Perm that referred to it as Penza—a city nearly a thousand kilometers away. In 2012, I listened as a scholar visiting Perm from Moscow embarrassed himself by boasting that not everyone—in contrast to himself—would fly "somewhere out beyond the Urals" to give a lecture. Perm is unambiguously, as his audience's murmurs and muted groans communicated, on the Moscow side of the Ural Mountains.

This concern with post-Soviet Perm's place in Russia and the world was not new. It became widespread after the city was opened to visitors in the late 1980s, and putting Perm back on the map, at least figuratively, has been an ongoing effort for regional politicians and businesses competing for federal and international investment since then. (Perm aimed, for example, to be Russia's Capital of Civil Society under Governor Iurii Trutnev in the early 2000s and a European Capital of Culture under Oleg Chirkunov, Trutnev's successor.) Indeed, looking back through my fieldnotes and interviews from what are now over twenty years of visits to the Perm region, one of the most enduring themes is the remaking of space: the perpetual resizing and renaming of territories and districts; the quality, direction, and construction of roads and pipelines; the relationship between centers and peripheries, capitals and provinces. Maps were everywhere during my fieldwork in the 1990s and 2000s, from the walls of contemporary art exhibits to the glossy covers of the endless stream of reports issued by state agencies and regional companies. A series of public discussions hosted by a journalist in 2009 invited local notables to talk about the "map of the world" that oriented their actions. A state official involved in financing cultural development projects in the region's small cities and towns impressed upon me that her task was to help people "find themselves in space."

Representations of the spaces of the Perm region were especially common at Lukoil-Perm, a regional subsidiary of the Moscow-based multinational oil company Lukoil. The rooftop at the company's headquarters in downtown Perm, for instance, featured a slowly revolving globe emblazoned with red "LUK" logos indicating the locations of Lukoil's main operations around the world. Inside the building below, the company's multiroom museum included a large interactive map of the Perm region's constituent districts and cities. Pressing an array of nearby buttons illuminated, by turns, districts that were home to production facilities, refining sites, gas stations, and social and cultural development projects that the company was sponsoring. Pushing all four buttons at once—and who among the museum's parade of school-aged visitors would not do just that?—set the region aglow from top to bottom. The museum's map of the Perm region was unapologetically corporate in its perspective, and it mimicked in key respects Lukoil's own organizational structure in the region, but it was not wholly inaccurate. Over the course of the 1990s and early 2000s, oil had displaced defense and metallurgy to become the Perm region's single most significant industry.

Scholars' map of Russia has likewise shifted drastically in the past two decades, and it now includes many more studies of frontiers, borderlands, edges, and margins than it once did. There is also much to be gained from attending to Russia's depths—to its *glubinka*. Glubinka generally describes locations that are considered out of the way, underdeveloped, hard to get to, and, depending on one's opinion of such places, either hopelessly backward or enviably traditional. Glubinka certainly describes many of the more remote areas of the Perm region where oil is extracted. From the perspective of many in Moscow and Saint Petersburg—perhaps including that visiting scholar—glubinka might describe the entire Perm region or even, at least in jest, the entire Urals. Like so many other spatial categories, glubinka was a term on the move in the 1990s and 2000s. Its shifts unfolded alongside transformations of Russia's depths in another, geological sense: the increasing centrality of subterranean oil and gas deposits to all dimensions of life.

These are the depths of Russia—locations often considered remote or peripheral and the increasingly significant deposits of hydrocarbons located beneath some of them—from which I take my title.

This book is, then, my own effort to put Perm, its region, and its oil on the map. For much of the twentieth century, the oil pumped from the Perm region's subsoil flowed through a socialist political and economic order, one that did not organize production, circulation, or consumption in the ways that have fed oil booms and busts—and shelves of scholarship about them—around the capitalist world. Despite the Soviet Union's high ranking among the world's major oil producers and exporters, Soviet citizens did not experience oil as directly associated with massive inequalities, destabilizing influxes of money, soaring expectations of rapid modernization, or grand cultural spectacles—all hallmarks of capitalist oil booms. Socialist oil was never the basis for the creation of an industrial-cum-financial elite that could influence, rival, or even take over agencies of a federal state. In these and other ways, oil's place in the Soviet Union differed significantly from its place in capitalist centers like the United States and the so-called petro-states of the Euro-American postcolonial periphery, from Nigeria to Saudi Arabia to Venezuela and beyond.

These differences did not simply evaporate with the end of the Soviet Union in 1991, or even with the boom years of the early to mid-2000s. Even as the Perm region's oil sector began to converge with the global oil industry in significant ways—including vertical integration and the adoption of corporate social responsibility programs—it did so against the backdrop of both Soviet-era practices and those of the ensuing transition from socialism. The resulting hybrids, layers, legacies, and innovations are fertile ground for theorizing the place of oil in the contemporary world more broadly—a necessarily comparative project that I pursue throughout this book. Oil companies rose to influence in the post-Soviet Perm region, for instance, not through the accumulation of money but in money's absence, in conditions of barter and surrogate currencies. By the early 2000s, a combination of Soviet-era expectations and post-Soviet conditions created an uncommonly tight relationship between regional state agencies and regional corporate subsidiaries, a relationship that powerfully shaped life throughout the Perm region in domains from infrastructure to health care and from ecology to religion. Of particular note in this state–corporate alliance was an effort to remake regional identities and meaningful worlds, what I term the post-Soviet cultural front. This cultural front drew residents' sensibilities away from the factories of Soviet Perm, with their long history as central administrative and symbolic sites in efforts to forge socialist citizens, and reoriented those sensibilities to the region's subsoil oil deposits and *their* symbolic possibilities for reimagining the Perm region and its residents. What it meant to live in the Perm region became ever more closely tied to the oil pumped from beneath it, even for those who sought to chart a trajectory for the region that was not based on hydrocarbons.

By itself, the entangling of oil and human imaginations and possibilities is hardly a unique story in global context. But the socialist past and ensuing postsocialist transformations have made the substance of these entanglements in the Perm region very different from the ones we know best on some scores, and more similar to them than we have yet appreciated on other scores. Plumbing the depths of Russia thus makes the study of oil more fully global than it has been to date, more fully attuned not just to the first and third of the Cold War's three worlds—and to the relationship between them—but to the still larger twentieth- and twenty-first-century orders in which oil continues to feature so prominently.

Acknowledgments

This book was made possible by the support of many institutions and many people, although I bear sole responsibility for the final result. The National Science Foundation Cultural Anthropology Program, the National Council for Eurasian and East European Research, and Yale University's MacMillan Center for International and Area Studies funded the fieldwork portions of this project, and I am grateful to their dedicated staffs and program officers for support and advice. In the Department of Anthropology at Yale, Mary Smith, Frank D'Aria, and Jennifer DeChello ushered me through the thousand administrative hoops, details, and forms that make the practicalities of research happen. The Radcliffe Institute for Advanced Study at Harvard University, where I was the Elizabeth S. and Richard M. Cashin Fellow in 2012–13, provided me with a year of time and a space apart to write, and the Office of the Provost at Yale made commuting to Cambridge possible. The Radcliffe Institute also conjured an inspiring crowd of scholars with whom to spend that year. Thank you to Bryna, Daisy, David, Henry, Judy, Katherine, Mariella, Rebecca, Renée, and many others for your encouragements and your companionship during my short weeks (but long days) in Cambridge. Your examples of smart and collegial scholarship nurtured this book more than you realize.

The Perm region has been the most hospitable and welcoming of places to do research for many years. I am indebted to the Political Science Department in the History Faculty at Perm State University, chaired by L. A. Fadeeva, for providing me with a local institutional affiliation during this project. The publications of the department's faculty and graduate students have been simply indispensable in writing about the post-Soviet Perm region, and I thank L. A. Fadeeva, P. V. Panov, and N. V. Borisova especially for their insights and unparalleled intellectual generosity. Other friends and acquaintances, among them G. N. Chagin, V. V. Kal′pidi, D. M. Kholmogorova, N. G. Nechaeva, A. V. Shakhaev, S. A. and T. A. Vetoshkin, and G. A. Yankovskaia, cheerfully provided me with more contacts, ideas, and delightful evenings than I knew what to do with. I am especially grateful to S. I. Fedotova for sharing her own work, published and unpublished, on the Perm region and its oil industry. M. G. Nechaev and S. A. Ponamarev, two directors of the Perm State Archive of Contemporary History (PermGANI), went above and beyond their ordinary duties in supporting the archival aspects of this project. Their dedication to carefully documented historical scholarship on the Perm region is nothing short of awe-inspiring, and deserves to be recognized as such far and wide. O. L. Kut′ev, who more than anyone else saw the potential of this project in its very early days, deserves separate thanks; the Perm region is not the same without him. One of the great pleasures of doing research for this book has been the opportunity to sneak out of Perm to visit the small town of Sepych, site of my earlier research and home to my oldest and dearest friends in the Perm region. This book is about a very different topic from the one I was working on when I lived in Sepych, but I think my friends there will understand that their own experiences in and after the Soviet Union were never far from my mind as I chased the intersections of oil, power, and culture in their part of the Urals.

I owe thanks, too, to the many colleagues and friends whose careful readings and comments on bits, pieces, drafts, and early ideas improved this book substantially, among them Mariella Bacigalupo, Margarita Balmaceda, Barney Bate, Dominic Boyer, Paul Bushkovich, Elizabeth Dunn, David Engerman, Kristen Ghodsee, Bruce Grant, Chris Hann, Erik Harms, Karen Hébert, Cymene Howe, Marcia Inhorn, Michael Kennedy, Mike McGovern, Serguei Oushakine, Oscar Sanchez-Sibony, Helen Siu, K. Sivaramakrishnan, Veli-Pekka Tynkkynen, Katherine Verdery, Michael Watts, and Susanne Wengle. Bill Kelly's wisdom has been essential these past years. Audiences at talks and workshops hosted by Miami University of Ohio, Brown University, Middlebury College, Yale University, the University of Pennsylvania, Rice University, New York University, Harvard University, and the University of Chicago pushed me to sharpen, and on several occasions reconsider, my arguments—as did many stimulating conversations with colleagues and students in my home department. Sara Brinegar, Peter Rutland, Oksan Bayugan, and Julie Hemment read and commented on the penultimate draft of this book. Julie's extensive and insightful comments, and an advance look at her own book on Russia in the 2000s, were particularly influential in the final stretch. Katherine Verdery deserves her own special category of appreciation. The further I go in this line of work, the more her example and her encouragement mean to me.

After the fieldwork and the writing came the book itself. I thank Roger Haydon at Cornell University Press for his enthusiasm, his careful readings, and his speed. I am likewise grateful to Bill Nelson for the maps, Kate Mertes for the index, April Wen for the sketches, and Kim Giambattisto and her colleagues at Westchester Publishing Services for the production. Il′ia Viktorov, Aleksei Ivanov, and Iulia Zaitseva granted permission to use Viktorov's wonderful "Pipeline" image on the cover. Portions of this book are sliced, diced, deepened, reframed, or upgraded versions of Rogers 2011, 2012, 2014a, 2014b, and 2014c; I thank Paradigm Publishers, the American Ethnological Society, the Wenner-Gren Foundation for

Anthropological Research, the Institute for Ethnographic Research, and Indiana University Press, respectively, for permission to reprint those portions here. I have had the good fortune to work with excellent research assistants over the past few years: Elizabeth Bronshteyn, Ian Convey, Nataliya Langburd, Leonid Peisakhin, Megan Race, and David Willey chased down numerous leads, read drafts, and commented instructively on matters large and small.

My deepest thanks, finally, to Carole, Leo, Melanie, Ella, Caden, and Anne for their flexibility, patience, encouragement, and good humor. Melanie and I decided some time ago to work on our ambitious projects in parallel rather than in sequence, and it has been one hell of a ride ever since. I can't imagine a better companion.

Note on Place, Corporate, and Personal Names, Translations, and Transliterations

The Perm region has expanded and contracted over the time period covered in this book, as well as being renamed several times. (For instance, Perm was rechristened Molotov from 1940 to 1957, and the surrounding region was the Molotov region.) For the sake of clarity, I use the current name, "Perm region," throughout, unless the name change is significant for the argument I am making. The same applies for cities, towns, districts, and other geographical locations. Corporation names are a more complicated matter, inasmuch as tracking the details of privatizations, mergers, acquisitions, and other kinds of restructurings—all causes for name changes—is central to many of the claims I make. Readers who become momentarily lost in the tangle of companies, subsidiaries, and corporate divisions should rest assured that many of my interlocutors did, too. The glossary and the appendix (which lists post-Soviet governors of the Perm region by years in office) have been designed with rapid reorientation in mind.

Many of the individuals quoted in this book were, at the time I spoke with them, state, corporate, or other employees speaking with me in their official capacities; with their permission, I use their real names in attributing those portions of our conversations that were "on the record." The vast majority of my interviewees, however, preferred that I withhold their names and other identifying details in print. In these cases, I use more general descriptions that are calibrated to obscure their identities: "one former employee," "an acquaintance," "a participant I spoke with," and so on.

All translations are my own, unless otherwise indicated. In transcribing Russian in the main text, I eliminate terminal soft signs and shorten soft endings (e.g., Perm rather than Perm′, Bashkiria rather than Bashkiriia) and use conventional English transliterations for well-known names (e.g., Yeltsin rather than El′tsin).

THE DEPTHS OF RUSSIA

Introduction

OIL, STATES AND CORPORATIONS, AND THE POLITICS OF CULTURE

I awoke at home on the morning of September 14, 2008, to National Public Radio reporting a plane crash in Perm. Eighty-eight people on the Aeroflot-Nord flight—all aboard—had died. Russian news sources did not have much more to say, other than adding some eyewitness reports, mourning the dead, and announcing state agencies' investigations. Rumors, of course, quickly began to circulate. The pilots were drunk, one said. A Russian military commander in the Chechen War was on board and rebels had taken their revenge on him by blowing up the plane, said another. One such rumor caught my attention in the summer of 2009, when I was next in Perm. Terrorists from the Caucasus had hijacked the plane, a taxi driver told me. They had planned to crash it into the headquarters of Lukoil-Perm, in the center of downtown Perm. But they never made it. The plane was shot down before it got there, he told me, either by the Russian military or, perhaps, by Lukoil's own security forces. This, he added, accounted not just for the crash but also for the ongoing secrecy of the investigation. When I inquired among friends and acquaintances in Perm, I confirmed that this was a fairly widely circulating version of events, although the alleged target shifted between Lukoil-Perm's administrative offices in downtown Perm and Lukoil's major refinery in the city's industrial district, closer to the site of the crash (and to the airport at which the plane was scheduled to land). An online forum included a lengthy discussion of whether intentionally crashing a plane into a refinery could, in fact, start a fire that would burn an entire city district to the ground.

I find this rumor interesting because it so clearly adopted and transformed the narrative of the September 11, 2001, attacks in New York City. Rather than the target being a symbolic center of American-led finance capital, it was a symbolic center (or processing center, in the refinery variant) of the Perm region's oil industry. Rather than al-Qaeda terrorists targeting the United States, it was terrorists from the Caucasus—presumably Chechen—targeting the Russian heartland. Rather than fast-acting civilians bringing down a plane before it could reach its destination, it was the military that acted swiftly to save the population—or perhaps the oil company itself, which, this version of the rumor assumes, possessed surface-to-air missile capabilities on par with the Russian state's. Some of the more religiously, mystically, or conspiratorially minded—as well those who preferred morbid jokes—would later link the plane crash with a horrific 2009 nightclub fire in Perm that killed over 150 people to offer theories of who was *really* controlling events in the Perm region. God, the devil, various Russian or international cabals, and competing tourist bureaus in the Urals were all candidates.

The Aeroflot-Nord crash clearly haunted many residents of Perm, and it continued to come up in conversations for years afterward. In the summer of 2010, a street art project, sanctioned and funded by the regional Department of Culture, installed a series of giant concrete blocks in the shape of the letters v-l-a-s-t-'—power, especially state power (fig. 1)—on the square in front of the Perm Regional Administration Building. The concrete letters were intended to be a somewhat playful materialization of that which is so often immaterial—a concretization, so to speak, of power. Here, mighty Russian state power could be used for something as mundane as a bench. *Power* was a small part of the regional government's sustained effort to rebrand Perm as a European Capital of Culture, thereby attracting tourists, diversifying an economy heavily reliant on oil revenues, and putting the city in the national and international news for something other than plane crashes and nightclub fires. Katia, an acquaintance of mine who worked in the ten-story administration building, was not, however, amused. As I looked down on the concrete letters from her office window a couple of days after the installation, she worriedly looked up at the sky. "Airplanes fly over here all the time," she said, "and those letters are big enough to see from way up there. You can probably see them on Google Earth. It's like someone wants to make state power into a target." If, that is, it was at least possible that unknown parties had aimed to crash a plane into the Lukoil-Perm offices just across the street, then should she not look out her window and worry as well?

Rumors, conspiracy theories, and anxieties are interesting in part because they reveal the preoccupations of a time and place.[1] Those that followed the 2008 plane

1. On conspiracies in Russia, see especially Oushakine 2009a and Borenstein 2014. Boyer 2006 provides an excellent overview of the utility of conspiracy theories for anthropology.

FIGURE 1 *Power*, by Nikolai Ridnyi, a public art installation erected in 2010 in front of Perm's Regional Administration Building, at one end of the central city esplanade. Lukoil-Perm's downtown headquarters is just out of the frame to the left.

Photo by author, 2010.

crash sought out and explained relationships among issues that were already on the minds of many residents of the Perm region in the first decade of the twenty-first century: the new and growing centrality of Lukoil to nearly all aspects of life in the Perm region; the company's close, if sometimes fraught, relationship with state agencies; the fashioning of new identities for the city of Perm and its surrounding region; the role of cultural production—including public art installations—in all these transformations; and, not least, a generalized feeling of vulnerability and uncertainty. These issues and the relationships among them are also the subject of this book. Although I draw on quite different sources and styles of explanation, I am just as interested as that taxi driver and my acquaintance Katia in giving an account of how these issues are connected to each other. My account is not so monocausal as "God" or "rival tourist bureaus," but neither is it as straightforward as "oil," "money," "Putin," or "the Soviet past"—each of which has anchored influential and decidedly mainstream scholarly studies of contemporary Russia.

The relationships I trace cross the length and breadth of the Perm region and extend into Soviet history. They run through museum displays and privatization battles, refineries and state bureaucracies, taxation policies and folklore festivals; at times, they stretch beyond the Perm region to Moscow and an array of transnational corporate and cultural connections. They have emerged from interviews, archives, libraries, participant observation, and immersion in the impressive Russian-language scholarship and journalism dedicated to the history, culture, and contemporary political economy of the Perm region. My understanding of these relationships draws on and contributes to interdisciplinary bodies of scholarship devoted to oil, to states and corporations, and to the politics of culture. All these disparate elements—ethnographic, historical, methodological, and analytical—revolve around a single issue that I return to again and again from multiple, intersecting vantage points: the making of the Perm region as an oil region.

Oil Regions

Reflecting on the historiography of Europe at the end of the twentieth century, Celia Applegate observed that the "great, hulking presence of nations" (1999, 1159) has long overshadowed the potential for studies of subnational regions. Much the same could be said for scholarship on oil around the world, which has generally privileged the nation-state—especially the "petrostate"—as its unit of analysis, despite the fact that oil is usually produced, refined, consumed, and reincarnated into money quite unevenly across territories within any given state.

This book joins a smaller number of studies of oil regions that use the details of region-level processes as basic building blocks for larger-scale analysis.[2]

Oil regions are generally assumed to be production regions—that is, subnational administrative and/or political units sitting atop oil deposits that are being actively exploited by oil companies. The Perm region has been an oil production region since the accidental discovery of oil in 1929. In the decades after World War II, it joined the neighboring regions of Tatarstan and Bashkortostan in making up the "Second Baku"—the Soviet nickname for the Volga-Urals oil basin that geologists and central planners hoped would replace the older and geopolitically vulnerable oilfields of the Caucasus (see Brinegar 2014, 264–66). The Perm region's oilfields peaked in output in the late 1970s, as the center of Soviet oil production moved to West Siberia. Today, around half of the Perm region's forty-two constituent districts are home to oil production or exploration operations, and new drilling and oilfield management technologies have revived production numbers, although they remain far below their 1970s heights.

But serving as the geographical home to oil production is not the only way in which an oil region can be constituted. Indeed, within the practical workings of socialist political economy, refining turned out to be a far more prestigious and influential dimension of the Perm regional economy than production, and Permneftorgsintez, Perm's massive refinery that began operations in 1959, played a bigger role in the making of the late Soviet Perm region than the regional production association Permneft. The importance of refining and circulation continued into the post-Soviet period, as the Perm region sought to extract itself from the deep economic crisis of the early 1990s by relying on long chains of bartered oil products emanating from its refinery, and eventually by introducing a surrogate currency backed by refined oil.

Attending to the making of an oil region offers much more than local color or details with which to fill in the larger-scale analytical frameworks advanced by studies focused on the nation-state or the petrostate. My claim is that a theoretically informed historical ethnography that extends outward from a single oil-producing region illuminates the larger picture of Soviet and post-Soviet oil in ways largely unexamined by the many studies that concern themselves primarily with the federal center, whether they focus on privatization policy, the role of oil revenues in the federal budget, or contests between oligarchs and the Kremlin.

2. See, for instance, studies of Saudi Arabia's Eastern Province (Jones 2010), Mexico's Gulf Coast (Breglia 2013), Nigeria's Niger Delta (Watts 2004, 2012), Venezuela's Lake Maracaibo (Tinker Salas 2009), and, in the United States, Oil Creek Valley in Pennsylvania (Black 2003), San Joaquin Valley and the Los Angeles Basin in California (Sabin 2004; LeMenager 2014), and the Louisiana Gulf Coast (Austin et al. 2006). In conceptualizing this ethnography as the study of a region, I have been influenced especially by Sivaramakrishnan 1999 and Ferguson 1999; for recent reviews of the anthropology of subnational regions, see Bitušková 2009 and Wilson 2012.

Scholarship that trains its attention primarily on the federal level is particularly inadequate for the 1990s, when central state power in Russia was at an ebb tide and region-level (and even district-level) decisions and processes had greater than usual significance in determining the overall course of transformations. Even when an assertive central state apparatus reemerged, beginning with Vladimir Putin's first two terms as president, regional processes in the oil industry continued to be far more crucial, diverse, and determinative than many existing studies are able to capture.

This book thus joins a robust line of scholarship about Russia and the Soviet Union that uses region-level processes to call into question representations of the central state apparatus as all-powerful, totalitarian, or consistently able to shape events in the provinces. We must see the Russian state, I will suggest, as constituted along a center–region axis in which regions are just as often the driving force as the federal center—a dynamic that is as true in the oil boom years of the early twenty-first century as it was in the imperial and Soviet periods, if also in some new and different ways. It is for a good reason, I show in chapter 6, that Lukoil-Perm has assertively claimed the mantle of the Stroganov family of nobles who controlled much of industry and agriculture in tsarist-era Perm province. Out in the Urals, that is, central state power has long relied on shifting alliances with regional notables, whose projects, profits, and politics divert and challenge the federal center as much as they support it.[3]

The Material Lives of Oil

The qualities of oil, along with the material and symbolic paths that it has traced through the Soviet and post-Soviet Perm region, form a central part of my argument. As an initial way to understand what I mean by this, consider the work of Russian conceptual artist Andrei Molodkin, who works in the medium of oil. Molodkin's installations, which began to receive international attention after he represented Russia at the Venice Biennale in 2009, often feature loud oil pumps and hoses that cycle crude oil through scenes, words, people, or objects arranged in provocative juxtaposition. They prompt critical reflections on the ways in which oil flows through history and politics—Russian and international, socialist and capitalist (see fig. 2)—and they invite us to trace what Dominic Boyer calls "energo-material transferences and transformations incorporated in all other

3. I defer my specific engagements with the literature on Soviet and post-Soviet regions to the substantive chapters; on the general topic, see especially Evtuhov 2012 and Vladimir Gel'man's wide-ranging work on local regimes (e.g., 2002, Gel'man et al. 2002, and Gel'man and Ryzhenkov 2011).

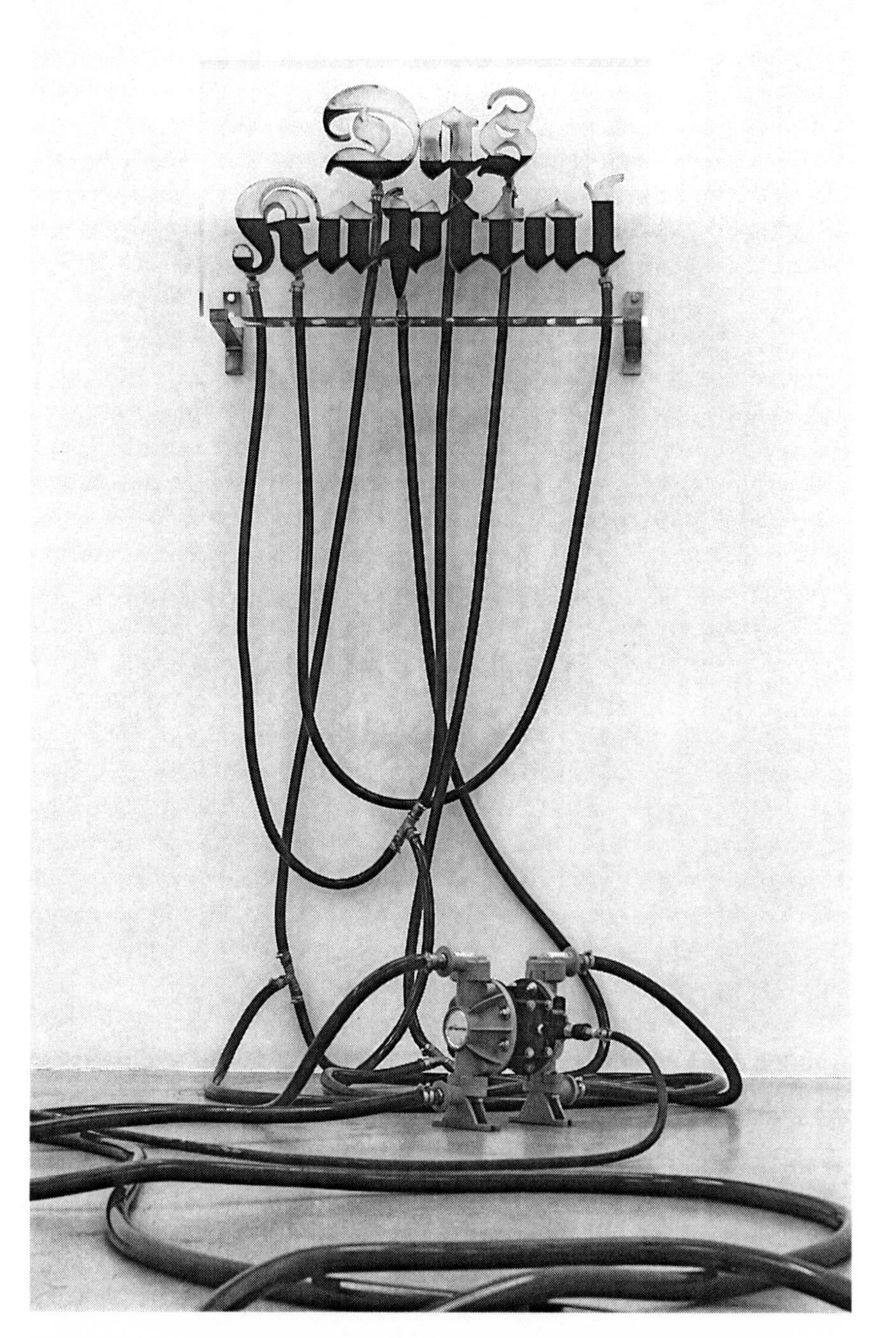

FIGURE 2 Andrei Molodkin's *Das Kapital* links Karl Marx's classic—a touchstone text for the architects of socialist societies and an enduring analytic framework for understanding capitalism—to the material lives of oil. Andrei Molodkin, *Das Kapital*, 2008. Acrylic block and plastic hoses filled with crude oil, pump, and compressor, 72 x 90 x 10 cm.

Courtesy of a/political, the Tretyakov Family Collection, London.

sociopolitical phenomena" (2014, 325). Molodkin's *Oil Evolution*, for instance, which toured European and North American galleries in 2009, featured a series of transparent hominid skulls partially filled with bubbling oil. Which ancient inheritance is really more important to the human condition, the installation seemed to ask, the genetic inheritance we commonly associate with human evolution or the transformation of millennia of organic matter into a fossil fuel? In *Liquid Modernity*, Molodkin displayed a rectangular cage, several meters in each dimension, constructed out of clear tubular bars. Oil flowed rhythmically through the bars from a nearby pump, its hisses and clangs filling the gallery space. What, the installation suggests, if we replaced Max Weber's iconic "iron cage" ([1904–1905] 1958, 181) with an oil cage as a central metaphor for the condition of modernity? Would the shift from iron to oil—from solid to liquid, perhaps, or from mining to drilling—make a difference for our theories of structure and agency, economy and society?

Molodkin's work gains its conceptual power by using some of the properties of oil as a substance, along with the oil industry infrastructure of pumps and pipes, to provoke reflection on the many relationships and transformations involving oil. His work draws attention to the technical systems that channel political and economic configurations and to the ways in which words and concepts and spaces can be colored, inflected, shaped, and filled by their associations with oil. It reflects on the nature of form and content, transformation and connectedness, time and space. Molodkin's art certainly touches on some common issues in the social science of oil—the relationship of oil, money, and global politics favored by political science (e.g., Ross 2012), or the technical infrastructures so central to strands of science and technology studies (e.g., Mitchell 2011; Barry 2013).[4] But it also draws attention to the materiality of oil in a more encompassing way, tying color and viscosity, for example, to transformations of humans and their environments that take place everywhere from extraction sites themselves (Rolston 2013) to representations of oil in museums, literature, or film (Yaeger et al. 2011; LeMenager 2014). Molodkin's art reveals many more, and more subtle, connections than do correlations between oil export levels and political systems, and many more ways of thinking about the relationship between humans and energy sources than tabulations of who produces or consumes how much oil.

It is instructive, for instance, to compare the wide range of connections, attributes, understandings, and transformations on display in Molodkin's work to the narrow focus of so much social science scholarship on a single transformation: that of oil into money. Michael Ross speaks for much of the social science of oil in this regard when he declares in the first pages of *The Oil Curse* that the

4. On this point, see also Richardson and Weszkalnys's (2014) ontological framing of "resource materialities" as always in the process of becoming.

book's subject is not really oil as such but the "unusual properties of oil revenues" (2012, 5).[5] This focus on the transformation of oil into money and oil as commodity predominates—although not nearly to the same degree—in a number of critical approaches to oil as well. "Oil, more than any other commodity," Fernando Coronil writes in his study of oil, money, and modernity in Venezuela, "illustrates both the importance and the mystification of natural resources in the modern world" (1997, 49). The Retort collective takes a similar line, opening their analysis of oil with the remark that, if "a commodity is the economic cell-form" of capitalism, then oil is a "perfect specimen" of that cell-form (2005, 38).

There is certainly ample reason to trace the relationship between oil and money and to theorize oil's special affinity with the commodity form, in Russia and elsewhere. One argument of this book, however, is that a more full-spectrum view of the material lives of oil, a view of the sort invited by Molodkin's installations, permits some new insights into the place oil has occupied in human social and cultural life.[6] Andrew Apter's study of Nigeria in the late 1970s, to give but one example, shows how the blackness of oil ("black gold") became part the same "matrix of value transformations" (2005, 83) as another sense of blackness: the pan-African black culture that the Nigerian state sought to revive with its oil revenues.[7] There is no reason, of course, to expect that oil's blackness would resonate in the same way in the Russian Urals, and indeed it did not, because which of the many qualities of an object become significant within broader "representational economies" (Keane 2007, 18–21) varies with a host of circumstances.[8] As Coronil puts it, "what is socially significant in the case of natural resources is how the material properties of these resources are made to matter by the network of social relations woven around them" (1997, 41).

This approach to the materiality of oil opens up possibilities at multiple scales of analysis, from the very local to the international. It is particularly instructive for understanding an oil region because it serves to bring together geological, geographical, infrastructural, and representational dimensions of the oil industry with associated social and cultural transformations in an ethnographically manageable way. Chapter 6, for instance, shows that it was not oil's quality of blackness—as in Apter's study of Nigeria—but its depth in subsoil deposits that became

5. For an anthropological take on the resource curse, see Behrends, Reyna, and Schlee 2011.

6. For work on energy in Eurasia that successfully and imaginatively integrates state and international policy, the politics of knowledge and culture, and materialities of the energy sector, see, above all, Kennedy 2008; 2014, 195–228, and Barry 2013.

7. Apter argues that the strand of semiotic theory most applicable to the Nigerian oil boom and its aftermath is Baudrillard's (1981), making his approach differ somewhat from the one I take here.

8. This approach to materiality derives from anthropologists' use of Peircean semiotics. See Keane 2003 for an especially instructive exposition that has influenced my approach and Munn 1986 and Meneley 2008 for analyses of the general sort that I pursue here.

highly significant in the Perm region, especially through Lukoil-Perm's efforts to cast itself as a socially responsible company by sponsoring historical and cultural programs. In scores of these programs, the depth of oil and the depth of historical and cultural tradition rubbed up against each other, constituting one of the chief ways in which residents of the Perm region rediscovered their pasts and fashioned new senses of cultural belonging in the early 2000s. It was, I argue, in large part through this language of depth that Lukoil-Perm's Connections with Society division, responsible for corporate social responsibility (CSR) projects, recast Soviet-era patterns of identity formation, cultural authenticity, and national belonging for a new and hydrocarbon-saturated age. The association of legitimacy, authenticity, and the subsoil that ran through Lukoil-Perm's cultural programming was strong and influential. Even a cultural organization that was largely critical of Lukoil-Perm—the independent Kamwa Festival, discussed in chapter 7—cast its own cultural projects in a similar manner, drawing attention to the material qualities of archaeological artifacts excavated from beneath the region and suggesting that they, too, were important sources of energy—the shamanic energy of the ancient denizens of the Permian lands.

Corporations and States

Tracing the material lives of oil permits new perspectives on the workings of and relationships between Russia's most two significant social and political actors: energy corporations and the state. As we look back from the distance of two decades, the enterprise privatizations of the Yeltsin era might equally be understood as processes of *incorporation*: the first steps in the making of new corporations, understood broadly as economic, social, cultural, and political collectivities. Many of those early post-Soviet corporations failed or disappeared, but the fate of a good many others—especially those in the energy and natural resource sectors—can be traced to the present day. These corporations have become tightly entwined with state agencies on multiple fronts, not just at the federal level but, and perhaps even especially, at the regional and local levels.

I take the study of corporations and states to involve two strategies simultaneously. In the first place, it is important to show how these powerful actors—"leviathans" in Alex Golub's (2014) recent and yet centuries-old framing—are not the unitary entities that they often present themselves, and are presumed by others, to be (Welker, Partridge, and Hardin 2011, S5).[9] Both corporations and

9. Ballard and Banks (2003) review a large anthropological literature, published between 1985 and 2003, that shows increasing overlaps among states, corporations, and local communities.

states are composed of diverse divisions, operations, offices, people, and practices that push and pull in multiple directions. The borders and boundaries of a corporation are always on the move, always internally and externally contested, always enacted rather than having an a priori existence (see, above all, Welker 2014). Taking all this into account considerably improves our understanding of the actual workings—and varieties—of corporations. The same is true for "the state," which Katherine Verdery and Gail Kligman instructively understand as "a contradictory ensemble of institutions, projects, and practices rather than an organized actor" (2011, 454). Much of the ethnography to come, therefore, follows the language and interpretive strategies of my interlocutors, who spent a good deal of their time—with me and with one another—sussing out the shifting networks, connections, teams, alliances, clans, friendships, employment histories, family relationships, ethnic backgrounds, generational cohorts, past business deals, places of birth, professor–student relationships, patronage circuits, and more that lie just beneath the abstracted labels of "corporation" and "state."[10] I am especially interested in showing how these varieties of social relationship can be distributed over space, morph over time, and congeal into divisions or subsidiaries within a larger corporation, state, or interlaced state–corporate field. I will show, for instance, that divisions and subsidiaries of Lukoil that specialized in finance capital, industrial capital, and corporate social and cultural sponsorship each grew out of specific and traceable networks of social relationships. These relationships, in turn, intersected in ever-shifting ways with those that made up regional state agencies, from the governor's office to the regional Ministry of Culture. At various times in the 1990s and 2000s, these state-corporate configurations ensured a supply of fuel for the Perm region, set social policy, or supported cultural institutions.

However, disassembling states and corporations into their constituent parts and practices comes with some analytical risks, chiefly the risk that we miss the significant social power and consequences that flow from the pervasive perception, indeed the common sense, that states and corporations are, in fact, unitary and coherent entities. Enormous social and cultural (and, in the extractive industries, environmental) consequences flow from the wide variety of actors—from judges to activists to scholars, and including state and corporate agents themselves—who speak and act in ways that conjure the unity of states and corporations out of their evident disunity (see, e.g., Gal and Kligman 2000, 20). The

10. In many ways, the first anthropological work to use this approach in the study of the oil industry was John Davis's ethnography of Libyan tribal politics in the age of "hydrocarbon society" (1987). For insightful recent examples, see Golub and Rhee 2013 and Steffen Hertog's terrific unpacking of the Saudi Arabian rentier state into the "institutions, societal groups, and social networks" (2010, 3) through which oil rents have unevenly flowed. In the post-Soviet world, some of the very best tracing of energy rents and their role in constituting elites of various sorts has been done by Margarita Balmaceda (e.g., 2013).

massive effects that corporations have had on all domains of life, especially in the past three decades of global capitalist transformations, cannot be accounted for exclusively by laying bare the networks and contingencies that compose them (esp. Kirsch 2014).

I therefore also attend to the considerable social and cultural work that goes into projecting and naturalizing the coherence of states and corporations—to what Timothy Mitchell calls the "powerful, apparently metaphysical effect of practices that make such structures appear to exist" (1999, 89)—and to the ways in which this apparent coherence has been consequential in the transformation of the Perm region. The Perm region over the course of the 1990s and 2000s is a useful context in which to see the processes of abstraction that are central to the making of both "state effects" (Abrams 1988; Coronil 1997) and "corporate effects" (Shever 2012, 197–99). Indeed, the circumstances of the postsocialist Perm region, where basic processes of state formation and corporate formation were taking place at the same time, made for an exceptionally tight relationship between the efforts of region-based corporate subsidiaries and regional divisions of the Russian state apparatus to project their coherence and power.

The project of using ethnography to historicize corporate forms and track their movements is well under way in the case of the global oil industry.[11] Suzana Sawyer's (2006; see also 2012) investigation of indigenous Ecuadorians' efforts to sue Chevron in a United States court illuminates how the oil company's legal teams used concepts of corporate personhood and limited liability rooted in a century and a half of U.S. jurisprudence to attempt to insulate Chevron from claims against it. Strikingly, similar reifications of the corporation as a person—such as indigenous Ecuadorian claims that Chevron must take responsibility for its actions as a true *patron* would—motivated the plaintiffs' contentions in the court case. Elana Shever likewise situates her study of the gendering of development activities and plans carried out by Shell in Argentina in a long genealogy of corporate personhood in the West, from the collective personhood of Roman law to the English concept of the "king's two bodies." Shever is able to show that these concepts, refracted over the centuries, inform even "everyday practices of framing liability, responsibility, and empowerment" (Shever 2010, 30) in Argentina. Indeed, the very notion of the modern capitalist corporation, including key concepts such as trusts, subsidiaries, and vertical integration, grew up hand in hand with Rockefeller's Standard Oil, as muckraking journalist Ida Tarbell showed in her late-nineteenth-century serialized reporting (see Tarbell 1904) and

11. On the broader extractive industry, and with particular attention to CSR, see Rajak 2011. The anthropology of the corporation is usefully outlined in Welker, Partridge, and Hardin 2011; Hann and Hart (2011, 155–59) envision the place of corporations in a revitalized economic anthropology for the twenty-first century in a way that resonates well with my own goals here.

as Daniel Yergin's *The Prize* (1991) has been only the most recent high profile study to document.

In the case of Lukoil in the Perm region, the relevant history of the corporate form is less that of Western legal and business practices exported around the world by multinational oil companies with Euro-American roots than that of Soviet socialist state-owned enterprises and their trajectories in the formative years of the 1990s. To be sure, Lukoil and other Russian oil companies have worked hard to emulate their Western counterparts and establish themselves as thoroughly multinational businesses, in legal and corporate structure and in other ways. Global accounting standards and other requirements for entry into Western capital markets demand no less. Western oil companies have also actively sought partnerships and joint ventures in the former Soviet world, in the process seeking to export their preferred corporate forms to Russia in ways broadly similar to other parts of the world. There is, in other words, ample room in the post-Soviet world for studies of the ways in which the global oil industry seeks to be, and often succeeds at appearing to be, "modular" (Appel 2012a)—entirely self-sustaining and cut off from the surrounding social and cultural contexts.

But other dimensions of the case I explore have led me to focus just as much on the trajectories of Soviet-era corporate forms into the 1990s and 2000s. First, Lukoil has been, more than many other Russian oil companies, the inheritor of the Soviet oil and gas industry; indeed, it is often portrayed as the primary post-Soviet home of the Soviet *neftianiki*—oil workers understood as a profession. It therefore invites an analysis attuned as much to the remaking of Soviet forms as the importing of Western ones. Second, although the Volga-Urals oil basin upon which the Perm region sits was the Soviet Union's production leader in the 1950s–1970s, it was declining rapidly by the 1980s. It was, therefore, never an item of interest for major Western oil companies, whose attention at the time was focused on the Caspian, Siberia, and the Russian Far North. The regional oil industry in the 1990s—the crucial years when privatization and its immediate aftermath were creating a diversity of new corporate forms, the topic of chapters 2 and 3—was thus far more in the control of homegrown interests and regional corporations than were, say, the oil and gas fields of Sakhalin Island or pipeline projects that extended into Europe. It is not a coincidence that Lukoil Overseas Holding—the international arm of Lukoil—was for many years based in Perm, and continued to be staffed by an outsized number of former Permian oil workers well after its central offices moved to Moscow.

It is also significant that the aspects of the formation and development of Lukoil-Perm as a corporation on which I focus here were not always those fashioned to appeal (or even to appear at all) to international observers or, for that matter, to federal authorities in Russia itself. Region-based coalitions and backroom privatization deals, extensive improvised barter chains, and the quite local organization and staffing of development projects—all topics of sustained discussion in the following chapters—were arenas where imported Western notions of the corporation were less salient and relevant than in Lukoil's other regions and in those domains of the company's own operations that were packaged for global circulation. Beginning from this appreciation of the internal and external diversity of the corporation points to ways in which we can tell the history of the postsocialist corporation without judging it by an assumed Western or universal standard. This, in turn, allows us to see how socialist histories and early postsocialist trajectories continue to be relevant for our understandings of Russian oil in the twenty-first century.

This history is particularly instructive, for instance, in understanding a major development of the 2000s in the Perm region: Lukoil-Perm's adoption and adaptation of the now-common practice of corporate social responsibility.[12] Chapters 4 and 5 trace the ways in which Lukoil-Perm's CSR initiatives, dedicated to all manner of cultural, economic, and environmental issues in the Perm region, spread in concert with the rebuilding of regional state agencies beginning in the early 2000s. This state–corporate collaboration replaced the civil society–building initiatives sponsored by international aid agencies in the 1990s, taking on many of their attributes (such as competitive grant writing) and adding others (such as including corporate image-making in the goals development projects).

By the mid-2000s, Lukoil-Perm's CSR projects stretched to cover everything from health care to ecology to sports to culture, and the company came to explicitly share with regional state agencies the roles of metacoordinator and authorizer of diverse fields of social interaction—roles that have more often been associated with "the state" in social theory than with the corporation (e.g., Mann 1986; Bourdieu 1999). The result was a field of state and corporate power in which it was, at least for a time, hard to distinguish state from corporation as they sought, separately and in concert, to remake the social and cultural order of the Perm region. It was only later, as the first decade of the twenty-first century came to a close, that Lukoil-Perm and the Perm regional state apparatus began to draw more noticeably away from each other, beginning to produce region-level state and corporate effects that could be more easily differentiated. Chapter 8 explores a key part of this process by focusing on Perm's concerted attempt to rebrand itself as a postindustrial European Capital of Culture.

---

12. On the wide diversity of projects often grouped under the rubric of CSR, see Auld, Bernstein, and Cashore 2008. For ethnographies and allied approaches that have influenced my approach to CSR, see Barry 2004; Watts 2005; Kirsch 2007, 2014; Welker 2009; Rajak 2011; and Welker and Wood 2011. On the curious absence of the state from many studies of CSR, see Gond, Kang, and Moon 2011.

A word is necessary, finally, on the thorny issue of states, corporations, and neoliberalism. Many of the topics I treat in this book, from privatization to CSR, from newly flexible labor regimes to efforts to transform Perm into a creative city of cultural entrepreneurs, might fairly be framed as aspects of neoliberalism or neoliberalization.[13] In each of these cases, it might be suggested, we can discern the retreat or reformulation of the state and the increasing significance of various forms of entrepreneurial or corporate capitalism—a serviceable enough capsule definition of neoliberalism. To substantiate this claim, however, would require explicit and integrative theorizing that is beyond the scope of this book, especially because the literature dedicated to neoliberalism has grown to such proportions that even analytically linking two elements—such as flexible labor regimes and CSR—requires significant ethnographic and theoretical lifting (see Smith and Helfgott 2010 and Benson 2012 for two studies that do this well). I prefer to simply stipulate that all the transformations traced in this book are linked, in one way or another, to global transformations in the nature of capitalism over the past four decades and to indicate, where appropriate in my account, how this has played out. At the end of the day, it is on the intersections of oil, corporations and states, and cultural production that I place the weight of my analysis. The insights thus generated would, I think, be obscured by an effort to unite everything under the theoretical sign of the neoliberal.

Cultural Politics between Socialism and Spectacle

"Culture" in the Soviet Union was, as Bruce Grant puts it, "something to be produced, invented, constructed, or reconstructed" (1995, xi). Indeed, all socialist states were centrally concerned with the production of culture, from the Bolsheviks' earliest efforts to create new, proletarian forms of theater and visual art to the national ideologies that, paradoxically, both helped to shore up the legitimacy of socialist regimes and became an important channel for resistance to those regimes. Culture specialists and intellectuals of all stripes thus played a significant role in the workings of socialist societies: it is no accident that one of the foundational studies in the anthropology of socialism—Katherine Verdery's *National Ideology under Socialism* (1991)—turned to competing intellectuals

13. Indeed, Sawyer's (2004) and Shever's (2012) ethnographies of the oil industry orient themselves precisely in this way, and I find persuasive the contention that looking in on the relationship between states and oil companies—which has provided excellent material for theorizing the place of states and corporations in earlier phases of capitalism—is again a useful place to embark on this sort of analysis in the present.

and their visions of the Romanian nation in order to illustrate the workings of socialist political economy.[14]

The transformations associated with the end of socialism, however, brought rapid and radical change to this configuration. Especially in the former Soviet Union, once stalwart intelligentsias were marginalized, and along with them many of the projects of official cultural construction—socialist, national, or hybrids thereof—that had been an important part of their turf (see, e.g., Oushakine 2009b). During his fieldwork on the island of Sakhalin in the early 1990s, Grant found cultural construction in "ruins" (1995, 163). In this context, studies of the intersection of culture and capitalism in the former socialist world shifted away from intellectuals and their efforts at concerted cultural construction to the newly relevant sphere of consumption, yielding studies of nation branding (Manning and Uplisashvili 2007), consumption and consumer goods as national signifiers (Humphrey 2002, 40–64; Shevchenko 2008; Klumbyte 2010), and even nations imagined as collective consumers (Greenberg 2006).[15]

Examples of all these phenomena could be found in the Perm region in the postsocialist years. Beginning in the early 2000s, they were joined—and often dominated—by an ascendant field of cultural production centered on oil, the oil industry, and oil-producing districts of the Perm region. The chapters of part III chart this reemergence of a cohort of culture producers and cultural managers, first at Lukoil-Perm, somewhat later in the Perm regional state apparatus. To understand these culture managers, however, we need to look beyond their socialist-era predecessors to some patterns of cultural production commonly found in oil-exporting states. Terry Lynn Karl once remarked that socialist states were, in important ways, akin to petrostates. She had in mind the key role that politics played in shaping national economies "when the state owns the central means of accumulation" (Karl 1997, 189), but the insight is instructive for cultural production and broader cultural politics as well.[16] Indeed, oil-exporting states

---

14. On the role of intellectuals and cultural producers in socialist systems and their early aftermath, see also Suny and Kennedy 1999, Kennedy 1992, and Konrád and Szelényi 1979; on intellectuals and nations more broadly, see especially Boyer and Lomnitz 2005. Technical intelligentsia worked largely in the redistributive planning apparatus, while cultural or creative intelligentsia worked in fields of art, literature, theater, and so on. Soviet cultural construction extended, importantly, into the domain of everyday life and the making of more "cultured" people (see, e.g., Kelly 1999, 2004).

15. This shift took place within a broader scholarship attuned to the fact that capitalism itself appears to be going through a phase that valorizes consumption over production: studies of the commodity image, advertising, and branding have become central to the ways in which experiences of culture are theorized, in the postsocialist world and elsewhere. Oushakine's *Patriotism of Despair* (2009a) is an exemplary account of some ways in which intellectuals and consumption could still be drawn into the same analytic framework in a study of the 1990s and early 2000s.

16. I leave discussion of whether the Soviet Union *was* a petrostate until Chapter 1; in brief, I think the issue is a red herring, answers to which obscure the crucial ways in which socialist oil differed from capitalist oil.

around the world have been particularly adept at the production of cultural and historical identities, in part because so many of them emerged at moments of postcolonial independence and/or the nationalization of their oil industries. As Karl notes, the subsequent growth of a domestically controlled oil industry as the leading sector of the economy coincided with initial processes of state formation, pushing to the fore the identity construction projects often at stake in cultural production and cultural politics. Moreover, the outsized revenues flowing into state coffers meant that petrostates specialized in producing not just garden-variety national culture, but elaborate spectacle—grand displays and festivals such as those analyzed by Apter (2005) and Coronil (1997) in the context of 1970s oil booms and that are epitomized in more recent times by the glittering cities of Dubai and Abu Dhabi and, on some readings (e.g., Retort 2005), the global order writ large.

Il′ia Kalinin, in a brief article titled "The Past as a Scarce Resource," beautifully illustrates some underappreciated ways in which oil and cultural/historical production in Russia began to wrap around each other in the twenty-first century.[17] Kalinin quotes, for instance, from a commentary that circulated widely on the Internet in Russia in 2012:

> History today, one might say, is something like a natural resource. And all around us are not only mineral deposits, not only gas and oil somewhere deep in the subsoil. Under our feet is a whole ocean of thousand-year history. The upper strata literally ooze it. (Kalinin 2013, 213)

Metaphors and similes linking the production of history to the production of hydrocarbons, Kalinin goes on, could also be readily found in President Vladimir Putin's speeches, such as one on the moral education of youth in 2012:

> As our own history shows us, cultural self-consciousness, spiritual and moral values, value codes—this is a sphere of sharp competition. . . . Attempts to influence the worldview of entire peoples, the attempt to bend them to one's will, to instill one's own system of values and understandings—this is an absolute reality, just like the struggle for mineral resources, which many countries, including our country, have encountered. (Kalinin 2013, 214)

Kalinin draws on Douglass North's neoinstitutionalism (as incorporated into North, Wallis, and Weingast 2009), coupled with Etienne Balibar and Immanuel Wallerstein's definition of economic rents as a domain in which profits are earned without one's own labor (1991), to argue that the Russian state's reassertion of

17. I thank Gulnaz Sharafudtinova for bringing Kalinin's work to my attention.

control over the oil industry has paralleled—and funded—a similar state attempt to monopolize the production of Russian history and culture. Both are domains in which resources that should be held in common become rent-generating enterprises for an elite in control of the state apparatus. Kalinin concludes by suggesting, ironically, that the future will bring the creation of a new state corporation, Rosistoria (on the model of the state oil company Rosneft), whose task will be to stop "orchestrated attacks" on the Russian past (Kalinin 2013, 214).

Kalinin's material would not be at all out of place in the Perm region in the mid-2000s; indeed, viewed in light of cultural production in and around Perm, Kalinin's examples from 2012 are rather late entries into an already densely populated discursive and material field. If he were writing about the Perm region in the mid-2000s, Kalinin would not even have needed his parting shot about the incorporation of history production under the auspices of Rosistoria, because the lead sponsor of historical and cultural revival in the Perm region by this time was none other than Lukoil-Perm, through its high-profile CSR projects and extensive collaboration with the regional Ministry of Culture.

Although I cover some of the same terrain as Kalinin, I approach the intersection of the oil industry and historical-cultural production in the Perm region not through North's neoinstitutionalism and Balibar and Wallerstein's notion of rents but by building on the two literatures already mentioned: socialist and early postsocialist cultural production and cultural spectacle in oil-exporting states. This expanded view enables me to show that the rise of Lukoil-Perm's CSR projects created a newly ascendant group of cultural managers that stretched from rural libraries and museums to the central public and private cultural institutions of the Perm region. Many of these managers were former Communist Party members—particularly from the Komsomol, the Communist Youth League—and former Soviet culture workers who had fallen on hard times in the immediate aftermath of the Soviet Union. If, as chapters 2 and 3 show, the 1990s saw a new configuration of regional political and economic elites based on an alliance of Soviet-era oil insiders and new financiers, then the 2000s saw this elite joined by a resurgent network of cultural managers whose task became, in no small part, to legitimize the processes of social stratification and exclusion that were becoming ever more noticeable.

By following the ways in which a new cultural front opened in the Perm region, we can also note departures from familiar patterns of oil-fueled cultural production. In contrast to postcolonial national identity and nation building, for instance, Lukoil-Perm's CSR campaigns closely followed socialist efforts to construct a "family of nations" by carrying out cultural programming across the diverse districts of the Perm region, from the largely Tatar and Bashkir populations in the south to the mostly ethnic Russian populations in the north. This was true both for the ways in which different national identities were presented as fitting together and in the very techniques by which this construction was carried out, many of which were recycled from the Soviet period.

By the end of the 2000s, the central weight of regional cultural production moved from Lukoil-Perm's CSR projects to full-fledged state spectacle, beginning with Governor Oleg Chirkunov's alliance with Moscow-based benefactors and cultural managers and culminating in a massive push to have Perm designated a European Capital of Culture. The Perm Cultural Project, which featured the opening of a new contemporary art museum in 2008, met with enormous resistance from the local cultural intelligentsia and, at least publicly, studied indifference from Lukoil-Perm. Tracing these debates and conflicts, chapter 8 shows the ways in which the cultural field remains tied in numerous ways, both official and unofficial, to the regional oil complex. The networks and circuits of exchange that grounded an oil-based sense of Permianness in the 1990s and early 2000s were, for example, precisely those that challenged the new projects of Moscow-based cultural managers and the new spectacles of culture they brought with them.

Many observers of the Russian political scene have drawn attention to the role of cultural producers in some of the innermost circles of Russian national politics in the Putin era: Vladislav Surkov, "gray cardinal" of Putin's Kremlin, is a novelist and trained theater director; Konstantin Ernst, general director of Russia's state-controlled Channel One, has orchestrated and designed many of the state spectacles of the Putin presidency, including the Opening Ceremonies at the Sochi Olympics; and Valerii Gergiev, artistic director at the Mariinsky Theater in Saint Petersburg, has been a prominent and influential supporter of President Putin and an architect of Russian cultural policy. Part of my purpose is to show that there are regional dimensions to this return of the cultural intelligentsia to Russian politics as well, and that these regional configurations should be seen not as simple extensions of federal state cultural policy or cultural politics, but as crucial dimensions of a larger cultural landscape that stretches across both center and region. As for states and corporations, so, too, for culture: the view from Moscow is at best part of the picture.

The Spaces and Times of the Perm Region

The Perm region is not a static canvas on which these intersecting issues have played out. As the rumors and conspiracy theories following the Aeroflot-Nord crash of 2008 briefly illustrated, the making and remaking of the Perm region has been an ongoing discursive and material process. Categories of space and time are some of the most significant arenas in which material, corporate, state, and

MAP 1 The Perm region, including major locations mentioned in the book.

cultural dynamics have combined and recombined. The wheres and whens of the Perm region, it follows, are less starting points than questions that I will explore throughout.

One of the reasons for the ongoing conversations about space—including but not limited to the pervasive maps and talk of maps that I have already noted—was that it kept changing. The 1990s and 2000s were a period of significant territorial reorganization across the Perm region. At the largest scale, the former Perm region (*oblast*) was united in 2004 with what had, until that point, been an entirely separate region, the Komi-Permiak Autonomous District (*okrug*). The result was a new entity called the Perm *krai*, although pretty much everyone spent months correcting themselves as they said "oblast." Districts at smaller scales were also being combined and divided at a healthy clip, and everywhere streets gradually, and not without controversy, shed their Soviet-era names and street signs. Governing a region larger than New York State, especially as centralization increased in the early 2000s, was an enormous technical undertaking. The elected heads of all the region's districts made their way by car to Perm every week for meetings—an eight-hour round trip for some of them. On the days that they were home in their district offices, they nearly always received delegations from one or another agency of the Perm regional government. This constant back and forth was one vector along which many of the exchanges, networks, and state projects that I discuss rode; indeed, I will argue that various networks associated with the oil industry helped to *create* these new spatial vectors over the 1990s and 2000s.

Time, no less than space, was a shifting category of experience in the post-Soviet Perm region. In the 1990s, in conditions of widespread demodernization and involution, most residents were as likely to think they were moving backward in history as forward (see also Verdery 1996, 204–28). By the early 2000s, however, the rise of the oil industry and the associated torrent of money contributed to a widespread sense that history was accelerating, at least for some—a phenomenon noted in other oil boom contexts as well. The majority of the fieldwork for this project took place in 2009–12, years in which the remarkable oil boom of the mid-2000s had stalled and the circumstances of global financial crisis led some to cast their minds back to the turbulent 1990s. Was Russia still "in transition" after all? Was it entering a new crisis? Or had rising oil prices finally put it back on the right path to a bright future? Given what were assumed to be declining reserves, for how long would the oil—and the social, cultural, economic, and political configurations informed by its labyrinthine path through the region—last?[18]

18. On the anthropology of resource temporalities, see especially Ferry and Limbert 2008 and Limbert 2010; Rogers 2015 includes a more detailed overview of scholarship on oil and temporality.

The production, circulation, and consumption of particular kinds of temporality also featured in a variety of other contexts. The geological time of the Perm region—of new interest as subsoil deposits of oil became ever more prominent in regional consciousness—began to feature in public histories and museum displays. The ground beneath the region offered up other grist for the historical imagination as well. Archaeological artifacts in "Perm Animal Style" began to play a central role in regional branding efforts; as I discuss in chapter 7, representations of the stylized figurines appeared everywhere from fashion shows to the redesigned facade of Perm's Central Store. At the same time, the Stroganovs—masters of the Permian lands in the Russian imperial period—enjoyed a new renaissance. Anniversaries were celebrated everywhere, inspiring a significant portion of the social and cultural projects I discuss throughout the second half of this book: 20 years of Lukoil, 75 years of Permian oil, 395 years of Perm, 5 years of the Perm krai, 60 years since the end of the Great Patriotic War, 580 years since the founding of Solikamsk.

Conceptualizing the temporality of the Perm region in this way allows me to gain some critical distance on the research paradigms of "transition" and "resource curse" that have been so prominent in scholarship on post-Soviet Russia, including that on oil. These paradigms embed their own temporalities, drawn less from experience than from the predictive models of social science (see esp. Weszkalnys 2011, 2014). When and how fast will Russia attain capitalism and democracy? Do particular policies or practices, when projected into the future, point to Russia following other countries along the resource curse path or managing to "escape" (Humphreys, Sachs, and Stiglitz 2007; Gel′man and Marganiya 2012)? Which areas or regions (of Russia, the former Soviet Union, the world) are moving faster or slower, or are more or less advanced?

In the 1990s, anthropologists' preferred categories of socialisms/postsocialisms accomplished the task of critiquing the temporal certainties of academic transitology. In more recent years, many have questioned the ongoing relevance of these categories, pointing to the fact that the experience of living after socialism is no longer as salient as it once was. This is certainly true. From my perspective, however, the research framework of socialisms/postsocialisms was always more analytically diverse than this critique allows: postsocialisms was never just about the experience of life after socialism. The utility of the socialist/postsocialist distinction depends, therefore, on the larger argument into which it is recruited.[19] I make a distinction between socialist oil and postsocialist oil, for instance, when my goal is to track decades-long historical trajectories in the state-corporate field and to

19. On this point, I join a number of other observers of ongoing postsocialist transformations, among them Hann et al. 2007, Chari and Verdery 2009, Creed 2011, 5–11, and Rogers and Verdery 2013.

situate them in global contexts. At other points, I leave this dichotomy behind, pegging my arguments to other relevant temporalities that emerged from my fieldwork: between the Yeltsin and Putin eras; among the governorships of Gennadii Igumenov, Iurii Trutnev, and Oleg Chirkunov in the Perm region; between alternating periods of monetization and demonetization; or in selective reclamations of historical periods that are designed to collapse or extend time.

The shifting times and spaces of the Perm region underscore the extent to which the story I tell in this book is a highly specific and contingent one. Like most anthropologists, I make no claim that the ethnography presented here is representative of anything in a part-for-whole sense. The relevance and broader significance of this sort of study lie along other analytical paths. In addition to the more theoretical discussions of oil's materiality, state and corporate forms, and cultural production introduced above, the particular conjunctures and contingencies of Permian oil illuminate some significant issues in broader Soviet and Russian history. For instance, the oil discovered beneath the Perm region in 1929 was the first truly Soviet oil. In contrast to the oilfields of Baku and Groznyi, which were developed in the imperial period and then Sovietized after the Revolution, the oil of the Perm region flowed, from exploration in the 1920s up through steady decline beginning in the 1970s, within a socialist political economy. There is, then, no better place than the Perm region to explore the questions of just what was socialist about the Soviet oil industry and how this socialist oil complex compares and contrasts with the capitalist oil industry of the same time period.

With respect to the post-Soviet period, I have often been told that I should have based this study in West or East Siberia, or in the Far North, all areas that are producing more oil—and more oil revenue—than the Perm region and that appear far more frequently in the strategic calculus of Russian and international oil companies and states. There is doubtless much of interest to be studied in all those regions (see, e.g., Thompson 2009; Stammler 2005, 2011), but the assumption that high-output oil production regions or regions almost exclusively dependent on oil revenues are necessarily the best place to study oil is inaccurate. In its most lucrative years, Lukoil-Perm's contributions to the Perm regional budget never exceeded 20–25 percent, a far from trivial but far from exclusive proportion. (Robert W. Orttung [2004, 53] terms the Perm region a "plural" region, home to a number of big businesses in addition to Lukoil-Perm.) Precisely because Lukoil-Perm does not fully dominate either Perm or the Perm region in the way that Lukoil (and other oil companies) dominate the oil districts and regions of Siberia, questions about how the oil industry relates to state agencies and other segments of society are on display with particular clarity in the Perm region.

One of my interviewees made this point nicely through a pun that contrasted the Perm region with the Khanty-Mansi Autonomous District in West

Siberia, home to massive oil production managed by Lukoil and other companies. There, he said, grinning, AO, or Avtonomnyi Okrug (Autonomous District) actually stood for Aktsionernoe Obshchestvo, or joint-stock company. The political district *was* the company. Not so in the highly populated Perm region, where the state–corporate field in the oil sector was far denser and more intricate. It was, for instance, in the Perm region—and not in the Khanty-Mansi AO—that Lukoil's policies and practices of corporate social responsibility first took shape in response to public critiques. CSR spread only later to corporate headquarters in Moscow and to other regions that were home to the company's operations. Similarly, the fact that, for some time in the late 1990s and early 2000s, one in five shares in all of Lukoil was held in the Perm region makes this region an ideal context for a ground-level study of processes of privatization and incorporation in the post-Soviet oil industry. In these and other ways, the ethnography of the Perm region speaks beyond itself not because the processes that unfolded there are representative, but because a number of large-scale processes converged in the Perm region to produce a particular set of outcomes; the narrative and analysis of this configuration can tell us quite a lot about those larger processes.

Fieldwork, Access, and the Study of Oil

In chapter 2, I introduce a former employee of the Perm Commodity Exchange who helped me understand a great deal about state and corporate interactions in the Perm region's early post-Soviet days. He began our first interview with what was a standard refrain whenever I let one of my interlocutors know that I would be asking, in part, about the oil deals and exchanges of the early 1990s: "Everything about that time is more or less shrouded in darkness for me. It's all on the level of rumors, incomplete thoughts, and so on. There are no documents to be found. Nothing is documented." He sounded exactly like certain of my academic colleagues in the United States. When I described this part of my research plan to a knowledgeable political scientist, for instance, his response was to tell me, gently, that no such research was possible. The deals were too opaque, the key players often assassinated, the memories purposely selective. Such were the conditions of 1990s "wild capitalism" and there is little to be done about it, especially, he said, out in the Russian regions. I believe I have reconstructed more than he thought possible, and, more notably, found analytic paths to theorizing the shape and significance of an emergent Russian oil corporation that do not require precise accountings of who mysterious stockholders were, how much money flowed through bank accounts in Cyprus or Switzerland, or how tax officials were or were not duped.

To a significant extent, however, both my interviewee from the Perm Commodity Exchange and my political science colleague were right, and I am sure that much has remained outside my view. After all, even in conditions much more stable than 1990s Russia, states and oil companies are among the most secretive of the world's institutions, and so the issue of "oilpacity," as Coronil called it (2011) in a discussion of political intrigue in Venezuela, deserves some special comment as a methodological and analytical issue. To begin with, it is worth recalling that ethnographic access is never complete. Having spent a year doing fieldwork in a very welcoming small town—a social context that would seem, at first blush, far more accessible than a major capitalist corporation in the extractive industry—I can say that it is not just when the research topic is oil that many things remain unsaid, unknown, offstage, inappropriate to share with a researcher or just in general.[20] The question, then, lies not so much in how close one comes to some imaginary ideal of access, but, rather, how the materials gathered in the course of fieldwork fit with the questions asked and arguments made in the overall analysis.

This project began to take shape at the suggestion of a long-time friend in the Perm region, Oleg Leonidovich Kut′ev. I first met Oleg Leonidovich in the late 1990s, when he was a mid-level official in the Perm region's Ministry of Culture. His long-running and influential research on the history of the Perm region, including on the Old Believer communities that interested me at the time, made him an important interlocutor as I developed the arguments of my first book. When I was visiting Perm in the summer of 2004, Oleg Leonidovich surprised me by telling me that he had switched jobs and now worked for Lukoil-Perm, where he oversaw grants in the company's growing program of sponsoring social and cultural projects. On a summer evening in one of the outdoor bars that crowded the Perm waterfront at the time, Oleg Leonidovich told me that I simply had to study the issues of oil, society, and culture that he was dealing with—this job was among the most interesting things he had ever done and he was gleaning new insights about Russia every day. There were some things he and others could not and would not tell me, and much to which he was not privy himself, he said, but scholarship focused on the company's innovative social and cultural projects would likely be welcomed by his superiors. I would get to know everyone of importance in the Perm region, he said, by paying close attention to the genesis, funding, and implementation of a year's worth of Lukoil-Perm's social and cultural projects. The cycle, he joked, was not so different from the agricultural cycle I had gotten used to in my earlier fieldwork.

By the time I had pieced together my own project centered on this suggestion, funding for fieldwork, and a research leave in which to do it—the space of a few

---

20. I am grateful to Stuart Kirsch for sharing his insights on this point.

years—Oleg Leonidovich no longer worked at Lukoil-Perm. He had been laid off in the aftermath of a leadership shuffle in 2006, in which an entirely new team of managers was brought in from a West Siberian division of Lukoil to run the company's operations in the Perm region. I will suggest in chapter 3 that this shakeup was likely an effort on the part of big Lukoil—the common nickname for the Moscow-based holding company of which Lukoil-Perm was a subsidiary—to inject some distance into relationships between the company and regional state agencies in the Perm region, relationships that had grown very cozy in the 1990s and early 2000s. With Oleg Leonidovich's job went much of the access on which my original fieldwork plan was based. Although I was a vanishingly small fish in the sea of connections at stake, the principle was the same: by 2006, relationships and connections built up over the decades in the Perm region were, for me and for many others, no longer the access route into the oil industry that they were even a couple years earlier. As things turned out, I had fairly minimal access to Lukoil-Perm itself during my primary fieldwork in the summers of 2008–12 and longer trips in 2009–10—something that many others in the Perm region, from journalists to academics to state employees, also noted in the aftermath of the management turnover. Even my tour of the Lukoil-Perm museum, an excursion taken by schoolchildren far and wide, was prefaced by a somewhat nervous plea that I not photograph anything or take any notes while the tour was in progress.

Although I still wish I had the access Oleg Leonidovich thought he could get me, the fact that I was not able to work directly with Lukoil-Perm sent me in other productive directions, directions that both showed me a wider field of interactions on the entangled fields of culture and energy (not just oil) and enabled me to trace a longer-running set of transformations of both state and corporation. In place of the embedded fieldwork that I had originally planned, I focused on interviews with former Lukoil-Perm employees and many others in the region who dealt with the company in one or another capacity. I relied on an expanded set of methodologies and focal points designed to get at other aspects of state and corporate activities, including archival and library research, and began to realize the importance of Soviet and early post-Soviet history to the larger issues I was beginning to understand. I started attending different and far more kinds of cultural events and productions than I had originally intended, doing everything I could to follow the Perm Cultural Project as it gathered proponents and critics with each successive visit.

I also began to talk much more than I had expected with state officials, who were going through their own reorganizations in parallel with the management restructuring at Lukoil-Perm. When it comes to state agencies and offices, the post-Soviet Perm region has a justly deserved reputation for comparative openness among Russian regions—a side effect, perhaps, of its efforts to put itself on the map after decades as a closed city. I found state officials at all levels ready and willing to talk, often at great length, about the projects that they were working on. Outside of their offices, and as I got to know people better, some spoke quite freely about the inner workings of Perm regional politics and economics, and gave their candid evaluations of all manner of events and intrigues. By comparison to state administrations in other regions, and especially during the governorship of Oleg Chirkunov (when my primary fieldwork took place), the regional state agencies in the Perm region were incessantly chatty: blogs, public discussions and forums, a habit of welcoming academic interviewers from Russia and abroad, and a seemingly endless back-and-forth with an often critical local press and local intelligentsia who were quite engaged in matters of regional governance. A flash drive for copying the growing output of public reports, pamphlets, and books that were no longer in print became an indispensable piece of fieldwork equipment.

This is not to say that the practices of masking often identified as mechanisms of state power were not present—transparency can be a terrific mask—but I do think that they took on a particular flavor in the Perm region. This flavor was nicely captured by one of my interlocutors, reflecting on an open forum held by Oleg Chirkunov's replacement as governor in 2012. He noted that he was not particularly happy with the new governor, but added immediately that at least he appeared to understand how things are done "the Permian way" (*po-Permskii*)—that is, not without engaging the public and respected members of the local elite in extensive conversation. I try to capture this aspect of the regional state in the ethnography that follows, especially by attending to the evolving self-conception of the Permian elite as it crystallized in the post-Soviet period through alliances that crossed state, corporate, and cultural domains (see esp. Fadeeva 2006, 2008). This strategy, I hope, serves as a counterweight to some of the assumptions of a unified, autocratic, and unresponsive state that inform many studies of Russia in the Putin era.

In the cases of both corporation and state, the readiness of employees and former employees to speak with me likely had something to do with the fact that I was asking, at least in the first instance, about things that they generally wanted to talk about: state and corporate projects aimed at improving social and cultural life in the region. My desire to discuss and analyze these projects by comparing them to others—at other times and in other regions, industries, and countries—meshed well with their own desires to market the Perm region and learn about what was being done elsewhere. Sometimes conversations ended on these topics; in many cases, they were a perfect springboard to talk about the ways in which social and cultural projects were tightly wrapped up in other aspects of the state–oil nexus in the Perm region—elections and politics, infrastructures, and the remaking of regional identities on multiple, intersecting axes. The ease with which conversations slipped from social and cultural projects to these other issues is itself evidence for the significance of what I call in part III "The Cultural Front."

These, then, are a few introductory examples of the kinds of methods and data that lie behind the claims of this book. If readers should not expect detailed descriptions of the inner workings of Lukoil-Perm, then they should also not expect descriptions of everyday life in the Perm region's oil producing districts, parallel to ethnographies of specific oil towns or smaller-scale districts elsewhere in the world.[21] One of the most frequent questions I have been asked when talking about this research runs as follows: All this ethnography of state and corporate projects is interesting, but how have ordinary people responded to all these attempts to remake them? This is an important question, but everyday life understood as a domain of practice is not the primary focus of this book. Especially in recent decades, it is far from the only canvas on which anthropologists have worked. Nowhere, then, will I claim that the projects and processes traced here have been received, understood, or responded to uniformly across the Perm region.

My position on this differs from that of many intellectuals and scholars in the Perm region itself. In an interview with the Sverdlovsk television segment *Accent on Culture* in 2011, renowned Perm-based novelist and culturologist Aleksei Ivanov was asked about his expeditions to far-flung areas of the Perm region and the Urals—expeditions that informed both his novels and his culturological essays. Did he spend much time asking local residents about history, or collecting folklore? No, he replied, this history "has been lost." Local residents are no longer "carriers" (*nositeli*) of identity or useful historical knowledge, and there was no longer any authentic folklore out there to be collected. They might know of good fishing places, or where to find firewood, but "book people . . . people who have read books about those places" were now the only remaining legitimate sources for describing the historical and cultural trajectory of the Urals.

I strongly disagree. I wrote a book about the ways in which residents of one remote town in the Perm region have a quite developed and intricate historical consciousness, one that often eludes passersby, outsiders, and so-called book people—especially those who seek it in a narrow expectation about authentic folklore (Rogers 2009). By shifting my focus in this book to the formation of regional elites, the interactions of state and corporation, and the ways in which social and cultural projects have been conceived with the goal of transforming local residents of all sorts, I am all too aware that I run the risk of replicating Ivanov's ignorance of the lives of those local residents, if not in intent or design then in practice. I have elected to run this risk in order to make a different kind of argu-

21. Book-length studies pitched at this scale include Reed 2009, Auyero and Swistun 2009, and Shever 2012.

ment. The practice of everyday life *was* more important in the 1990s and early 2000s—a time when structures and institutions lay largely in ruins and everyday practice often had large-scale consequences in ways that it rarely does (see Burawoy and Verdery 1999, 2–3; Humphrey and Mandel 2002). In those contexts, anthropologists quite appropriately focused most of their attention on everyday life in order to show how it had far more implications for postsocialist trajectories than the large-scale, highly abstracted plans and projects envisioned by the architects of transition. But things have changed. The story I tell in this book is, in good part, about the ways in which the practices and projects of an emergent state–corporate–cultural elite came to have ever more pronounced consequences for the reshaping of an entire region. I do not doubt for a moment that residents across the length and breadth of the Perm region did some combination of embracing, diverting, and resisting the agendas and projects of this new elite—and the larger forces they channeled—at the level of everyday practice. My claim, however, is that the conditions of possibility for embracing, diverting, and resisting have shifted substantially in the past two decades. Accounting for the role of oil in this shift is one of the major projects of this book.

Part I
FROM SOCIALIST TO POSTSOCIALIST OIL

1

THE SOCIALIST OIL COMPLEX

Scarcity and Hierarchies of Prestige in the Second Baku

Long live the oil of the Urals,
rushing from the depths of the Earth
to the aid of workers and peasants!

—Banner at a political rally in Verkhnechusovskie Gorodki in 1929

It is instructive to begin with a pair of oil regions that were central to the capitalist world in the twentieth century.

In July 1929, oil was discovered near Darst Creek in Guadalupe County, Texas, east of San Antonio. There was nothing particularly special or notable about the Darst Creek oilfield itself, especially against the backdrop of nearly three decades of oil strikes in Texas. The reservoir lay beneath a one-mile by six-mile stretch of the Gulf Coastal Plain, and was developed by a group of oil companies, including Humble Oil and Refining (eventually part of Exxon) and the Texas Company (forerunner to Texaco). By 1930, 186 wells had been completed, a network of pipelines ran to Texas refineries, and the field was rated at a daily potential of just over 155,000 barrels.[1] However, by order of the Texas Railway Commission (charged with the regulation of the state's oilfields), Darst Creek was permitted to produce only a fifth of that amount. In fact, it was the first oilfield in this area of Texas to operate under voluntary proration—a hotly debated system of production quotas designed to avoid saturating the market and driving down prices.

Voluntary proration is emblematic of a crucial aspect of oil production and consumption in conditions of capitalism. In the long term, humanity is rapidly exhausting finite reserves of oil: there is an absolute scarcity of oil relative to current consumption patterns. But, at any given moment in time, the capitalist oil industry has usually struggled to *limit* the amount of oil extracted from the

Epigraph: Abaturova 2003, 11.
1. On the Darst Creek oilfield, see Smith 2010.

subsoil, the better to keep market prices—and therefore both profits and state taxes on those profits—as high as sustainable. In other words, the capitalist oil industry has survived and thrived in key part by producing scarcity. In Texas during the Great Depression, when plummeting demand exacerbated an already acute crisis of overproduction, voluntary proration turned out to be an inadequate means of keeping oil in short supply. By 1931 Governor Ross Sterling had declared martial law and dispatched the National Guard to the giant East Texas oilfield in order to ensure that the oil remained in the ground, and the question of how to ensure the scarcity of Texas crude quickly became a matter of national concern as well (Huber 2013, 27–59).[2]

There was good reason that Texas was at the center of efforts to forestall overproduction in the early 1930s. The Texas–Oklahoma region of the United States had been central to the world oil economy since the first gusher at Spindletop, near Beaumont, in 1901 (Yergin 1991, 82–87). It remained so through the Great Depression and the New Deal and into the decades of suburbanization, expanding consumption, and automobility that followed World War II. In these postwar years, as the world center of oil production moved to the Persian Gulf and the battle against overproduction moved from the Texas Railway Commission to the OPEC cartel, Texan oil became increasingly associated with mythologies of (and nostalgia for) the American West. The booming postwar fields were in west Texas, and, as Karen Merrill (2012) shows, the men who made their fortunes there embraced a legacy of the nineteenth-century American West, complete with sprawling ranches and cowboy boots, ten gallon hats, and art collections featuring the work of Frederic Remington. They became iconic in the American imagination of the time, and they—and their money—played an increasingly prominent role in U.S. politics. By the summer of 1980, much of the country—and, Merrill notes, the world—was focused on an urgent question at the intersection of oil, money, and Texan intrigue: Who shot J. R.? (J. R. Ewing, fictional head of

---

2. Other well-known chapters in the long struggle against overproduction include Standard Oil's efforts to establish and maintain a monopoly under John D. Rockefeller and, decades later, the founding of the Organization of Petroleum Exporting Countries (OPEC) cartel. Much of the "special relationship" between the United States and Saudi Arabia in the Cold War era and beyond rested on the fact that the Kingdom's oil reserves and production capacity were so massive that they made it a swing producer, able, at least in theory, to intervene in crises of overproduction by adjusting its output to move global oil prices in one direction or another (see, for instance, Mitchell 2011, 200–30). Pervasive discourses that predict the depletion of the earth's oil in the near- to middle-term future—such as those that circulated in the early 1970s oil crisis or in recent worries about global peak oil—also feed the sense that oil is scarce and help to keep profits up. Gary Bowden (1985) argues that shifting oil industry reserve estimates in the 1970s tracked closely with, and indeed supported, industry strategies for capital accumulation; see Bridge and Wood (2010) for a study of corporate strategies in relationship to current discussions of peak oil and Retort (2005, 38–77) for a sophisticated account of oil's relative scarcity and industry accumulation strategies in the twenty-first century. Labban 2008 theorizes the relationship between the production of oil scarcity and space, and Sabin 2004 provides a view on these issues from another U.S. oil region—California.

Ewing Oil, was a character in the popular television series *Dallas*; his mysterious shooting at the very end of season three had audiences debating which of his many enemies pulled the trigger.)

In 1929, following a remarkable boom over the previous decade, Venezuela became the largest exporter of oil in the world, and the second largest producer (after the United States). Venezuelan oil production began in Mene Grande in the west in 1914, and spread around the shores of nearby Lake Maracaibo throughout the 1920s. Foreign companies, chiefly subsidiaries of Standard Oil and Royal Dutch Shell, ran this oil production, and their expanding operations quickly began to transform Venezuela. At the national level in the 1920s, the companies' enormous profits and search for new land on which to drill set in motion processes of state formation in what had previously been a highly decentralized political order. The Venezuelan state, often indistinguishable in practice from the person of President Juan Vicente Gómez, became the increasingly wealthy landlord to foreign corporations—a central dynamic in "nature-exporting" states, including petrostates (Coronil 1997). At the subnational level, Venezuela's oil-producing regions hosted a vast number of oil camps, populated by foreign supervisors and Venezuelan or migrant laborers. Miguel Tinker Salas (2009; see also Coronil 1997, 109–10) argues that these oil camps were the drivers of social and cultural change far beyond their heavily patrolled borders. The racial hierarchy of the oil camps—with white managers at the top and Venezuelan and migrant workers arrayed in ranked jobs beneath them—began to transform concepts of race and identity throughout the country. Venezuelans learned the rhythms of capitalist labor and time discipline at the oil camps, where foreign managers taught and expected efficiency and timeliness. "[The oil camps'] presence," Tinker Salas writes, "dramatically altered the panorama of the Venezuelan countryside and even major urban areas, inaugurating new residential prototypes, consumption patterns, and forms of social organization, influencing fashion, leisure, sports, and diet" (2009, 170). These new patterns and practices spread, albeit unevenly, from the oil camps around Lake Maracaibo to the remainder of Venezuela. They had echoes and refractions in the twentieth-century "oil encounter" (Ghosh 1992) between wealthy international corporations based in the United States and Europe and local populations around the world.

In April 1929, oil was discovered in the Perm region of the Soviet Union. A team led by geologist P. I. Preobrazhenskii from the Geology Faculty at Perm State University had discovered large deposits of potash in the north of the Perm region in 1925 and was searching for new deposits in similar geological structures near the town of Verkhnechusovskie Gorodki, some fifty kilometers from Perm. With no results after drilling down to the level at which they expected to find potash, Preobrazhenskii and his team pushed deeper for exploratory purposes and to test the limits of their equipment (Kurbatova 2006, 84). On April 16, oil gushed to the surface from a depth of 328 meters. By May 5, *Pravda* was proclaiming the "Enormous Promise of a New Oil Region"; on May 7, the Supreme Soviet of the National Economy (VSNKh) noted the major significance of discovering oil in a region that was already home to significant metallurgical factories; and on May 18, the VSNKh ordered the creation of a special bureau, Uralneft, to oversee and intensify the quest for oil in the Urals (Vikkel′, Fedotova, and Iuzifovich 2009, 25–26).

The oil that flowed from Verkhnechusovskie Gorodki was the world's first fully socialist oil: discovered, developed, and flowing through a set of political, economic, social, and cultural configurations that differed radically from the capitalist oil regions of Texas–Oklahoma in the United States, Lake Maracaibo in western Venezuela, and, indeed, all other oil regions discovered to date.[3] The Soviet Union was, of course, heir to the massive oil deposits of Baku and Groznyi that had been developed in the prerevolutionary period (Tolf 1976), but oil from the newly discovered Volga-Urals basin fed the world's first oil region that was built up from scratch in the encompassing context of socialist political economy and socialist social and cultural construction.[4] As the VSNKh's swift actions following Preobrazhenskii's discovery indicate, oil was understood to be of major significance to the socialist project writ large. Without oil, neither the Soviet Union's signature transformations of collectivization and industrialization nor its aspirations to military might—nor, decades later, its expansionist aims in Eastern Europe (which had little native oil resources)—could be realized. As Lazar Kaganovich, Stalin's deputy and People's Commissar of the Oil Industry from 1939 to 1940, put it at the Eighteenth Congress of the Communist Party in 1939, "Everyone understands that without oil there is no tractor. And if there is no tractor there is no wheat or cotton. Without oil there are no cars, no aviation" (quoted in Slavkina 2007, 44).

Kaganovich's declaration notwithstanding, the most remarkable aspect of socialist oil was its comparative lack of significance in the Soviet Union, relative both to other sectors of the Soviet economy and to the capitalist oil industry. Compared to the political influence of Texas oilmen and Soviet heavy industry,

3. In choosing an analytic strategy that contrasts capitalist and socialist oil complexes, I am mindful that stories emphasizing commonalities might also be told to great effect—an analytic strategy pursued nowhere better than in the work of Kate Brown (2013) on Soviet and U.S. atomic cities. I do not deny that important similarities might be explored in the case of oil as well, ranging from technological practices to imaginations of modernity to the role of oil revenues in federal budgets. My claim is not that there were no similarities but, rather, that recapturing some of the differences and distinctions of the socialist oil complex considerably improves our understanding of both twentieth-century oil in global contexts and, most crucially for my argument here, the transformations of the 1990s and 2000s in Russia itself.

4. The labor movement in the Baku oilfields had helped launch Joseph Stalin's career as a revolutionary in 1907–8 (Suny 1972), and a significant number of former oil workers from Baku went on to noteworthy careers in the Soviet Union (Sara Brinegar, personal communication).

socialist oil production associations had little clout in regional or national affairs—especially when, as was the case in the Volga-Urals basin, they were not the only industry around. Compared to the prestige and exemplary status that accrued to Venezuela's oil camps and Soviet factory labor, the Perm region's oil workers were held in low esteem. Recalling her Soviet-era opinion of the oil industry for me one day, an acquaintance who had worked in a prestigious Soviet armaments factory in Perm wrinkled her nose with disdain: "An oil worker, an *oil worker*? Who's that?" (Her phrasing also accurately captures the place of oil production in the social and cultural history of the Soviet Union and Eastern Europe until very recently, even as scholarship on the heavily oil-dependent projects of collectivization, industrialization, and militarization fills entire library sections.)[5]

In order to unravel the riddle of the critical Soviet industry that nevertheless enjoyed little political clout and not much prestige, this chapter moves through a number of aspects of what I will call, adapting a term from geographer Michael Watts, the socialist oil complex.[6] I include within the socialist oil complex elements that range from production plans issued by Moscow in the late 1920s to ecological activism directed against the oil industry at the very end of the Soviet period. A lone chapter covering some sixty years in a single region cannot hope to contribute more than signposts for a more comprehensive social and cultural history of oil in the Soviet Union. Rather, as my opening examples from the United States and Venezuela indicate, my aims are analytical and typological. I use a series of targeted examples from the Soviet Perm region in support of the overall claim that, despite important differences among time periods and across regions, it is useful to speak of a single socialist oil complex that had a number of interacting elements and that differed in fundamental ways from the capitalist oil complex. Charting this socialist oil complex both expands our catalog of the ways in which oil has entered human social, cultural, economic, and political configurations and serves as crucial background for my analysis of the 1990s and 2000, when the oil industry *did* become politically powerful and prestigious in the Perm region.

Giving an account of the socialist oil complex from today's vantage point requires reading both with and against a large number of recent histories of oil in the Perm region produced by local scholars, journalists, and culture workers.

5. The major exception is Sara Brinegar's work on oil in revolutionary and early Soviet Azerbaijan (2014). See Slavkina (2007, 13–36) for one of the best overviews of the Russian-language historiography related to oil.

6. See, for example, Watts 2004. In more recent publications, Watts writes persuasively of the "oil assemblage" as a way to link elements of the oil industry itself, including wells, flow stations, tankers, and so on, to "regimes of life and death in the postcolonial South and the advanced capitalist North" (2012, 440); see also Watts, Appel, and Mason 2015. Although this is close to what I have in mind for the socialist world, I stick with the earlier and less marked "oil complex," in part because using "assemblage" would necessitate substantial engagement with burgeoning anthropological uses of this term (e.g., Ong and Collier 2004) that would, however instructive, distract from my purposes here.

I read with these new histories—memoirs, oral histories, archival publications, and other documents—because they provide an enormous amount of instructive information about oil in the Soviet Perm region. I read against them because I understand most of them to be, as I show in more detail in chapter 6, artifacts of a concerted and well-funded effort on the part of Lukoil-Perm (beginning around 2001) to create a far more prestigious history for oil in the Perm region than would have rung true in the socialist period itself. The first step in unraveling this chapter's riddle is to be skeptical that oil really was as important in the Soviet Perm region as it is today. For reasons that have to do with the workings of socialist political economy and its accompanying cultural sensibilities, it was not.

Oil Production and Socialist Shortage

The amount and pace of oil production in the Perm region depended crucially on socialist central planning, which generated binding production targets for each brigade and each deposit as well as for the Perm region and the Volga-Urals oil basin as a whole. These plan targets were never far from the consciousness of oil workers, their managers, and higher-level party-state representatives. Some sense of the urgency of fulfilling the plan is conveyed by the recollections of two oil workers in Chernushka, in the south of the Perm region, in the early 1950s: "The country demanded oil. The demand on oil workers was enormous," recalled one. "It was not always possible to correctly develop a deposit because we were always being rushed. They demanded so much oil from us . . . 'The plan—at all costs,' " said another (Bondarenko 2003, 50). In principle, these plans aimed to coordinate activity across all economic sectors of the Soviet Union in ways that would advance the party-state's overall goals and strategies, which were set out in five-year plans. Despite the considerable attention devoted to planning, however, there never seemed to be enough oil. Oil shortages were pervasive, not only at production sites like Chernushka but at refineries and distribution centers as well. Socialism, that is, produced its own kind of oil scarcity, one quite different from the succession of corporate-cum-state strategies to avert crises of overproduction that have characterized the history of capitalist oil. Attending to the dynamic between socialist planning and socialist shortage illuminates a number of important dimensions of the socialist oil complex.

Mobilizing for Oil

The accidental discovery of oil in Verkhnechusovskie Gorodki in 1929 came only months before Stalin's "Great Break": an end to the hybrid state capitalist system of the New Economic Policy (NEP) of the 1920s and an all-out effort to build socialism through collectivization and industrialization. The tantalizing possibility of new oilfields close to the heart of a region that was central to the Soviet Union's industrialization plans—and far from the oilfields of the Caucasus, with their geographical vulnerability and thoroughly capitalist history—meant that Permian oil was rapidly incorporated into the mobilizations of the time. At political rallies in Perm, placards proclaimed, "We will wake the sleeping subsoil" and crowds waved small bottles containing samples of the first oil in the Urals (Vikkel′, Fedotova, and Iuzifovich 2009, 27). Stalin devoted considerable personal attention to developing oil in the Urals and was in the habit of making personal inquiries to ensure that oil exploration in the region received everything it needed from central planners (Slavkina 2007, 59). Plans called for forty to eighty wells within a year (Igolkin 2005, 132–33), and, by 1930, rail lines were extended to Verkhnechusovskie Gorodki to speedily deliver additional drilling infrastructure—some of it ordered from as far away as the United States—and to transport the extracted oil away for refining. Over seven hundred specialists in oil prospecting and drilling were brought to the Urals from Baku and Groznyi, including a number of the most experienced oilmen and high-level party members, quickly bringing the number of employees in the newly formed Uralneft trust to more than three thousand (Markelova 2004; see also Vorob′eva 2000). Some concentrated on drilling additional wells in and around the initial strike at Verkhnechusovskie Gorodki, and the number of wells rose to fifty by 1932. Others fanned out into dozens of brigades seeking oil throughout the area between the Volga and Kama Rivers. The new oil region also received its first plan from Moscow: five million tons of oil extracted by the end of the First Five-Year Plan, or by 1933 (Gasheva and Mikhailiuk 1999, 11).

This figure was ambitious from the start (indeed, it was close to the plan target for the well-established oilfields in Groznyi). The high number is likely due not just to federal ambitions but to regional lobbying as well. Economic administrators in the Urals were eager to win federal investments and to outcompete other mining-metallurgical regions (especially Ukraine), and overstated their own potential, especially during the First Five-Year Plan (Harris 1999). The plan faltered quickly when the oil deposit at Verkhnechusovskie Gorodki turned out to be shallow, low pressure, and low quality. In its most productive year, 1933, the deposit produced only fifteen thousand tons, and production declined steadily from there until it was abandoned for good in 1945. In the early 1930s, some additional small oil deposits were found south of Perm, but there was considerable doubt about whether the Urals basin would, in fact, turn out to be the major oil region that some geologists had predicted. New hope was kindled in July 1934, when another accidental discovery lit up telegram lines, this one in the city of Krasnokamsk, fifty kilometers to the west of Perm, where an artesian well being drilled to support a new paper factory struck oil at a depth of 190 meters. As was the case in Verkhnechusovskie Gorodki, specialists and new equipment immediately poured into the city, with drilling teams often housed in temporary barracks or boarding with peasant families on the outskirts.

Again, though, the first oil at this location proved to be less extensive and of much lower quality than had been hoped. By late 1934 and over the course of 1935, many geologists reached the conclusion that Krasnokamsk was another false lead, perhaps even proof that major oil deposits would never be found in the Urals. Moscow ordered a halt to drilling operations around Krasnokamsk in early 1936, but Nikolai Gerasimov, head geologist, decided that his team would nevertheless keep drilling on a single well—Number Seven, one of the deepest planned for the area (Abaturova 2003, 12–13)—and they struck a larger and higher-quality oil deposit at 934 meters on April 2, 1936. This find was taken as decisive proof of new and plentiful oil in the Urals, and by 1938, so many new wells were exploiting this deposit, with more planned, that high-level officials briefly considered moving the entire city of Krasnokamsk to make room for oil production facilities, planned factories, and small refineries (Vikkel′, Fedotova, and Iuzifovich 2009, 38). Instead, planners settled on a series of wells and derricks arrayed along some of the city's main streets. The Soviet base for oil exploration and production operations in the Volga-Urals basin moved to Krasnokamsk and remained there for decades.

In 1939, with production steadily increasing in Krasnokamsk, the Eighteenth Congress of the Communist Party officially prioritized the development of the Volga-Urals oil basin, with V. N. Molotov proclaiming that, "Creating a new oil base—a Second Baku—in the region between the Volga and the Urals is considered a first-order and urgent state project" (cited in Igolkin 2005, 155). An article in *Pravda* at the time captures some of the intensity of the campaign to develop oil in the Urals:

> The country will not spare strength and resources to create, at a pace that old capitalist Baku never knew, a mighty new oil base in the east, essentially in the center of the country. It is hard to overvalue the enormous political, economic, and defense implications [of this effort].[7]

Indeed, oil operations in the Volga-Urals basin intensified just in time to offer a boost to the Soviet effort in World War II, by providing both a reserve (more hoped for than actual) if Baku should fall into German hands and, as the war stretched on, modest additional supplies to the front. As evacuees from the west-

7. "Vtoroe Baku." *Pravda*, August 7, 1939.

ern provinces flowed into the Perm region, refined oil flowed out, much of it from the ever more intensively exploited deposits in and around Krasnokamsk. With the help of evacuated drilling experts from Baku and Groznyi—who arrived with four thousand tons of drilling equipment transported by steamship (Vikkel′, Fedotova, and Iuzifovich 2009, 46)—Krasnokamsk-based teams set to breaking records for the number of meters drilled. The prominence of Volga-Urals development continued after the war, with the Fourth Five-Year Plan (1946–50) placing its development as a higher priority than any other oil-producing region (Slavkina 2007, 55).[8] In a major speech in 1946, Stalin set the goal of 60 million tons of oil production a year by 1960 as one of the conditions for the Soviet Union begin able to feel itself safe from all unexpected events (production was, at that time, at only 19.4 million tons, down considerably from a prewar high of just over 30 million tons). In the postwar period, the management of the country's oil production was placed in the portfolio of Stalin's Deputy Premier L. P. Beria, a graduate of the Baku Polytechnicum, which specialized in preparing workers for the oil industry (Slavkina 2007, 60).

The first decades of oil exploration and drilling in the Perm region, then, unfolded at an astounding pace, driven by ambitious domestic plans and external threats, with the fate of the Soviet Union's socialist experiment understood to be hanging in the balance. Capitalist efforts to establish control over oil deposits, whether to secure access or to forestall crises of overproduction, have often been accompanied by state and/or corporate violence—at times in defense of oilfields themselves (recall the National Guard summoned to the East Texas oilfields in 1931); at times in violent clashes between management and labor such as those portrayed in Upton Sinclair's muckraking *Oil!*; and at times in broader insurgencies, interstate wars, and simmering international conflicts.[9] In the Stalinist thirties, the questions of whether, when, and where more oil would be found—one dimension of socialist-style oil scarcity—could be the difference between freedom and imprisonment, and in some cases life or death, for everyone from geologists to drilling engineers.

---

8. For a region-by-region account of the development of the Volga-Urals basin that situates the history of the Perm region within the context of exploration and discovery in the Tatar and Bashkir ASSRs, see Trofimuk 1957, 97–142. The Tatar ASSR was home to the largest oilfield in the Volga-Urals basin, the Romashkino field, discovered in 1948. See also Igolkin 2005, 146–54 and Slavkina 2007, 48–94.

9. On oil and violence in the capitalist world, see, for example, Watts 2001, 2008 and Huber 2011. Among the most thought-provoking understandings of the relationship of oil and violence in capitalist contexts is Nitzan and Bichler's *The Global Political Economy of Israel* (2002), which tracks the interaction between armaments corporations and oil corporations (the "weapondollar-petrodollar coalition") in the global capitalist system, and theorizes its relationship to Middle Eastern wars in the second half of the twentieth century.

The fate of many geologists, for instance, depended on the discovery of oil in the Urals. In the 1920s and early 1930s, the Soviet geological establishment was divided, like many academic disciplines, between the established elite of the prerevolutionary academy and a new generation of revolutionaries.[10] One of the sharpest divides had to do with the question of where in the Soviet Union it was most promising to seek out new oilfields. On one side of this issue stood Ivan Gubkin, a staunch revolutionary and deep admirer of Stalin, whose preferred geological theory predicted that the Volga-Urals basin would yield an enormous bounty of high-quality oil. Other geologists were skeptical, viewing exploration in the Urals as a waste of time and money. As the seesaw of reports from Verkhnechusovskie Gorodki and Krasnokamsk filtered up to Moscow between 1929 and 1936, these groups fell in and out of favor. Gubkin proclaimed each new discovery as vindication of his theories, whereas his opponents pointed to each failed well as evidence that what oil was to be found in the Volga-Urals basin was "dead" oil and that exploration efforts should be directed elsewhere.

The series of high-quality oil discoveries around Krasnokamsk eventually vindicated Gubkin's position, and he was appointed head of a special commission charged with making recommendations for the future development of oil in the Volga-Urals basin. Gubkin died before he could complete work on the commission's report, but a portion of his late writings were published by some of his colleagues and students in 1940. The introduction to these papers was unsparing in its criticisms of Gubkin's opponents and shows the geology of oil clearly caught up in the nexus of early socialist planning and state violence:

> The historic decision of the Eighteenth Congress of our Party, which decreed that the creation of a 'Second Baku' is the decisive task for Stalin's Third Five-Year Plan, reached Ivan Mikhailovich [Gubkin] on his deathbed. The decision brought an end to the titanic battle that Ivan Mikhailovich, along with the best of the Soviet geological sciences, fought against the many skeptics and enemies of the people who sought to discredit the presence of oil in the Volga-Urals basin with all of their strength. (Gubkin 1940, 5)

The stakes were high for those lower down the hierarchy as well. By early 1936, Gubkin's opponents had persuaded Soviet leadership that additional exploration outside of Krasnokamsk was fruitless. The Perm-based geologist Nikolai Gerasimov, who had been placed in charge of the site, was ordered to cease operations and report immediately to Moscow. Even as he traveled to Moscow to appear

---

10. The Geological Commission had already been rocked by a series of arrests and investigations in 1929–30, known as the Geolkom Affair, in which a number of leading geologists were arrested and sentenced to lengthy prison terms (see Zabolotskii 1999).

before the Council of People's Commissars, Gerasimov allowed drilling to continue on the Number Seven well outside of Krasnokamsk, reasoning that a last-minute oil discovery was his only hope of avoiding arrest and trial. Gerasimov brought an assistant with him to Moscow and stationed him near the telegraph station closest to his hearing room. Astonishingly, the telegram announcing that the Number Seven well had struck oil arrived in the middle of Gerasimov's grilling. The assistant burst into the meeting room with the telegram in hand and, he later recalled, "Gerasimov was saved" (see Vikkel′, Fedotova, and Iuzifovich 2009, 37–38; Abaturova 2003, 12).

Gerasimov had every reason to have been worried. The second year of the Second Five-Year Plan and the height of the Great Terror, 1937, was fateful for much of the central leadership of the oil industry, as it became clear that the actual production figures would not come close to the ambitious plans set by central and regional planners. Dozens of geologists and bureaucrats at Glavneft were arrested and tried (Igolkin 2005, 97–115). In Krasnokamsk alone, 750 arrests were made in 1936–37, with most of those arrested for counterrevolutionary activity or wrecking—catchall terms that often amounted, in the Urals, to scapegoating for the underfulfillment of plans (Harris 1999, 170–71). Analysis of the secret police files of those arrested in Krasnokamsk when they were opened in the post-Soviet period, however, yielded an interesting result: not a single oil worker was, in the end, executed. Nearly all returned to their homes and work within months or years—a testament, it seems, to the value that was placed on their skills in alleviating what continued to be a pervasive sense that there was not enough oil (Abaturova 2003, 18–19).

Plenty and Scarcity in the Second Baku

Although wartime production concentrated on exploiting the known deposits around Krasnokamsk as thoroughly as possible, attention returned to exploratory drilling shortly after World War II. By 1948, a dozen geological-exploratory parties were searching for oil up and down the Perm region and neighboring regions (Vikkel′, Fedotova, and Iuzifovich 2009, 56). Following discoveries in the Tatar and Bashkir Autonomous Soviet Socialist Republics (ASSRs), the plans for drilling in the Perm region called for deeper wells that would tap older, Devonian geological formations. (The region's earliest discoveries had come from shallower and smaller Permian and Carboniferous deposits.) A number of former armaments factories in the Urals were tasked with producing the new materials and equipment—especially steel pipes and casings—that would be necessary for the oil industry to drill deeper and deeper wells (Igolkin 2009, 158–60). These wells struck deposit after deposit in the southern reaches of the Perm region in the

1950s—in and around Chernushka, Osa, Kueda, and other small towns in what would soon become the Perm region's major oil production centers. Between 1945 and 1969, seventy-two oil deposits were discovered in the Perm region, and they were tapped by over eight hundred wells. Production soared, reaching a yearly peak of nearly twenty-six million tons in 1977—up from just over two million in 1960—and helping to realize Ivan Gubkin's dream that the Volga-Urals basin would become the Soviet Union's Second Baku.

But still there was never enough oil because socialist shortage was a matter neither of absolute scarcity fixable by discovering more oil nor of overly optimistic and impossible-to-meet plans. It emerged, rather, from the systemic practical workings of centrally planned socialist economies, where everyday strategies devoted to realizing, faking, negotiating, or otherwise coping with the plan resulted in bottlenecks, hidden resources, and constant readjustments and extra-plan exchanges across and among all enterprises and all sectors. These processes were systemic, neither consciously initiated by nor stoppable by the efforts of any single actor at any point or any level. In Hungarian economist Janos Kornai's famous phrasing, socialist economics *was* the economics of shortage (Kornai 1980).[11]

One of the most often-noted ways in which central planning shaped socialist-style oil shortage was in plans and incentives for the drilling of new wells.[12] After negotiations between central and regional planners arrived at a total desired output for oil, numbers were set for how much of that amount would come from established wells and how much from newly drilled wells, given estimates of known productivity and required depth of wells in a certain region. These numbers were then broken down further and assigned to individual oil production units. Out in the oilfields in practice, this method of planning produced results that exacerbated shortage even as they increased production. At the level of the drilling brigade, plans were set in number of meters, which incentivized brigades to maximize the distance they drilled. Teams often drilled a large number of shallow wells, racking up meters in the layers of the subsoil that were easiest to move through and ignoring, deferring, or abandoning deeper, more demanding wells. On other occasions, wells were overdrilled in order to accumulate meters rather than spend time moving equipment to another location. Brigades often did not stop drilling to test or properly case their wells—a time-consuming process that would reduce their total meters drilled per plan period.

---

11. In a very large literature on socialist shortage, see Berliner 1957 and Burawoy and Lukacs 1992 for examples of firm-level plan negotiations and manipulations of the sort that characterized oil production in the Soviet Union. On the thoroughly political nature of planning, shortage, and their workarounds, see, for instance, Rutland 1985.

12. The paragraphs below draw especially on Campbell 1968, 87–120; Campbell 1976, 14–25; Gustafson 1989; and Moser 2009, 109–14.

Drilling for oil thus took on the characteristic rhythms of socialist production. Teams raced to meet plans at the end of the month and year ("storming"). They kept a reserve of production by underreporting meters (and even entire wells) drilled; these meters could be called up in case of unexpected deficits in future planning periods ("hoarding"). The hoarding of meters drilled to apply against future plan targets was particularly useful, given that factories were poorly incentivized to produce the high-quality steel pipe necessary for deep drilling into Devonian deposits, and often provided pipes with major imperfections—or no pipes at all (Campbell 1968, 92–93; Goldman 1980, 36–37). Analogous strategies prevailed in other brigades—specializing in repair, production, and so on—with the net result that it was impossible for planners to get a good fix on how much oil there was. This difficulty was likely even more acute in the oil industry than in the rest of the energy sector, for predicting and charting the location of oil deposits—as the parade of dry wells around the Perm region in the 1930s attests—was an inherently harder and less predictable business than, for instance, mapping the extent of a coal deposit or predicting the energy output of a hydroelectric dam. The permanent contingency of oil production was a constant headache for socialist planners.[13]

Given all this, it is not surprising that the negotiation and improvisation of plans make up a significant part of the recollections that Soviet-era oil workers provided to local historians. One oil worker recalled the conditions of work in the early 1950s at the time of the discovery of the first oil near Chernushka:

> Moscow determined the regions for drilling—in the Ministry there was an enormous geological division. From there came the orders: "Here's a square near the town of Demenovo. Drill!" And you look at the spot: here's a swamp, there's a forest, it's far from water. How to drill there? So we picked the spot ourselves, and then sent it to Moscow for approval at the Ministry and the personal signature of Stalin. Yes, Iosif Vissarionovich. All of the land was collective farm land, and the basis for transforming it into oil prospecting required the personal signature of the highest authority. (Bondarenko 2003, 26)

If drilling brigades could use their own local knowledge to adjust some of the precise details of drilling, the oil industry had a much harder time bargaining its overall plans for the amount of oil it was to extract. No sooner had production levels hit twenty-six million tons in 1976 than a target of twenty-eight million tons appeared for 1980. According to one account, a candidate for the Politburo who was visiting Perm then looked over the production and planning

13. I am grateful to Peter Rutland for this phrasing.

figures, pronounced, "that number isn't so round," and bumped the target up to thirty million tons (Gasheva and Mikhailiuk 1999, 179). In some ways, the Perm region was a victim of its own success. The massive new finds of the 1950s and 1960s convinced some central planners that quotas throughout the 1970s could be extrapolated from past decades. By this time, however, production in the Perm region had already peaked and begun to decline, and local geologists and oil industry specialists were unable to convince their higher-ups that slower production was possible or, given the status of the fields over the longer term, wise.

The constant drive to drill new wells and open new fields, coupled with unceasing plan bargaining and shortages, had also created a maintenance and repair nightmare. Of the 2,200 wells in the Perm region in the mid-1970s, only half were actually operational (Vikkel′, Fedotova, and Iuzifovich 2009, 81). In 1978, not a single oil-producing division in the Perm region hit its plan target. A report from the Communist Party leadership at the Krasnokamsk division explained things this way:

> The plan for 1978 was fulfilled at 83.3 percent. The main reasons for the nonfulfillment of the state plan were the waterflooding of wells and the high level of water pipe and oil pipeline accidents . . . an enormous quantity of wells are idle, and the necessary drilling liquids are lacking. We are losing 360 tons of oil every day. (Abaturova 2003, 113)

As total production figures in the Perm region tapered off in the middle of the 1970s, many experienced oil workers were fired for not meeting plans. As one party organizer attached to the Chernushka division put it, recalling what he viewed as the unfair firing of a talented division head:

> Permneft [the regional production association] simply overloaded the production situation . . . major new deposits were not being opened, and the old ones were being filled with water. And the plan was being raised! Thirty million tons of oil was required. Such an insane number should never have been chosen. Thirty million! Where is it? There was never that kind of oil here. It was impossible to pump thirty million. (Kurbatova 2006, 430)

At best, it was possible to retroactively ensure that the shortage was blamed for the lack of physical reserves in the ground, rather than on the oil workers themselves. V. D. Viktorin, head of geology and oil deposit development at Permneft described his efforts to do just this in the late 1970s:

> The state plan was raised. But we couldn't change the plan, as it had the force of law. The one thing we could do was establish the assignment

> (*zadanie*) lower than the plan, in order to evaluate everyone, including prizes for workers, in relationship to the assignment. But for that you needed the permission of the Council of Ministers of the USSR. . . . I was at the Council of Ministers for three days. I was told: "You have five lines to summarize your point." And I created a proposition that ultimately went into the decisions of the Council of Ministers of the USSR: "In connection with the non-confirmation of reserves of oil, permit the Gosplan and the USSR Ministry of the Oil Industry to set the assignment for Permneft below the state plan for oil extraction." So at the end of the five-year plan, the oil workers did not fulfill their plan, but they did fulfill their assignment. (Kurbatova 2006, 431)

Socialist-style shortages of oil thus grew out of millions of daily interactions that revolved around the negotiations of central planning. They cascaded quickly, and massive inter-enterprise and interpersonal networks of exchanges grew up out of the necessity of locating and trading oil and oil products. This was socialism's famous "second economy," where goods and favors of all sorts moved among friends, contacts, and business associates in lengthy and elaborate chains. All that I have described here—from the various machinations involved in increasing meters drilled (with little regard to the productivity of the resulting wells) to the fine distinctions between not fulfilling a plan versus not fulfilling an assignment—appear not just inefficient but downright perverse to capitalist eyes. But summoning the Texas National Guard to *stop* the flow of crude from East Texas oilfields would have appeared no less incomprehensible from the perspective of socialist planners in the all-out mobilizations of the late 1920s and 1930s. Mid-twentieth-century capitalist and socialist oil complexes both produced a relative scarcity of oil, but did so in very different ways.

Oil and Allocative Power

With all this plan bargaining, hoarding, and extra-plan activity, it should be evident that socialist party-state power did not work through the top-down, "totalitarian" control of the economy. A different way to conceptualize party-state power and its implications for the socialist oil complex is therefore necessary. Building on several discussions of the workings of socialist systems in practice, Katherine Verdery (1991, 74–83) argued that the central drive of socialist political and economic organization was toward the *maximization of the capacity to redistribute* (see also Konrád and Szelényi 1979). She termed this "allocative power" and distinguished it from both a drive to maximize profit (characteristic of capitalist systems) and a simple drive to maximize the quantity of resources under control. The field of allocative power was the primary terrain of struggle in socialist societies, both within the planning bureaucracy itself—as planners and enterprise directors at different levels and in different sectors sought to increase their capacity to allocate relative to one another—and between the party-state as a whole and its citizens, who often registered discontent that not enough was being allocated in their direction.

To accumulate allocative power was not, it is important to note, the same as actually allocating; indeed, it often meant hoarding as much as possible and only selectively alleviating shortage. Moreover, because allocative power was about the capacity to distribute, it congealed with special density in particular sectors such as heavy industry, where the means of production were themselves produced. In other words, more allocative power could be amassed by controlling the capacity to produce machines that produced other goods than by controlling the capacity to produce finished goods. And there was not much power at all to be accumulated at the points of distribution and consumption; after all, as soon as something was consumed, it ceased to exist on the field of struggles for allocative power. What, then, of the oil sector, which would seem to be involved in turning out a crucial means of production—essential, as Lazar Kaganovich said, for everything from tractors to trucks to airplanes? This question has three interlocking answers that take us close to the heart of the socialist oil complex and its relative lack of power and prestige.

First, oil production is highly dependent on specialized machinery and equipment, particularly for the exploitation of deeper and/or depleted deposits. In capitalist contexts, acquiring this specialized equipment is generally not a problem. Oil companies pay top dollar for it, and so, therefore, do consumers, through prices at the pump or through tax breaks for oil companies. The Soviet oil production industry, however, depended on Soviet heavy industry for the provision of steel pipe, drilling bits, motors, and many other items of basic industrial equipment—recall that shortages of high-quality steel pipe plagued the opening of the deep wells of the southern Perm region in the years after World War II. With far greater stores of allocative power located in heavy industry, planners and procurement specialists in the oil sector competed with their counterparts in other sectors to push the items they needed higher on the list of what would be produced. Would steel mills and machine shops produce drilling pipe or tractor frames? In this context, the frequent proclamations by the oil production industry about the fundamental importance of oil to Soviet goals should be understood less as descriptive statements and more as political gambits, as efforts to shake loose those pipes and drill bits—the means of oil production—from the more powerful and stingy stewards of heavy industry (see also Hewett 1984, 11–12).

Second, considering allocative power draws our attention not just to production figures—how much oil was produced—but also to the important questions of where the extracted oil was headed and what segments of the socialist planning apparatus made those determinations. From the 1950s to the end of the Soviet period, a portion of the Perm region's oil was exported abroad, either to refineries in Eastern Europe, especially through the Druzhba pipeline network, or to the capitalist world.[14] Indeed, Western scholars have most often considered Soviet oil production precisely for its effects on global markets and its importance to negotiations among states within the Soviet bloc and the broader global socialist diaspora.[15] For my purposes here, what is important about this exported oil is that its allocation was determined from the very top of the Soviet planning bureaucracy, in interministry negotiations in Moscow. These negotiations could certainly be influenced by political and economic coalitions based in the regions (see esp. Chung 1987), but the determination of the overall export/domestic mix of oil was far out of the hands of Perm's regional production association. All exports ran through a single entity, Soiuznefteksport (see esp. Goldman 1980, 72–73, 186–91), and the proceeds realized from international exchanges—whether bartered goods or hard currency—went into the federal budget and did not return directly to Permneft. However important they were for the Soviet budget at the federal level—and many observers claim they were increasingly crucial as the twentieth century wore on (e.g., Kotkin 2001; Gaidar 2007)—these exports controlled by Moscow simply fell out of the struggle for allocative power at the internal regional level. They certainly affected Permneft indirectly, as central planners pushed for more oil to export by raising plan targets and sought, at least in the official plan, to provide the equipment that would enhance production. But, as we have seen, the official plan was a weak mechanism for generating outcomes, and none of this gave the leadership of Permneft much in the way of influence in the Perm region where it really mattered: in the ongoing struggle for the capacity to allocate the means of production. Soviet central planning and allocative power, in sum, created a fundamentally different relationship between national center and oil region than we find in the capitalist world, one in which exports *diminished* the power of regional producers by taking oil out of their hands with little in return save for increased plan targets (see also Sanchez-Sibony 2010).

14. Soviet oil exports ran at about 70 percent crude and 30 percent refined oil products during the late Soviet period (Chadwick, Long, and Nissanke 1987, 32)—meaning that, when it came to export, refineries within the Soviet Union were, by and large, not included in the planning equation. On the provision of oil and other energy sources to Eastern European countries as part of larger trade within the Soviet bloc, see Stone 2002.

15. See, for instance, Gustafson 1981, Hewett 1984, and Stern 1987. Rogers 2014a suggests that the standard assumptions of this literature do not account very well for the fact that much of this Soviet oil trade took place on barter terms.

This brings us to the third dimension of oil and allocative power: the movement of oil within the socialist system, from production fields to the engines of Soviet tractors, trucks, airplanes, and furnaces. As M. V. Slavkina (2007, 95–142) shows, based on careful economic and statistical analysis, refined oil from the Volga-Urals basin made possible some of the major transformations of the post-Stalin decades, among them increasingly mechanized agriculture in Khrushchev's turn from collective to state farming and the integration of different regions of the Soviet Union through truck and bus transportation. Although the information remains classified, it is clear that Soviet military operations consumed a gargantuan share of oil produced in this period as well (Slavkina 2007, 46–47). Here, too, however, Permneft as an oil production association was in a comparatively weak position. Even though Permneft's oil was critical to all these aspects of the socialist project, the crude it extracted from the subsoil was fairly useless before it was refined into various types of fuels and other petrochemical products. There is simply not much that can be done with crude oil by itself, and so the capacity to produce crude oil allowed the accumulation of much less allocative power than the capacity to refine that crude into oil products. These refined products, slated for allocation across the socialist world, were the true means of production emerging from the oil sector. In sum, within the socialist oil complex as it worked in practice—at least domestically and especially at the regional level—the greatest potential for accumulating allocative power lay in refineries. Perm was home to a big one.

Socialist Refining and Distribution

In the second half of the nineteenth century in the United States, Standard Oil's empire was built not in oil production but in refining and transport. Unable to deal with persistent problems of overproduction and consequently declining prices by controlling who extracted oil from the ground and how much—there were simply too many independent producers drilling in too many locations—Standard Oil concentrated on acquiring as many refineries as possible, using all the tools in its expansive arsenal to drive competitors out of business and acquire their assets. Daniel Yergin reports in his account of the American "oil wars," that, by 1879, Standard Oil controlled fully 90 percent of the refineries in the United States (1991, 43). This control, in turn, enabled Standard Oil to fight overproduction simply by refusing to accept more oil for refining than it deemed necessary. Acquiring a near-monopoly over refining was one part of Standard Oil's broader strategy of vertical integration: uniting all aspects of the oil industry, from production through refining to sale, into the same corporate structure. In the century and a half since the American oil wars, vertical integration has become standard in the global oil industry.

In the socialist world, where problems of overproduction did not obtain and the scarcity of oil had quite different causes, vertical integration never became a standard practice. Indeed, although the Baku oil industry nationalized by the Bolsheviks in 1920 closely united production, refining, and distribution (see Tolf 1976) and the early 1920s NEP period generally reproduced this structure as Bolsheviks sought international markets for their oil products, the trend over the course of the Soviet period was toward the greater institutional and administrative separation of production from refining and distribution. In the world of capitalist oil, the fact that the same corporations have generally controlled production and refining through complicated subsidiary and holding company arrangements has enabled analysts to speak of a unified oil industry and to place most of their emphasis on production, reserve size estimates, and so on. But vertical integration is a contingent rather than necessary form of oil sector organization. The socialist oil complex demands greater recognition of the fact that oil is, in fact, produced twice—at the point of extraction of crude and again at the point of refining—and that these sites can be very differently situated within broader political and economic structures (cf. Huber 2013, 61–96).

Refining Socialist Oil

In the Soviet 1920s, the basic model of enterprise organization for crucial economic sectors was the state-controlled trust. Trusts contained a number of enterprises within them that focused on a particular commodity, and they owed a great deal to the model of a vertically integrated holding company (see esp. Shearer 1996, 36). There were four oil trusts in the Soviet Union in the NEP era, based in the four oil-producing areas of Baku, Groznyi, Kuban, and Emba (on this period, see Igolkin 1999a; 1999b). The discovery of oil outside of Perm in 1929 called for the creation of a new trust based in this region—Uralneft.

Uralneft (and its successor Prikamneft) included various small enterprises focused both on production and refining, with refineries situated quite close to the oilfields themselves. A small refinery was built in Verkhnechusovskie Gorodki in 1933, with a capacity of only eighty tons of oil a day (Dement′ev 1967, 18). A major reorganization in 1942 resulted in another unified structure in the Perm region (at that point named the Molotov region), called Molotovneftekombinat and reporting to the People's Commissariat of the Oil Industry. Divisions of Molotovneftekombinat included both drilling operations—divided into groups dedicated to exploration, drilling, transport, repair, and other tasks—and refining operations, especially in Krasnokamsk (Abaturova 2003, 42; Kurbatova 2006,

243). Under the Molotovneftekombinat umbrella, a larger refinery was built in Krasnokamsk beginning in 1943, in part with evacuated equipment from the Berdiansk refinery in southeast Ukraine, to help with increased wartime production and the new wells that continued to come online in Krasnokamsk.

The discovery of deeper and more abundant oil in the Volga-Urals basin throughout the 1940s and subsequent decades demanded vastly increased refining capacity. In 1949, the Soviet Council of Ministers announced the construction of a new network of nine major refineries to be built across the Soviet Union. This plan followed a global trend of siting refineries closer to points of intended consumption rather than production (see Sagers 1984, 1–8), and several major refineries were constructed in Eastern Europe at this time as well, along with an extensive pipeline network connecting the Volga-Urals fields to both Eastern Europe and the heavily populated Moscow and Leningrad regions.

One of these new refineries was planned for Perm, and construction began quickly, initially under the supervision of the head of the small Krasnokamsk refinery. Although the original plans called for a new refinery capable of processing three million tons of oil a year from the Perm region and the Tatar and Bashkir ASSRs (including the massive Romashkino field, which began production in 1950), the rapid pace of discovery in the early 1950s convinced planners to double the expected capacity of the refinery a mere three years later. Construction took place throughout the 1950s, in part with the labor of the Perm region's many political prisoners from the Gulag system (Kolbas 2002, 8–9) and, later, Komsomol volunteers from Perm and around the Soviet Union (Kurbatova 2006, 288–89). The first delivery of oil from the Tatar ASSR arrived through a newly constructed pipeline in the fall of 1958, and Perm's new refinery was soon turning out kerosene and diesel as well as fuel for cars and airplanes (Sverkal′tseva 1998, 3–6). As oil production in the Second Baku increased in the region through the 1950s and into the 1960s, Perm's refinery grew in size, technological sophistication, and number of products in which it specialized (see esp. Iuzifovich 2008, 193–318). It became a center not just for refining oil into fuel products but also for the entire petrochemical industry of the Urals.

Another institutional reorganization took place in 1976, with the establishment of Soiuznefteorgsintez, an all-Union industrial association that combined oil refining with the production of all manner of synthetic organic materials. The Perm Oil Refinery was renamed Permneftorgsintez, or PNOS for short, a title it retained into the early post-Soviet years. It was located within the new Ministry of Oil Refining and Petrochemicals—an entirely separate ministry from the Ministry of Oil Production. Following this reorganization, PNOS began adding still more sections and operations, eventually including the production of butanol and other industrial solvents and mineral fertilizers. Fuel oil (*mazut*) remained a primary fuel for Soviet boilers well into the 1980s, constituting about a third of refined petroleum output (Sagers and Tretyakova 1985, 2). Although the oilfields of the Volga-Urals basin had reached their peak output in 1976, PNOS continued to operate at full capacity, refining oil from West Siberian fields that arrived by pipeline. The refinery went from employing 2,583 people in 1959 (Kurbatova 2006, 289) to nearly 15,000 at the end of the Soviet period, and its operations became central to the agricultural, defense, and industrial sectors of the Perm region.

The trend over the socialist period, in sum, was toward greater institutional separation between oil production and refining, with coordination between the two industries taking place—at least on an official level—largely at the highest levels of planning and interministry coordination, and decreasing interconnection among directors and employees at region-level sites of production and refining. Both Permneft and PNOS received their production and processing plans from their parent ministries, where there were often tensions and debates about where and how Soviet oil should flow. Refineries like PNOS were, of course, supply constrained, affected by the unpredictability, socialist rhythms, and shortages in the production fields that fed them and dependent on heavy industry for their own industrial machinery. But their ability to turn crude oil into an increasing number of products that were central to nearby industry, agriculture, and heat and power generation—as well as more distant transport, military, and aviation—made them far more central than production associations to regional networks and power dynamics that gathered around the drive to accumulate allocative power.

Distribution Networks and Consumption

Distribution networks for gasoline, fuel oil, and other oil products destined for consumption were still another matter, managed by still other aspects of the party-state apparatus, and divorced from production and refining from the earliest days of the Soviet Union. The early Soviet oil trusts of the Caucasus, as I have noted, combined production and refining operations. But, as the NEP period gave way to the full-scale building of socialism, the transport and sale of oil products—with its potential for bourgeoisie profit-making—was split off into other organizations more tightly controlled by the party-state bureaucracy (Brinegar 2014, 248–50). This organization continued through to the end of the Soviet period.

In the later Soviet period, the State Committee for the Distribution of Oil Products, Goskomnefteprodukt, managed distribution through rail, some pipeline, and tanker truck networks. It was divided into geographical subsidiaries—

including, in the Perm region, Permnefteprodukt—each of which included its own distribution centers (*neftebazy*) and, especially as car ownership increased in the later Soviet period, gas stations and service garages, or *avtozapravochnye stanstii* (AZS) (Iuzifovich 2008, 3–4; on the Perm region, see Lukoil-Permnefteprodukt 2003). Given the overall socialist economy's focus on production, allocation, and acquisition through planning structures, producers and refiners were even less likely to know or to be particularly concerned with the state-controlled distribution networks through which oil products flowed out of PNOS for public consumption. Indeed, many oil products would have leaked directly out of PNOS through less-than-formal channels before they even got to the official distribution networks.

With public consumption near the bottom of the distribution hierarchy, fuel shortages were acute as well. Even as car ownership—and with it "car culture" (Siegelbaum 2008)—continued to grow, the official state network of gas stations remained far from sufficient. In contrast to crude oil, which was comparatively hard to peddle, gasoline was one of the most common items moving in the late Soviet second economy, siphoned from state-owned vehicles into privately owned cars—and just as likely on again to the cars of neighbors and friends—along the shoulder of every highway and byway. In a study of gasoline in the later Soviet period, Michael Alexeev cataloged the many ways in which the second economy for gasoline worked: through inflated mileage and weight reports filed by official state drivers (reports that purported to have overfulfilled their plans and therefore generated fungible coupons); through gas station attendants' willingness to look the other way when private car owners presented them with state coupons; and through widespread forged and stolen documents. He estimates that, by the mid-1980s, private cars' technically illegal fuel consumption constituted fully 68 percent of all private car fuel consumption, or 7.5 billion liters of gasoline a year (Alexeev 1987, 18). Although we do not have good data, we can assume that gasoline and other oil products were in equally high demand in the developed inter-enterprise second economy as well. Indeed, the large amount of gasoline that moved from state enterprises into private gas tanks meant that enterprises—from state farms to factories—were always looking to acquire more gasoline than the official plan allocated to them.

These aspects of distribution and consumption underscore a crucial aspect of the socialist oil complex: neither production associations like Permneft nor refineries like PNOS were the direct recipients of any of the money paid for the consumption of their crude oil or oil products. Although there were occasional monetary transactions involving oil in the overall circuits I have described—both on international markets and through internal distribution—these transactions were firmly partitioned off from oil sector production associations and refining enterprises. They were decidedly peripheral to the accumulation of allocative power. Socialist oil, that is, never circulated through Soviet society *as money*, and there was no cause to coin a term such as "petrorubles" by way of analogy to "petrodollars." There were, it follows, no oil boomtowns or cities or regions, at least as they were known in the capitalist world. This aspect of socialist oil would become enormously significant for the course of post-Soviet transformations, when the accumulation of money did, slowly, become central to the Perm region.

In the popular Western imagination during the Cold War, stories of socialist shortage were epitomized by long lines for goods and products rather than by difficulties in obtaining quality steel casings for oil wells—that is, by tales of shortage at the point of consumption rather than at the point of production. Scholarly theories of socialism, by contrast, initially emphasized production (Verdery 1991; Burawoy and Lukacs 1994) and have only more recently foregrounded consumption practices (e.g., Fehérváry 2013; Oushakine 2014). Although these more recent approaches to socialist societies illuminate much that was hidden from view in earlier studies, they are far from a replacement. Understanding Soviet oil, I have argued, requires appreciating the ways in which production, refining, *and* distribution/consumption differed from their capitalist counterparts, separately and as integrated elements of a distinctively socialist oil complex. The next sections take up additional dimensions of this chapter's riddle of the crucial socialist sector that nevertheless lacked both political power and broader prestige in the Perm region.

Regional Oil in the Soviet Energy and Industrial Sectors

All the attention lavished on the first discoveries of oil in the Volga-Urals basin and on PNOS notwithstanding, oil was far from the primary energy source of the early Soviet Union. Indeed, the Soviet Union made use of a much more diverse set of energy sources—and for much longer into the twentieth century—than the capitalist world. A key decision for the early Soviet leadership, for instance, was: what kind of fuel would power Lenin's signature goal of electrifying the Soviet countryside? Already in 1920, the priorities were set in the following order: Moscow coal, peat, Urals coal, Donbass coal, firewood, and, only in last place, oil (Igolkin 1999b, 120; see also Brinegar 2014, 85). This ranking of priorities established the tone for energy development for the entire NEP era and, Igolkin suggests, contributed to the lack of oil exploration outside the known oil deposits of the Caucasus until the accidental discovery of oil in Verkhnechusovskie Gorodki in 1929. The Soviet Union was a net importer of oil (much of it from Romania) until the 1950s (Campbell 1976, 11).

This hierarchy of early socialism's energy sources translated into increased emphasis on the coal industry in the Perm region, based around the prerevolutionary coal-mining towns in the Kizel coal basin—the "Furnace of the Urals"—to the northeast of Perm. The first Soviet efforts to industrialize and electrify the Urals thus proceeded not on the basis of oil but through Kizel coal, especially as the electrification of the mines themselves by the mid-1920s allowed for significantly increased production.[16] Even the expansion of the regional oil industry following discoveries in Krasnokamsk and throughout the south of the Perm region in the decades after World War II did not fundamentally change regional expectations about oil's low ranking among regional energy sources. Other, nonoil sources of energy remained crucial and, at times, central to the Perm region for the remainder of the Soviet period, including both the continued exploitation of the Kizel coal basin well into the late Soviet period and hydroelectric power, especially the construction of the Kama Hydroelectric Station, in 1948–59 (the same years as PNOS) to supply the increasing energy demands of Perm's factories. This energy profile tracked well with broader Soviet trends, and the Soviet Union lagged behind the West considerably in embracing oil as a primary fuel source.[17] Oil's place in the overall Soviet energy mix was thus another reason that its producers and refiners were more remote from political power than their counterparts in the West, and for much longer. By 1960, coal continued to compose over half of Soviet energy production, with oil at 29 percent and natural gas at 8 percent. By 1982, following major investments and development initiatives, oil and gas had risen to 70 percent of the Soviet energy output (Hewett 1984, 34), and there continued to be pressure to produce more oil and gas for export (Dienes 1985). However, by this time—the time when the Soviet Union as a whole was most dependent on hard currency earnings from the export of oil—the Volga-Urals oilfields were in steep decline, replaced by the new and highly productive West Siberian oilfields.[18]

---

16. On the history and development of the Kizel coal basin, see Dedov 1959. There was no overall Ministry of Energy in the USSR, only the ministries associated with each energy source itself (oil, coal, gas, nuclear, etc.). At the very end of the Soviet period, from 1989 to 1991, the ministries in charge of oil and gas production were combined into a single federal ministry—the Ministry of the Oil and Gas Industries.

17. On the Soviet energy complex, see also Dienes and Shabad 1979, Campbell 1980, and Hewett 1984.

18. Timothy Mitchell (2011) argues that the capitalist world's shift from reliance on coal to oil as an energy source over the course of the twentieth century was bound up in efforts to roll back the welfare guarantees secured by organized labor in the nineteenth century. Those guarantees had often been won by strike actions directed at the chokepoints of a coal-based economy, especially at railway lines and labor-intensive coal mines that allowed possibilities for collective organizing. In Mitchell's view, oil's profile as an energy source—requiring much less labor and being far more transportable

A Factory Elite

In the context of socialist-style allocative power, in fact, the true pride of the Perm region was never energy of any sort. As in the rest of the Urals, it was metallurgy, defense, and heavy industry. The earliest five-year plans for the Perm region saw the construction and reconstruction of a number of factories in Perm and the surrounding region: a paper factory on the banks of the Kama and a second near Visher in the north (both using the region's abundant forest resources); a metallurgy factory in Chusovoi; a motor factory in Perm; and the intensive development of potash and magnesium mining around Solikamsk and Berezniki in the north (see Kurbatova 2006, 67–140). The construction of a new armaments factory beginning in 1934 continued Perm's prerevolutionary tradition of munitions production.

Although World War II and the attendant threat to Caucasus oil did bring a new emphasis on the Perm region's recently discovered oilfields—despite their still-small production figures—wartime was an even bigger influence on the developing factories and, as a consequence, on the city's labor profile, political elite, and developing self-image. The Motovilikha factories continued to be leading producers of artillery, and newer factories played a major role in Soviet wartime aviation, producing M-82 engines for fighter planes. The Kirov munitions factory became the first of several Soviet factories specializing in mortars and rocket launchers that went by the generic name "Katiusha" (Fedotova 2009, 33–37, 55–70). An additional 124 factories were evacuated to the Perm region from Moscow in the early war years, with 64 of them relocated to the city of Perm itself (Kurbatova 2006, 200–205). In his wide-ranging account of this era in the history of Perm, O. L. Leibovich shows that "the wartime political economy first and foremost served the interests of enterprise directors" (2009, 17)—primarily directors of factories. This was, he goes on to argue, a highly compartmentalized field of industrial, factory-based production, and it was still too early at that time to call Perm an "industrial center" (42–43). That status came later, with continued infrastructural development and postwar consolidation of the region's political elite around these factories.

With an increasing number of major factories, each answering to its own Ministry and negotiating (and fudging) its own plan, the role of region-based Communist Party committees and leaders became crucial. As Jerry Hough's

---

than coal (especially by tanker and pipeline)—helped to facilitate the flow of political power away from labor and back to state–corporate alliances. We do not yet have a study of socialist energy that is comparable to Mitchell's for the capitalist world; existing studies focus on policy decisions or economic outcomes, but provide little insight into the sociotechnical arrangements that are central to Mitchell's argument.

classic study showed, these "Soviet prefects" were among the chief greasers of the Soviet shortage economy at the regional level—cutting deals, pressuring the center for more resources, and mediating among regional enterprises (Hough 1969; see also Rutland 1993). This was the heart of regional allocative power—and neither Permneft nor PNOS were part of that inner circle. In *Wheel of Fortune*, Thane Gustafson notes the lack of vertical organization in the Soviet oil industry—he terms it a "stovepipe system" (2012, 32)—and goes on to ask how, in this kind of organization, any successful coordination among ministries at the regional or local level took place at all. His answer is the same as Hough's and Rutland's: the Communist Party, which was organized into a parallel structure, with representatives in each enterprise and each ministry, as well as comprising powerful regional and national committees. At any point, the regional head of the Communist Party could pick up a direct telephone line—the so-called *vertushka*—and seek to broker a deal among regional enterprises or between regional enterprises and Moscow.

Consider, however, an anecdote about the vertushka that I heard repeated in several contexts during my fieldwork. In the Perm region in the 1960s and 1970s—the heyday of the Second Baku, before the mid-1970s drop-off in production—the head of the Perm Regional Communist Party had a list of numbers next to the telephone on his desk. The numbers would connect him at a moment's notice to the directors of the Perm region's factories of national significance. The list never included Permneft. This is not to say, of course, that the deals brokered by regional Party heads never included the region's oil production association. Surely they did. But the widely circulating story about the absence of Permneft on the vertushka list effectively underscores a point that is crucial to understanding the socialist oil complex: for all the various ways in which oil production was important for the socialist system, the broader workings of that system, especially at the regional level, meant that this importance never translated into prestige or political influence for oil production companies in the regions of their operations. In recounting the vertushka anecdote from the perspective of the post-Soviet oil boom, a time when oil production *had* become absolutely central to regional political economy, one post-Soviet corporate history concludes incredulously: "Imagine that, oil! It's not some D-30K engine" (Vikkel′, Fedotova, and Iuzifovich 2009, 65). Yet a hierarchy in which the regional engine production factory was significantly more prestigious and well connected than the regional oil production enterprise is precisely what we must imagine in the case of socialist oil.

Indeed, none of the leaders of the Soviet era's regional Party committees hailed from the oil industry. Most had risen through the ranks at the Perm region's metallurgical and defense factories or through internal Communist Party structures, with some from the agricultural sector, which employed a significant percentage of the region's population. The major Soviet and immediate post-Soviet political leaders in the Perm region likewise emerged from the region's factories and nonoil mining enterprises. B. V. Konoplev, first secretary of the Perm Regional Committee of the Communist Party from 1972 to 1988, rose through the Party ranks by playing a significant role in the construction of the Kama Hydroelectric Station. B. Iu. Kuznetsov, the first post-Soviet governor of the Perm region, was an engineer, former director of steamship operations on the Kama River, and Communist Party insider. Gennadii Igumenov, who followed Kuznetsov in 1996, started working in the coal mining towns around Kizel. The one partial exception is Iurii Trutnev, Igumenov's successor, who began his career in the oil industry in Polazna; however, by his own and others' accounts—about which more in later chapters—Trutnev's rise to political power began precisely when he left the oil industry for the leadership of the regional Komsomol.

The Metallurgical Subjects of Socialism

The symbolic dimensions of the Soviet project—the ways in which labor, industry, subjectivity, and the country's path to a utopian future were represented in everything from art and film to everyday life and popular culture—followed suit in generating little in the way of prestige or exemplary status for oil and its industry. Images of steel and iron—and not oil gushers and rigs—were everywhere, constituting some of the most basic visual and literary imagery of the early Soviet period and providing an enormous wealth of metaphorical material for the transformations of human beings and society that the socialist project imagined and sought to implement. Rolf Hellebust argues that, "In Soviet literature and culture as a whole, the essential symbol for communist transformation is the metallization of the revolutionary body" (2003, 39). This may overstate the matter, but there is no denying that some of the central imagery of the socialist modernizing project was associated with the transformation of metals rather than hydrocarbons—a sharp contrast to, for instance, Venezuela, where, "Through the old alchemic magic of money . . . oil pulled off the trick of putting 'primitive' Venezuela into a hat and taking it out in the form of an 'oil nation'" (Coronil 1997, 111).

If there was a leading example in the Urals of what Soviet socialism and its laboring subjects would look like after the Great Break, it was to be found not in the Perm region at all but further east, in the ironworks of Magnitogorsk. "Magnitogorsk was no mere business for generating profits," Stephen Kotkin writes, "it was a device for transforming the country: its geography, its history, and above all its people. Magnitogorsk was the October Revolution itself, the socialist revolution, Stalin's revolution" (1995, 71). Kotkin goes on to note that the significance of Stalin's own chosen name—it meant "Man of Steel"—was lost on no one. Neither was the name of Stalin's protégé and deputy Viacheslav Mikhailovich Skriabin, who took the surname Molotov, derived from *molot*, hammer. In 1940, Perm was renamed Molotov, and it was the center of the Molotov region, a simultaneous homage to the Old Bolshevik himself and a telling indication of which segment of local industry was most identified with the socialist project. (The city reverted to its original name in 1957, following Khrushchev's Secret Speech and Molotov's marginalization.)

Soviet metal imagery faded by the later Soviet period, but the pride of place given to the factories that it had once held up as exemplar was reproduced. By comparison to Perm's vaunted metallurgical factories, the regional oil production industry was grubby, dirty, lower-prestige work, carried out by small crews in remote, inhospitable locations that required living in rural areas. Even petrochemical plants like PNOS, although certainly closer to the urban and industrial ideal than remote oil wells, did not stack up to metallurgical factories. As Zsusza Gille (2007, 84–87) shows in a fascinating comparison of Hungarian waste practices in metallurgical and chemical factories, the very nature of industrial chemistry operations was somewhat out of synch with a central planning system that was designed with factory labor and assembly lines in mind. Chemical production, in addition to requiring little direct human muscle of the sort that went on in steel and ironworks, was hard to break into targets for individuals, making it difficult to organize socialist competitions, reward individual workers, or cast chemical or petrochemical plants as industrial leaders. Checking gauges, confirming pressures, and adjusting temperatures was not the stuff of the most heroic socialist labor.

Nor did the rising dependence of the entire Soviet Union on oil revenues substantially change this symbolic hierarchy at the regional level; as I noted earlier, the socialist oil complex partitioned off sales from production and refining quite strictly. One acquaintance of mine, a journalist, summed this up nicely for me in an interview that touched on her recollection of the Soviet-era oil industry:

> I remember when I enrolled in the History Faculty [at Perm State University] in 1984, my friend enrolled in the Oil Faculty at the Polytechnical Institute. And she got in because there was only a little bit of competition, you could even say no competition. That is, where I enrolled there were five people for every spot, and for them, I think, maybe two spots for one person. . . . Oil workers—they're out there wallowing in the mud somewhere. And their salary was very low. My father worked at PNOS. He received about 120 rubles [a month]. He had gone to PNOS from being a bus driver, and as a bus driver had gotten 360 rubles . . . at [Perm] Motors they received 500–600. . . . That is, a normal

> salary. Oil workers . . . really only 120 or 140. Maybe, only somewhere out there on a drilling rig, 400.

Although she spoke in shorthand salary terms, the real difference was in prestige, connections, access to housing and vacations, and other dimensions of socialist-style inequality. The grand heroes and exemplars in the Perm region—those who became iconic of the region as a whole and constituted its political and economic elite—were most often to be found tempering steel or assembling rockets. In the early Soviet period, they were perhaps mining coal. If they were pumping oil, it was in a very specific, geographically bounded context that was not exemplary beyond itself: in the oil towns and cities that grew up, especially in the southern Perm region, in the 1950s–1970s.[19]

Oil Towns and the Soviet Social Sphere

I have thus far written of Permneft, the Perm region's oil production association, as a single unit. In reality, Permneft was composed of a shifting number of production divisions, called oil and gas production administrations (*neftegazodobyvaiushchie upravleniia*), or simply NGDUs. NGDUs were geographically organized by oilfield or cluster of oilfields within the Perm region, and headquartered in population centers—usually small cities—near to the oilfields they tended. Each included a full complement of brigades necessary for oil production, initial transport, well maintenance and repair, and other services that supported production, including a central office staff. In the last decades of the Soviet period, there were NGDUs based in Krasnokamsk, Chernushka, Osa, Polazna (see fig. 3), and Kungur; these were the oil towns of the Soviet Second Baku.

The oil towns and encampments of the capitalist oil complex—from Oil Creek Valley in nineteenth-century Pennsylvania to the boomtowns atop the Bakken formation in North Dakota's Badlands today—are much studied and generally well understood (e.g., Hinton 2008). They have often been places of overt and dramatic class inequality, with highly paid industry specialists supervising the extraction of oil in the midst of desperately poor local communities. These distinctions have, moreover, often been overlain by distinctions of race: the first American oil camps of Saudi Arabia in the mid-twentieth century, Robert Vitalis (2006) shows convincingly, were little short of Jim Crow enclaves modeled on the American West. Versions of these dynamics played out in the Perm region's oil

19. Historian Lennart Samuelson (2011) argues that the tractor and tank factories of Cheliabinsk, further east in the Urals, made the city (sometimes known as Tankograd) into a "Mirror of Russia in the Twentieth Century." The same could not be said of the oil cities of the southern Perm region.

FIGURE 3 Socialist and postsocialist representations of the oil industry on a central square in Polazna, an oil company town just north of Perm. The sign on the mural to the right reads, "We are Lukoil!"

Photo by author, 2010.

towns as well. Given oil's different place within the socialist mode of production and the party-state apparatus, however, they were far less dramatic and far less determinative of other outcomes. For instance, initial oil discoveries around the Perm region were facilitated with help from outsiders—mostly specialists and evacuees from Baku and Groznyi—but this divide between transplants and locals did not last long. Regional academies for training oil workers were set up in fairly short order, and it was not long before local residents took the lead in developing the Perm region's oil deposits. In the south of the Perm region, this meant Tatar and Bashkir communities were heavily represented, right up the leadership ranks of local NGDUs and Permneft by the later Soviet period.

Soviet-style socialism aspired to coordinate all aspects of human life in ways that would build toward the bright socialist (and eventually communist) future.[20] These aspirations failed in numerous, ongoing, and ever-morphing ways, to be sure, but they nevertheless created distinctive kinds of organizational structures, built environments, and generalized expectations that set the conditions for everything from high politics to everyday life. Socialist company towns, in which much of the population was employed in a single industry, were primary sites for

20. For a recent and innovative take on company towns and Soviet social modernity that gives an important place to energy infrastructure, see Collier 2011.

all this activity. In these towns, the local enterprise served as the central space not only for oil production activities narrowly construed but also for party politics, housing construction, education, culture, and more. Consider the case of Osa.

From its founding in 1591 until the 1960s, Osa, some 125 kilometers to the south of Perm, was a largely agricultural town known most famously for events in the late eighteenth century, when it was home for a time to Emel'ian Pugachev and the peasant rebellion he led. Even as other areas of imperial Perm province became factory centers under the influence of the Stroganov and Demidov noble families, Osa remained largely an agricultural area. This began to change rapidly with the full-scale opening of the Volga-Urals oil region. The first oil flowed from the subsoil beneath Osa in May 1963, and within a few short years Osa became a "city of oil workers" (Gasheva 2007, 44). It quickly became one of the primary bases of operations for oil production in the southern Perm region, a process that included the development of both an oil industry infrastructure and an entire social and cultural infrastructure to go along with it: the city's first multistory apartment buildings, a Palace of Culture, new kindergartens and schools, a hotel, and, by the late 1970s and early 1980s, a full set of asphalted roads running through and around the city, connecting each oil deposit to Osa and to the major regional cities of Perm and Chaikovskii. By the end of the Soviet period, nearly five thousand people were employed in the oil industry in Osa, working on ten separate oil deposits.

Permneft's Osa NGDU administered nearly all these aspects of life. *A Chronicle of Osa Oil, 1963–1993* (Kalimullin 1993) moves division by division through the constituent parts of the Osa NGDU; the book's structure thus provides some sense of the close entanglement of oil production operations and the rest of life in Osa. The book begins with the oil production divisions, arranged by the deposits to which they attended, and then catalogs the various divisions that supported oil production, ranging from well repair and reconstruction divisions to those tasked with drilling artesian wells. (Enormous amounts of water were pumped into depleted deposits to keep the oil flowing.) Fully half of the compilation is dedicated to divisions servicing aspects of life in Osa that extended out of the oil industry, strictly defined. Electricity generation as well as steam and hot water cogeneration, for instance, served both oil company operations at production sites and the urban apartment blocks that were home to oil workers and their families, stores, schools, kindergartens, a sanatorium, and more. The Oil Worker Palace of Culture, completed in 1985, included a main stage, an auditorium, numerous dedicated practice rooms, and even a square outside featuring a fountain. It hosted not just the standard Communist Party activities and assemblies but also dance and orchestral ensembles and a theater troupe. Similar configurations characterized a number of cities and towns in the Perm region during the years of the Second Baku. By the end of the Soviet period, Permneft's balance sheet included more than a million square meters of apartment space, forty-four kindergartens, and six sanatoriums, and it had primary responsibility for many of the major roads that connected oil towns like Osa to the regional center (Vikkel', Fedotova, and Iuzifovich 2009, 85).

Similar processes unfolded around PNOS in Perm itself. What would eventually become the industrial district of Perm started out as a collection of small villages on the edges of the city. Following the decision to build a refinery in 1949, a number of small factories, dormitories, and other facilities had been erected to provide the materials and labor required for basic construction purposes. Later, as the refinery began operations in the 1950s and early 1960s, these facilities were replaced with amenities for new employees, including extensive apartment space, a Palace of Culture, and a museum in 1965 (Sverkal'tseva 1998, 23). They joined schools and a hospital complex that had already begun to rise in the early 1950s. In 1972, the area around the refinery had grown to such an extent that it was designated a new city district: the industrial district.

As in Osa and other company towns, cultural and leisure activities were closely integrated with the enterprises themselves. A history of the industrial district framed this entanglement of enterprise and cultural production as a rhetorical question: "Is this a lot or a little for a single district: two Palaces of Culture, two musical and one visual arts school, a cinema, and six libraries? Are they able to 'enrich' the population of such large district?" (Kolbas 2002, 109).

The answer stretched over dozens of pages, with a lengthy catalog of the industrial district's many cultural services, distinguished culture workers, and signal accomplishments. The work of enriching the population that went on in the Soviet Union's network of enterprise-run clubs took on various guises. It ranged from the most boring of Communist Party–sponsored lectures—events that were generally poorly attended and largely ignored—to far more popular celebrations of workplace accomplishments and holidays. If oil workers took a backseat to factory workers in the grand, metallic symbolism of socialist cultural production, then this was not the case in their own clubs and houses of culture. Here, in these limited contexts and in the hands of trained Soviet culture workers employed in special cultural divisions of NGDUs and PNOS, oil and the labor of production and refining were celebrated, chronicled, and linked to local and national histories. They were, in a word, "recognized" (Donahoe and Habeck 2011, xi; see also Grant 2011), often in ways that exceeded the official party line.

Looking back on this era from the perspective of the post-Soviet years, Nikolai Kobakov, one of the leaders of OOO Lukoil-Permneft, characterized the oil industry's company towns in the region as follows:

> What was Permneft? A natural economy, where there was everything, from scientific and technical institutes to agricultural farms and greenhouses. We explored for oil ourselves, drew up the plans for developing deposits ourselves, drilled the wells ourselves, completed the infrastructure around a deposit ourselves, produced oil ourselves, did the repairs ourselves, grew the food ourselves, educated our children ourselves—in our own kindergartens, organized leisure for ourselves . . . (Bondarenko 2003, 126)

Ecology, Environment, and the End of the Socialist Oil Complex

A final significant element of the socialist oil complex was its relationship with the surrounding natural environment. For much of the Soviet period, "communist environmentality" (Snajdr 2008) contained little space for "Nature" as an autonomous or active agent worthy of protection—as opposed to exploitation in the service of socialist projects of human betterment. It is rather surprising, then, that by the end of the Soviet period, environmental activism emerged as one of the chief vectors of political critique.[21] In the Perm region, both production and refining sites were home to oil company practices that, in the wake of Chernobyl and in the broader conditions of Gorbachev's glasnost and perestroika, became crucial mobilizing events.

The Fallout of "Peaceful Nuclear Explosions"

As Soviet planners, geologists, and engineers sought to maximize the productivity of oil deposits in conditions of socialist shortage, they turned again and again to technological innovations. Some of these were new drilling techniques. The Perm region was home to some of the first Soviet experiments with diagonal and horizontal drilling (Abaturova 2003, 39) and it was in the Volga-Urals basin that the turbodrill became a standard tool (Campbell 1968, 108–20). Other techniques sought to transform oil deposits themselves, chief among them waterflooding, the practice of pumping copious amounts of water into declining wells in order to raise the pressure and extract more oil.

21. On the rise of late socialist environmental activism, see Harper 2006, Snajdr 2008, Henry 2010, and Weiner 1999, 429–40. For cases broadly analogous to those explored here, see Tsepilova 2007 and Bolotova 2007.

In the 1960s, these efforts extended to include the use of underground nuclear explosions that, it was thought, would reshape subsoil oil reservoirs and increase production output. The first two underground nuclear explosions in the Perm region were carried out twenty kilometers from Osa, in 1968, each at a depth of just over 1,200 meters and at a force of 7.6 kilotons. Although teams from the Perm region were involved in the drilling of the dedicated wells into which the nuclear devices were inserted, the Soviet military and the Ministry of Atomic Energy, following approval at the level of the Politburo, carried out the actual explosions. They were part of the Soviet Union's broader Peaceful Nuclear Explosions for the National Economy program that envisioned using nuclear devices for everything from increasing oil and gas output to changing the shape of rivers and mountain ranges (Kurbatova 2006, 401–2). (The United States pursued similar projects in Operation Plowshare, including the use of underground nuclear explosions to spur natural gas production in the American West. These efforts were on a considerably smaller scale than the Soviet program.)

The explosions near Osa did not appreciably increase oil production from the surrounding wells, although they did apparently rattle houses in the Osa district enough that most brick stoves had to be rebuilt (Vikkel′, Fedotova, and Iuzifovich 2009, 72). By 1976 it was clear that a significant amount of radiation was seeping into the surrounding groundwater, as well as into the extracted oil itself, including the infrastructure used to transport it for refining (Kurbatova 2006, 401). Although the specific wells that had been identified as having unacceptably high levels of radiation were closed, the Osa oil deposit as a whole remained in production. These results did not, however, stop the Soviet Union's efforts to use underground nuclear explosions in the effort to increase oil production. An additional five devices were detonated in the subsoil of the Perm region in the 1980s, all of them in the far north, at the Gezh oil deposit, eighteen kilometers from the town of Krasnovishersk (Sokolova 2003, 4). The result was the same as in Osa: no discernible increases in oil production, but increasing radiation throughout the local environment.

The final two underground explosions near Krasnovishersk in 1987 took place at a time of rising ecological consciousness in the period of Gorbachev's glasnost and perestroika, and a local drilling geologist led a group of concerned Krasnovishersk citizens in demanding more information about the potential health and ecological dangers of the underground explosions. Although they received no significant additional information, their demands appear to have contributed to Gorbachev's moratorium on underground nuclear testing and other explosions, which came only one day before the seventh planned subsoil explosion outside of Krasnovishersk (indeed, the warhead was already in the ground) (Kurbatova 2006, 402).

Clouds Overhead

The mobilization at Krasnovishersk had a much larger counterpart in Perm itself, also directed at the oil industry. On May 15, 1987, the newspaper *Evening Perm*—not an official organ of the Communist Party—published an article that was still being discussed and cited during my primary fieldwork nearly twenty-five years later. The article was titled "Clouds Overhead" (*Tuchi nad golovoi*), and it detailed the ecological and human health hazards created by emissions from PNOS in Perm's industrial district.[22] Before long, photocopies of the article were plastered on walls and in public spaces around the city, and the city's first environmental movement began to coalesce. *Evening Perm* was deluged with letters to the editor, and the paper ran a series of responses and additional interviews throughout the remainder of the year. This kind of open exchange and debate in a newspaper was previously unknown in the region; decades later it was often remembered as the moment in Perm's collective history when perestroika began to feel real. It features as a pivotal point in the memoirs of nearly all the notable politicians of the era (Sapiro 2003; 2009, 70; Igumenov 2008, 286).

The running series of articles in *Evening Perm* detailed specific instances of failures to abide by existing standards and ecological regulation as well as cases of negligence on the part of supervisors charged with monitoring these issues. One article, for instance, covered the dumping of phenol acid—a common petroleum derivative that can cause burns and rashes even in relatively small quantities—into the Kama River, despite procedures and regulations stipulating that any substances dumped in the river after processing must be of drinkable quality.[23] Even Communist Party members who worked in PNOS itself felt free to write letters to the editor calling for a comprehensive evaluation and mitigation of the enterprise's ecological impacts and harms in the name of the health and welfare of Perm's residents. The paper summed up this response by encouraging still more engagement: "But times are changing. Now there are no zones that are closed off from criticism, which means that intervening in and correcting situations in a businesslike, concrete way has become possible."[24] It was not long before the issue reached the highest levels of the Perm region—the Regional Committee of the Communist Party, which, by the summer of 1987, had ordered an investigation into ecological

---

22. Sergei Zhuravlev, "Tuchi nad Golovoi," *Vecherniaia Perm'*, May 15, 1987. Zhuravlev later recounted that the office of the state censor had judged his first title, "Clouds over the City," too provocative, but left the content itself unchanged. See Sergei Zhuravlev, "Pravdu! Nichego Krome Pravdy . . ." *Zvezda*, November 18, 2005.

23. A. Korabel'nikov, "Zhivaia Voda ili Mertvaia?" *Vechernaia Perm'*, May 28, 1987.

24. N. Golovanova, "Tuchi nad golovoi," *Vecherniaia Perm'*, May 26, 1987.

practices and conditions at all factories and other industrial operations throughout the entire region.[25]

In early June, the city-level Communist Party organization (*gorkom*) of Perm called a meeting to gather opinion from specialists and respond to the public outcry. At the meeting, PNOS director Veniamin Sukharev pronounced the charges enumerated in "Clouds Overhead" to be "fair and timely" and noted that the production demands on the refinery had outstripped its capacities to protect the local environment in recent five-year plans. He announced major repairs, new procedures for enforcing existing regulations, and the intensification of efforts to create a sanitary zone around PNOS by resettling city residents whose homes had been irreversibly polluted. He noted that such resettlements had taken place since the 1960s to some extent, but that they would now extend to include Pervomaiskoe Street, with one hundred families slated for relocation in 1987 and the remainder by 1990.[26]

The Party's official response, published in the paper later that week, acknowledged the severity of the situation, laid partial blame on the fact that ongoing demands for production and the crucial importance of PNOS to the regional economy had resulted in an unacceptably low emphasis on ecological safely issues, and promised a program of capital improvements.[27] An accompanying piece written by the acting director and Party chair at PNOS noted that the association's ecological footprint had been minor and laudable through the 1970s, when it was focused on oil refining. But, he said, as PNOS extended its work to include other aspects of petrochemical and chemical refining, including the production of butane, urea, and ammonia, ecological management became more complex, and the plant was not provided with the resources it needed to fulfill the ecological protection side of its plans. The current construction of new facilities that would enable even deeper refining of oil products would, he promised, unfold with full attention to environmental protection.[28]

The public outcry did not stop, and, in the months that followed, PNOS fought to stay open and pursued all manner of strategies for compromise. For example, because activists had focused much of their attention on a new refining facility being built at PNOS, the refinery paid for a number of leaders in the local environmental movement to travel to another city where a similar facility was already operating, in an attempt to demonstrate that it was safe. That trip produced no compromise, and all the while a series of public meetings and newspaper articles

25. "V Obkome KPSS," *Vecherniaia Perm'*, June 24, 1987.
26. L. Zherebtsov and A Korabel'nikov, "Okhrane Okruzhaiushchei Sredi—Rabochuiu Zabotu!" *Vecherniaia Perm'*, June 11, 1987.
27. I. Shchelgov, "Zona Osobogo Vnimaniia," *Vecherniaia Perm'*, June 11, 1987.
28. V. Shuberov and V. Vlasov, "Nash Dolg," *Vecherniaia Perm'*, June 11, 1987.

drew attention to an ever-widening set of ecological dangers: toxic waste heaps, river contamination, smoking chemical flares, and more. Although PNOS did survive, work on the new Pareks refining complex was brought to a halt until new funding for ecological modernization and mitigation could be extracted from the Ministry of Oil Refining and Petrochemicals.

A decade and a half later, Veniamin Sukharev, who had just become head of PNOS the year before the "Clouds Overhead" article was plastered all over Perm, recalled:

> At the time, no one called PNOS anything but a monster, destroying nature and human health. Ecological problems became for Perm the impulse for the processes that we now call democratization. And it just so happened that we fell into that big political-economic conflict.[29]

A post-Soviet corporate history acknowledges the episode in a broadly similar way, taking some umbrage that it was PNOS, and not one of the city's other major polluting factories (of which there was no shortage) that took the brunt of public ire:

> Every plant, every factory, that had a negative impact on human health attracted the close attention of the socially concerned part of the country. In our city, you don't have to walk far to find an example. Despite the fact that the Ordonikidzhe factory and its bright yellow cloud of smoke was located right in the center of a relaxation and leisure zone . . . the city rose up in arms against [PNOS]. "Clouds Overhead" had an enormous public resonance: residents of Pervomaiskoe Street demanded resettlement, and the whole city, if it didn't demand outright closure of the "primary monster," then at least demanded a stop to construction. (Gasheva and Mikhailiuk 1999, 57)

In addition to its geographical centrality in the city's industrial district, there is likely another good reason that PNOS and not some other industrial operation found itself at the center of environmental mobilization: it was *not* the most prestigious or politically connected of the city's industrial operations. It was *not* closely identified with the very top of the regional political order or with the military, and thus made an ideal target for fledgling, still tentative environmental mobilization. Even, that is, at the very end of the Soviet period, when an element of the socialist oil sector did come to occupy a central place in the regional imagination, it attained that position in part because of its comparatively low status in the regional pecking order.

---

29. "Veniamin Sukharev o vremeni i o sebe," *Zvezda*, March 31, 2006.

In the early 1990s and in subsequent years, there was considerable debate about whether PNOS actually improved much in the near term, and ecological challenges continued into the 1990s and 2000s. But perhaps the most enduring aspect of the controversy surrounding "Clouds Overhead" was that it showed how ecology and human health concerns could be a channel for widespread, indeed, essentially revolutionary, political change. Long before there were press secretaries, public relations departments, corporate social responsibility programs, and international standards of ecological protection, the controversy that swirled around PNOS in the late 1980s taught the region's oil executives and politicians their first lessons about how ecological critique could lead to political reform. The lesson stuck. Indeed, averting another similar incident was high on the list of those executives' and politicians' priorities as the oil industry became central to the region after 1991.

2

CIRCULATION BEFORE PRIVATIZATION

Petrobarter and New Corporate Forms

The disintegration of the Soviet Union over the period from 1985 to 1991 threw all the elements discussed in the previous chapter, as well as their integration into a distinctively socialist oil complex, into disarray. In contemplating the transformations that ensued, it is crucial to keep in mind that the nature and pace of post-Soviet changes varied tremendously—by region, by sector, and by element within any given sector. In the Perm region's oil industry, issues of official cultural-ideological construction in company towns and concerns about pollution, ecology, and environment—those aspects of the Soviet oil complex that had become most closely associated with the Communist Party's proponents and critics, respectively—quickly faded into the background. They would stay there for the better part of a decade. At the same time, questions about the reorganization of socialist enterprises in the oil sector and about how oil and oil products would circulate in new political and economic conditions became matters of utmost urgency, and they rapidly began to resonate much more widely throughout the Perm region than they had in the Soviet period. These two interwoven strands—the making of socialist enterprises into new corporations and new kinds and patterns of circulation—make up the story of Permian oil in the 1990s that I relate across chapters 2 and 3.

With respect to enterprise restructuring, the Russian oil industry after 1991 faced two daunting challenges: first, the Soviet system of ministries and their constituent units was quickly crumbling, and second, the Communist Party was no longer around to fulfill its previous role of mediating among different interests in the central planning process. Vagit Alekperov, deputy minister of oil and gas at the end of the Soviet Union, phrased these challenges succinctly: "the Russian oil and gas industry [at the end of 1992] included nearly two thousand uncoordinated associations, enterprises, and organizations belonging to the former Soviet industry" (2011, 324). The key word here is uncoordinated: the elementary matter of how oil would move through the various units involved in production, refining, and distribution—an intricate if more or less known quantity in the political economy of socialism—was suddenly up for grabs. It was not even clear anymore what the various associations, enterprises, and organizations that Alekperov's ministry helped oversee were, or who controlled them. It was especially unclear what they would become as the importance of socialist-style allocative power faded.

The standard narrative of the making of new Russian oil companies out of this initial post-Soviet chaos features a high-stakes battle over production fields, with all its implications for who controls the most oil, the largest reserves, and the greatest opportunities for short- and long-term production and profits. In this narrative, the overarching issue is the privatization of state-owned assets in the oil sector, including the infamous loans-for-shares deals of the mid-1990s and other machinations in Moscow. The view from the Perm region, from just a handful of the two thousand uncoordinated units mentioned by Alekperov, allows us to revise and expand this standard narrative in a number of ways.[1] To follow the winding route by which the Perm region's Soviet oil complex became a group of subsidiaries within the new, private, vertically integrated oil company Lukoil, for instance, is to trace the significance of processes that were under way well before privatization proper began in 1993, including the seldom-appreciated importance of refineries (as opposed to production fields)—an extension of Permneftorgsintez's (PNOS's) greater regional significance in the Soviet period—and the domestic circulation (as opposed to export) of oil products. These chapters also take an expansive view of what makes a corporation, situating the allocation of shares through privatization within an array of other crucial dimensions of incorporation: spatial organization within the Perm region; relationships with regional- and district-level political networks; and the reshaping of labor and Soviet company towns in new circumstances. They also treat a range of new corporations—among them the Perm Commodity Exchange and the Perm Financial-Productive Group—that were not technically part of the oil sector, but that nonetheless powerfully shaped its trajectory and the broader ways in which oil remade the Perm region.

1. For additional perspectives on oil in the Russian regions in the Yeltsin era, see Kellison 1999 and Glatter 1999 and 2003, all of which concentrate on oil-rich West Siberia. One regional issue that has been noted a great deal in existing scholarship is the dual-key provision of the 1992 Law on Mineral Resources, which required that licenses to drill for oil receive approval both in Moscow and the local region—a provision that allotted enormous power to coalitions of regional geologists, politicians, and oilmen (see, e.g., Gustafson 2012, 68–69).

Vertical integration here appears less as a type or an endpoint and more as an ongoing process, a strategy of accumulation that, although particularly mobile and chaotic in early 1990s Russia, continued long after.

Running in and out of the story about corporations in these chapters is a second story, about shifting patterns of exchange and circulation: the means by, terms on, and paths along which oil and oil products were and were not transacted in the 1990s. Contrary to many expert predictions, monetary exchange and markets did not replace Soviet shortage and rational redistribution overnight. Oil in the 1990s Perm region was therefore less an item of monetized exchange than of situational barter, and in 1995–98 it even anchored a regionally circulating surrogate currency. The rapidly shifting contours of these exchanges significantly influenced both regional corporations and regional politics. Indeed, I will suggest that the materiality of bartered oil and oil-backed surrogate currencies—the ways in which these modalities of exchange linked oil as a substance to other substances, objects, places, and institutions—was a crucial factor in shifting operative notions of regional identity from the defense-industrial sector to the oil sector.[2]

Only gradually and fitfully did barter wane and a solid relationship between oil and money take root in the Perm region. I want to emphasize at the outset that, in my view, tracing the emergence of this relationship over the 1990s is evidence for its contingency, rather than its inevitability or naturalness (as many economists and other theorists of post-Soviet "transition" would likely argue). Elsewhere, I use examples from a wider range of times and places to show that barter relationships have long been important in the global oil industry, and that they often exist alongside—and in intensely political relationship with—the monetized exchange of oil (Rogers 2014a). That I trace a more limited temporal and spatial arc here, one in which money did indeed come to predominate over barter as the 1990s wound down, should not be assumed to mean that this must always happen, that it is a natural progression, or that it will not be undone in the future. Indeed, it would be just as legitimate to argue the opposite: that the sustained and elaborate nonmonetary oil transactions of the 1990s Perm region are good evidence that oil and money are not at all natural partners.

To put this in still broader historical and comparative context: One of the claims of part I of this book, a claim illustrated more clearly by Soviet and early post-Soviet oil than by any other case around the world, is that the close relationship between oil and money that has so often been central to modern global capitalisms is far from natural or all-encompassing. It must be made, and then it must be perpetuated; it is always open to challenge and reformulation. In the Perm

2. On the economic struggles of Soviet defense-sector cities (including Perm), to adjust to post-Soviet circumstances, see especially Gaddy 1996.

region, the decade-long fashioning of a tight relationship between the circulation of oil and the circulation of money is inextricable from the fashioning of new corporate forms, of a new regional elite, and of powerful new senses of regional identity. The story begins where it ended at the close of the Soviet period: at Perm's refinery, PNOS.

PNOS, the Ministries, and Lukoil

Veniamin Sukharev, general director of PNOS, had a problem, or, more accurately, a whole nest of problems. In the late 1980s, his management of the "Clouds Overhead" environmental crisis had not only kept Perm's sprawling oil refinery and petrochemicals plant open, but speeded his promotion from chief engineer to general director. But the early 1990s presented a much bigger set of challenges to the continued existence of PNOS than ecological activism had. Without the interministry links and Communist Party-facilitated connections that had moved socialist oil from production through refining and on to consumers, PNOS found itself squeezed on both upstream and downstream ends at the same time. On the production side, the regional oil production association Permneft was on the verge of falling apart into its constituent units, many of which were making new and exorbitant demands on their own central office and on refineries seeking to acquire crude. On the distribution and consumption side, PNOS's biggest state customers, including the Soviet military, were drastically cutting back their orders or unable to pay. There were new and tantalizing possibilities for export abroad for hard currency that could—in contrast to the Soviet era—accrue directly to PNOS's own accounts, but much of the refinery's output was not yet up to European standards and would need major upgrades to get there. There was an important technical dimension to PNOS's problems as well: unlike production wells and car, truck, and airplane engines, PNOS's refining equipment needed to run around the clock at a minimum capacity in order not to break down. Temporary shutdowns were thus not an option for dealing with widely and unpredictably gyrating levels of supply and demand. PNOS was coming uncomfortably close to those minimums.

One place that Sukharev could look to solve his problems was to PNOS's parent ministry in Moscow, the Ministry of Oil Refining and Petrochemicals, which was still attempting to coordinate the refining of oil across the country's major refineries—dealing, in essence, with PNOS's problems on the scale of the entire Soviet system as it came apart at the seams. But Sukharev quickly ran into difficulties there, for PNOS was not especially well placed in the internal hierarchy of the Ministry of Oil Refining and Petrochemicals. Through regional and personal connections, the leadership of the ministry was most closely entwined with the refineries in European Russia and West Siberia; PNOS and the Perm region, with their combination of provincial location and declining regional oil production, simply had not accumulated the kinds of connections that would allow preferred treatment from ministry officials to solve its problems. Sukharev later recalled his concerns about which of the Soviet system's refineries might be forced to close first:

> Omsk? The head of the [ministry's] Central Board is from there. Iaroslavl? That's the former domain of the deputy minister. Angarsk? The minister himself made his career there. Moscow? Don't make me laugh. That means we sinners in the Urals will be the first to fall on hard times.[3]

In other words, the Ministry of Oil Refining and Petrochemicals would not be extending PNOS much of a lifeline.

State Concern Lukoil

In addition to the negative dimensions of its Soviet inheritance, however, there were also new possibilities for PNOS. In the multiparty elections of the fall of 1990, Sukharev was elected as a representative from the Perm region to the Supreme Soviet of the RSFSR (the Russian Republic within the Soviet Union), having run largely on a campaign of continuing to win additional resources for renovating PNOS in order to comply with promised higher environmental standards. This position placed him in close contact with the leadership of the Perm region at the time—which was no more eager than he to lose PNOS's fifteen thousand jobs in Perm's industrial district—and also afforded him a position in Moscow from which to pursue opportunities for PNOS outside its parent ministry. Sukharev first explored a variety of emerging consortia among refineries and producers, but none seemed to provide many new opportunities for PNOS.[4] In November 1991, he sought out Vagit Alekperov, a deputy minister in the Ministry of Oil and Gas Production and a longtime veteran of the Soviet oil industry.

At the time, Alekperov was leveraging his position to embark on a new experiment: the creation of a vertically integrated oil company that would be modeled after, and eventually able to compete with, Western oil companies. Alekperov had gained control over three significant oilfields in West Siberia—Langepas, Urai,

3. "Veniamin Sukharev o vremeni i o sebe," *Zvezda*, March 31, 2006.
4. S. Fedotova, "Vremia L—1991 g.," *Permskaia Neft'*, February 22, 2011.

and Kogalym—the initial letters of which would form a new company's name: Lukoil.[5] Sukharev offered Alekperov a crucial piece of the vertically integrated puzzle: a major refinery. Due to the honeycombed nature of the Soviet oil industry, however, Alekperov had little influence at the Ministry of Oil Refining and Petrochemicals, so it fell to Sukharev and his board of directors to officially extract PNOS from its parent ministry (see also Vikkel′, Fedotova, and Iuzifovich 2009, 141–144). This they did, although not without some difficulties and accusations of being a traitor. When Lukoil was officially established in November 1991, it included the three production fields in West Siberia, PNOS, and a second, somewhat smaller refinery in Volgograd.[6] Lukoil was, at that time, still a state concern (*kontsern*); privatization in the oil industry did not begin to take shape until the fall of 1992.

From Alekperov's perspective in Moscow, the acquisition of a major—if somewhat antiquated—refinery was a major step in Lukoil's effort to compete with Western oil companies on their own terms. Back in the Perm region, PNOS's entry into the new Lukoil structure seemed to offer solutions to some of Sukharev's problems. PNOS would be guaranteed a steady supply of oil to refine from the highly productive West Siberian fields, and would, therefore, no longer need to rely so heavily on the unpredictable and often fly-by-night operators peddling tankers and rail cars of crude to refineries around Russia. A place in State Concern Lukoil, with its openly global ambitions and roots in the Soviet oil establishment, also offered new possibilities for foreign and domestic investment and, increasingly, for export. With help from Lukoil, PNOS was quickly able to secure loans, credits, and grants that would enable technological upgrades designed to bring refining to a standard acceptable for international export, including a $500,000 grant from the United States Agency for International Development in early 1993. The refinery quickly entered into partnerships with Italian, French, and Japanese firms, and refining specialists from Shell and British Petroleum consulted on equipment upgrades, as did the U.S.-based refining services companies Bonner & Moore, Lummus Global, Foster Wheeler, and others. The parade of international specialists became so extensive that the company built a special office building for meetings with foreigners (Sverkal′tseva 1998, 64). By the mid-1990s, PNOS had increased the depth to which it was able to refine oil by nearly 20 percent, doubled its output of high-quality motor oils, and increased fivefold its output of high-

5. On the founding of Lukoil, see also the accounts in Kryukov and Moe 1991; Lane 1999, 26–30, 111–14; Alekperov 2011; and Gustafson 2012, 98–144.

6. For a period of months, State Concern Lukoil also included refineries in Lithuania and Bashkiria, but these firms left Lukoil under political pressure from their national and regional governments, respectively, shortly after the end of the Soviet Union.

quality gasoline products, all while decreasing harmful emissions (Kurbatova 2006, 514–15).

The entry of PNOS into State Concern Lukoil in the fall of 1991 made the refining segment of the Perm region's oil complex part of Russia's flagship vertically integrated oil company—a considerable step up from PNOS's position, only weeks earlier, near the bottom of the pecking order within the Ministry of Oil Refining and Petrochemicals. This move was only the first step, however, in the consolidation of oil operations in the Perm region, for both production and distribution organizations remained outside of the Lukoil umbrella, and would stay there for some years yet.

In step with State Concern Lukoil's plans for vertical integration, PNOS quickly began the process of spinning off several of its constituent petrochemical-focused enterprises into their own companies in order to focus more narrowly on oil refining. Enterprises concerned with metalworking fluids, mineral fertilizers, and other products that had been part of PNOS during the decades of the Ministry of Oil Refining and Petrochemicals were now on their own. (The fact that those now-independent companies had been responsible for much of PNOS's infamous pollution considerably bolstered the refinery's claims to be rapidly decreasing its environmental impacts.) Yet even as this older model of Soviet industrial integration was giving way to plans for vertical integration in the oil industry, PNOS was simultaneously acquiring and expanding into other sectors that would seem to some eyes much further afield. A renowned former collective farm in the Kungur district, renamed "Agroassociation Labor" in the post-Soviet period, joined PNOS in 1992 with the promise of providing a regular supply of milk and meat to the refinery's employees. (PNOS had an older, although less reliable and successful, relationship with other collective farms.) The refinery soon began construction on a sausage division for Agroassociation Labor as well (Sverkal′tseva 1998, 66).

Although the story of Lukoil's rise as Russia's first vertically integrated company has been told often, it is seldom noted that vertical integration at the national level proceeded in part through increasing cross-sectoral integration at the regional level. A consideration of early post-Soviet PNOS's place in regional patterns of exchange, including both its acquisition of agricultural enterprises and its growing role in the circulation of consumer goods, points to some key early transformations at the intersection of corporation and state, emergent elite, regional identities, and the materials and means of exchange involving oil—all of this before privatization and before the integration of the Perm region's oil production operations into a major oil corporation.

PNOS and Petrobarter: A Suzerainty with a Difference

PNOS's expansion into the agricultural sector was more norm than exception among early post-Soviet firms, for with the end of socialist central planning came an age of "parcelized sovereignty" (Verdery 1996, 208): collectivities of all sorts, from geographical and political regions to former collective farms to factories like PNOS, came increasingly to serve as total social institutions for their residents and employees. The early postsocialist landscape was, that is, dotted with disconnected suzerainties, in Humphrey's apt terminology (2002, 5–20), with little in the way of centralized state power or working markets to control and coordinate their interactions. Many of these suzerainties issued their own currencies and various sorts of rationing tickets or identity cards in efforts to cope with pervasive shortages. With many foodstuffs increasingly hard to come by, getting into the agriculture and sausage business was a way for PNOS to ensure the health and well-being of its employees and their families—a considerable portion of the industrial district of Perm—without depending on unpredictable outside sources and without needing to spend cash.

Parcelized sovereignty empowered local bosses, farm directors, and other low-level administrators within the fiefdoms they controlled. It also participated in a major shift in the nature of relationships *among* enterprises: with prices rising rapidly, inflation galloping, and socialist rational redistribution in tatters, inter-enterprise exchange took place more frequently on barter terms than through the medium of money. A number of studies have seen this parcelization of sovereignty and the associated flourishing of barter relationships as diagnostic of federal loss of control over nearly everything in the immediate post-Soviet period—from violence (Volkov 2002) to the money supply (Woodruff 1999) to deeper expectations about governance as a whole (Rogers 2006). All this applies in the case of PNOS, but PNOS was also a suzerainty with a very important difference. As an enterprise within the oil industry—and one situated within the new federal State Concern Lukoil—PNOS had one crucial advantage over nearly all other major firms in the Perm region at the time: it could exchange its products on international markets.[7] Although this potential was limited both by the quality of its oil products and by federal law, even a small amount of export was highly significant in the context of the Perm region's deepening economic crisis. Indeed, the very early 1990s transformation of PNOS's refined oil into many other kinds of goods through circuits of exchange that extended into the international arena was the opening wedge in the process by which oil came to undergird the means

7. The only other firm with this possibility at the time was Uralkalii, the major potash mining enterprise in the north of the Perm region; see chapter 3.

of regional exchange in the post-Soviet period—something that it had never done in the Soviet period.

Consider the account of a former Lukoil-Perm employee, whom I asked in 2011 to recall the place of the oil industry in the Perm region in the immediate aftermath of the Soviet Union. She began by focusing on PNOS:

> It was all done by barter—there was no money. . . . So the Chusovoi metals factory traded its product for sugar or other things. And thanks to that some sugar appeared in the region. It wasn't only the Chusovoi metals factory, but also PNOS. Yes, they also worked on barter, I think with the Chinese. They imported down jackets, and well-made ones. [PNOS employees got them] and just went down to the market and sold them for crazy [i.e., a lot of] money—there was nothing to wear, nothing in stores. . . . PNOS got those jackets for their workers, in exchange for oil products. It was outright barter, because there was no money.

PNOS employees, that is, received internationally bartered items like down jackets in lieu of cash salary, as bonuses, by dint of connections, or by other means. A significant portion of these items then made its way to street-corner markets and into the region more broadly (Sverkal'tseva 1998, 65). This pattern of exchange had its roots in the last years of the Soviet period. In 1989, taking advantage of perestroika-era reforms that gave increased autonomy to firms, PNOS opened a new division that was permitted to execute its own international oil sales. The earliest deals took place with partners that included Japan, Korea, Finland, and Hungary. Only a small portion of these international exchanges were for cash because PNOS could only keep up to 27 percent of profit from monetary sales abroad (the remainder being shunted into the central Soviet budget as the federal state struggled to maintain control of the planning and allocation apparatus), whereas the refinery was permitted to keep all the goods for which it bartered. Late Soviet and early post-Soviet PNOS also used this opportunity to barter extensively for technology and construction materials that would help modernize and repair its aging facilities (Sverkal'tseva 1998, 61).

As the Perm region's industrial base steadily eroded and PNOS entered State Concern Lukoil, the refinery's international cash sales and barter deals expanded and began to include all manner of fashionable items. It became ever more central to regional exchange. On one memorable occasion, for instance, an international barter deal between PNOS and Japan brought nearly one hundred brand new Toyotas to the Perm region. Those with close connections to the refinery snapped up the new cars immediately: PNOS had, in contrast to its Soviet days, "unexpectedly become a prestigious place to work" (Sverkal'tseva 1998, 62). In the very early 1990s, that is, a new regional elite was beginning to coalesce and spread outward from PNOS, and internationally bartered oil was central to the exchanges in which it participated. PNOS's petrobarter was relatively small-scale at this point, notable for its international dimension but only beginning to be distinguishable from widespread barter across the regional economy. That would soon change.

In the words of a well-connected acquaintance who recalled those years to me, it turned out that what Veniamin Sukharev of PNOS had really done in November 1991 was to "join Lukoil to the regional economy." To understand the next stage in this process, and its ever-growing implications for the remaking of regional identities and practices, we need to consider another of Sukharev's projects: joining forces with political leaders and the directors of other major enterprises in the region to launch of a new and, at first, controversial enterprise: the Perm Commodity Exchange.

The Perm Commodity Exchange

In the summer of 2011, I interviewed one of the earliest employees of the Perm Commodity Exchange (Permskaia Tovarnaia Birzha, hereafter PTB). "There was," he recalled of the early 1990s,

> a certain sense of general euphoria. The collective, the people who joined, were young and beautiful and raring to go. At the [commodity] exchange there was the consciousness that in a year or two we would build a capitalism in which we would have everything, there would be a middle class, everything would be great . . .

He trailed off, the unfinished sentence marking the distance between that initial euphoria and the reality of the intervening twenty years.

Such nostalgia for 1990 and 1991 was common among those whose post-Soviet careers started at the Perm Commodity Exchange. Andrei Kuziaev, who had headed the PTB from shortly after its opening in 1991 until its shuttering in 1993, was the figure most closely associated with the sweeping transformations of regional exchange patterns that it enabled. In 2011, Kuziaev was serving as a Lukoil vice president and president of Lukoil Overseas Holding (responsible for all Lukoil's operations outside Russia), and he returned to his native Perm from Moscow to join celebrations for the twentieth anniversary of the founding of the PTB. "Back then we all believed in the fairy tale that a market economy would make us all happy," Kuziaev recollected for a local newspaper correspondent, "but it turned out that only we can make ourselves happy."[8] In other interviews, how-

8. Svetlana Voronova, "Finansovaia Skazka," *Mestnoe Vremia*, March 10, 2011.

ever, Kuziaev was the first to admit that, his own prodigious work ethic notwithstanding, he had simply found himself in the right place at the right time. He had never expected that the PTB would turn him and many of his earliest collaborators into oilmen and send them rising through the ranks at Lukoil—first in the Perm region, later in Moscow, and eventually hopping among high-stakes negotiations around the world, from Central Asia to the Middle East to Africa. Indeed, when the PTB first opened in 1991, oil was not among the items commonly transacted on its trading floor. By 1993, however, it was, and so the story of the rise and fall of the PTB—and the ways in which it reshaped exchange patterns in the Perm region—is a crucial phase in the movement of oil from a peripheral and not particularly prestigious corner of the Perm regional economy in the Soviet period to one of its central pillars.[9]

"The Market Calls"

In the fall of 1990, as shortages deepened and spread, members of the Perm region's political elite—composed largely of the management of the region's major factories and enterprises—were wrestling with a larger-scale version of the problems faced by individual households and firms: how to coordinate the flow of inputs and outputs through the region as Soviet supply chains and their familiar workarounds disintegrated in front of their eyes? In December 1990, Evgenii Sapiro, an economist, former university professor, and, as of the previous May, deputy chairman of the Perm Regional Communist Party Executive Committee, gave a speech titled "The Market Calls." Sapiro's speech, later published in the regional party newspaper *Zvezda*, suggested that the Perm region should replace its division of the Soviet state supply agency (Glavnoe Snabzhenie, or Glavsnab) with a commodity exchange, where registered brokers would represent clients in the buying and selling of products on an open market. There was already a commodity exchange operating in Moscow as of May 1990, and others were beginning to spring up around the Soviet Union ("like mushrooms," one of my interlocutors recalled), but Sapiro still had to convince a number of skeptics.

The first of these was the leadership of Perm's division of Glavsnab itself, which was arguing, along with its leadership in Moscow, that new market conditions required not open commodity exchanges, but new intermediary companies, staffed by Glavsnab employees, that would work on behalf of enterprises to sell their products. Sapiro's position, however, was that the region needed not Soviet state supply bureaucrats in a new organization, but, rather, a "young, intelligent

9. For instructive overviews of the commodity exchange movement in the former Soviet Union, see Frye 2000, 84–106; Davis 1998; Wegren 1994; and Yakovlev 1991. On the Perm Commodity Exchange in particular, see Fedotova 2006, 40–44.

fartsovshchik" with a market mindset.[10] (Fartsovshchik was a Soviet slang term for someone who made a living buying and selling in the shadows of the official socialist economy.) With the powerful leadership of the region's factories and production associations, among them PNOS's Sukharev, on his side, and with the regional division of Glavsnab near the bottom of the regional party-state hierarchy, Sapiro's view soon prevailed. On March 1, 1991, the PTB was incorporated with the support and participation of more than one hundred regional factories and enterprises. Its first trading day was scheduled for April 11, 1991.

In a nod to the debates that accompanied the PTB's founding, Victor Gulia, the former head of Perm's Glavsnab, was named as inaugural president of the exchange. As his initial opposition suggested, this move was something of a demotion—an acknowledgment that state supply agencies had been dealt the worst possible hand in the exit from socialism and were far less in control of their own fate than the inheritors of other Soviet-era structures. As manager of operations (*upravliaiushchii*) at the PTB, Sapiro installed Andrei Kuziaev, a twenty-five-year-old graduate of Perm State University and, at the time, a graduate student in the Economics Department at Moscow State University. Although he was more student than denizen of the less-well-lit corners of the Soviet second economy, Kuziaev was to be Sapiro's "young fartsovshchik" at the commodity exchange. (He had also been Sapiro's former student at Perm State University.) Kuziaev's role at the PTB expanded rapidly, and by early 1992 he took over the presidency from Gulia.[11]

As Fedotova also notes (2006, 41), it is likely that only a young outsider with patronage connections in the regional Communist Party hierarchy such as Kuziaev could have taken the reins of the commodity exchange at this early stage. Nearly everyone occupying a position of note in the regional elite was looking for ways to mobilize Soviet state resources already under their control in new ways. They were unwilling to give up these Soviet inheritances to manage something entirely new and unpredictable. Indeed, the very existence of the PTB was evidence that Victor Gulia and others in the Soviet state redistribution offices had just failed in their effort to repurpose their own Soviet inheritance by starting new companies specializing in supply. It was Kuziaev's early genius—or, at least, so subsequent glowing biographies would put it—to recognize that the position of manager and then president of the PTB would soon give him and his associates access to the resources not just in any one existing enterprise or sector of the regional economy, but to all of them as they interacted with each other through new modes and patterns of regional circulation.

10. Voronova, "Finansovaia Skazka." On the Komsomol as a training ground for the new generation of entrepreneurs like Kuziaev, see Yurchak 2001.

11. V. Bel′tiukov, "Sdelan pervyi shag," *Zvezda*, March 13, 1992.

Sapiro, Kuziaev, and their collaborators also needed to convince a skeptical public that the Perm Commodity Exchange was a good idea, not an easy task given the official opprobrium that markets and "speculation" had received over the course of the Soviet period. In a framing that was to be repeated and amplified many times over the coming years, they did this with appeals to regional identity, pride, and independence—another face of the multilevel and multidomain parcelization of sovereignty that characterized those years. An early argument for the importance of establishing a commodity exchange in Perm, for instance, suggested that, "Residents of the Perm region have something to offer the all-Union market—timber and fertilizer, metals and machine-building products, paper and mineral natural resources. An exchange will be a possibility to really shift from words about economic independence to actions."[12] Six months later, mere weeks before the Perm Commodity Exchange was to open, S. Finochko argued for its legitimacy and importance (still not taken for granted by everyone in the region) in the pages of *Zvezda*:

> In conditions of unprecedented shortages, inflation, and coming changes in the rules for trade and setting prices, only the concentration of all the resources of the region into one "fist" [*kulak*] will allow the economy of the Perm region not to fall apart, but to preserve its stability.[13]

In these calls for a regional commodity exchange, we find some of the earliest evidence in the Perm region of what Susanne Wengle (2015, 6) terms "regionally specific developmental bargains" that unfolded in various ways across Russia: deals struck between government officials and new industrial or financial conglomerates that both enriched an emergent regional elite and gave state agencies the funding, expertise, and capacity to carry out their own projects.[14]

The PTB was formed as a joint-stock corporation with shareholders who had voting rights and rights to dividends—160 of them in February 1991. In addition to setting up the physical and procedural infrastructure to facilitate trading, the PTB needed—within the space of months before its scheduled opening in April—to hire and train two groups: brokers who would represent clients on the trading floor and the exchange's own staff. Many of the first brokers came from the region's factories and other established organizations, serving essentially as delegates from one or another enterprise or set of enterprises to the exchange. Independent brokers quickly joined them, and brokerage houses specializing in various

12. A. Vaniukov, "Nuzhna li Permi torgovaia birzha?" *Zvezda*, August 8, 1990.

13. S. Finochko, "Bez Birzhi ne Vyzhit'," *Zvezda*, February 16, 1991.

14. A theory of regional developmental bargains, Wengle argues very persuasively, better illuminates the interaction of state officials and industrial/financial conglomerates at the time than do existing theories of state capture by oligarchs, corruption on the part of state officials, or state predation (see 2015, 36–57).

products or types of sales and purchases cropped up as well. By early 1992, there were more than 1,400 brokers and 730 registered brokerage houses in Perm. Vsevolod Bel′tiukov, one of Andrei Kuziaev's key deputies charged with staffing the exchange, later recalled, "We hired young people, because there was no conservatism in them, just the desire to plow ahead."[15] Bel′tiukov himself was sent off to the New York Stock Exchange for a crash course in exchange management.

Similar hiring criteria applied to the staff of the exchange itself. One of the PTB's first auctioneers, whose job was to call up items for bidding from a precirculated list and then manage the incoming bids through to a sale, told me that his own background was a good example of who came to work in the exchange. He had spent his early career working for one of the Perm region's major factories as a *snabzhulik*, slang for a fixer (literally a rogue supplier) skilled at acquiring whatever was necessary in the informal nooks and crannies of the planned economy. Having tired of this work by the late 1980s, he enrolled in the Perm Institute of Art and Culture, hoping to realize his dream of becoming a theater director. But then, in 1991, "the most interesting period began . . . and I saw that everything had changed . . . wild monetary reforms and all that, and I had to change my life." On a whim, he responded to an ad in the paper for a job as auctioneer (*vedushchyi torgov*) at the Perm Commodity Exchange, and easily beat out the competition:

> It was the combination of two things, maybe, [that got me the job]: work, first of all, in material-technical supply [but *not*, note, Glavsnab], plus preparation as an actor-director. The synthesis of those things in a broker-trader is, I guess, optimal . . . that is, I actually know what I'm hawking!

Others I spoke with recalled his booming voice announcing finalized deals from the front of the trading floor. If everyday economic life for most people in the Perm region in 1991 played out increasingly on street-corner markets, through barter, and in petty trade—all strategies for basic survival—then the PTB was set to become the new exchange mechanism for the much smaller number of residents intent on becoming the vanguard of a new and specifically regional economic order.

Rise: A New Center of Regional Exchange

On April 11, 1991, and on Tuesdays thereafter, the Perm Commodity Exchange opened for trading in the meeting hall that had once been home to the regional

15. No author given, "Kapital, Chto Dorozhe Deneg," *Kommersant*, March 14, 2006.

Komsomol. Several days before the opening of each trading session, brokers representing sellers submitted descriptions and prices to the exchange staff, who circulated them in a newsletter. The items were called up one by one, and brokers representing buyers made bids until a deal was reached or the item for sale was taken off the floor due to lack of interest that day. As soon as a sale was made, the brokers would meet with PTB staff to negotiate the precise terms and sign a contract. "We had an [IBM] 286 and a laser printer!" the auctioneer with whom I spoke recalled. The setup was one of the first in the entire Perm region. Most PTB trades were spot contracts, meaning that the exchange of goods would take place immediately. A smaller number were short-term forward contracts, specifying delivery at a designated future date. (The PTB did not offer derivatives contracts such as options and futures until 1993.) The exchange made its own money by charging brokers fees for their seats on the exchange and by collecting a small commission on each transaction.

The first trading days at the PTB brought crowds of onlookers and considerable media attention. Practically anything was for sale, dredged up from the hoarded supplies of the region's major enterprises: from helicopters to construction materials, from timber to televisions, from office paper to trucks and cars, from 105,000 packs of cigarettes to 1,513 tons of washing-machine powder.[16] With almost no competitors and most enterprises in the region using its services at least occasionally, the PTB quickly accumulated more money in commissions than it easily knew what to do with. Indeed, with barter and rationing proliferating nearly everywhere else, large-scale exchanges through the PTB were among the only transactions on which enterprises were willing to spend scarce cash reserves in the very early 1990s. Many of the items acquired then made their way to street-corner markets or into other, off-exchange, inter-enterprise deals.

The income from operations in its first months allowed the PTB to rapidly expand the kinds of trading it facilitated and to present itself as a wellspring of regional know-how about new economic relationships. At its height in 1992, the Perm Commodity Exchange incorporated trades in six locations simultaneously (Perm, Berezniki, Izhevsk, Sverdlovsk, Cheliabinsk, and Maikor). Using technology adopted from the Soviet defense sector's "Spark" system, simultaneous trading took place over open telephone lines amplified into the trading hall in Perm; multiple private lines then allowed one-to-one negotiations to finalize deals.[17] In mid-1992, the PTB was the former Soviet Union's fifth largest commodity exchange by volume of exchange, following only commodity exchanges based in Moscow and Saint Petersburg. The PTB also entered into contractual relationships

16. "Permskaia Tovarnaia Birzha: Dinamika tsen i predlozhenii na torgakh PTB s 18 maia po 22 maia," *Zvezda*, May 5, 1992.

17. K. Petrov, "Poslednee slovo—za prodovtsom," *Zvezda*, April 10, 1992.

with other exchanges across the former Soviet Union, enabling trades that extended outside the region, and these trades ultimately accounted for nearly a third of the PTB's business. This longer-distance trade was made possible, in part, by the fact that the Perm regional government was generous in granting the special permits that were, at the time, required to move quantities of goods out of the region.

Already by August 1991—months before the official end of the Soviet Union—the PTB had opened a real estate section, looking forward to a near future, its architects said, when apartments and garages would be bought and sold. In early December of that year, it hosted a major conference dedicated to carving out a role for commodity exchanges in the upcoming privatization of property. As part of the conference, the PTB brought attendees to the commodity exchange floor to show them how the trading process worked, and how the already up-and-running trading processes at the PTB could serve as the most appropriate context for real estate transactions.[18] The PTB also opened a new securities section dedicated to trading shares of new companies. Securities trading took place every other Tuesday, a schedule in keeping with the relatively low numbers of trades on that exchange at the time, but with the assertive goal of being prepared to host the many new companies and privatization auctions that were on the horizon.[19] The exchange also ran a series of seminars aimed at training brokers, and provided an array of fee-based services as well, including everything from legal advice to packets of prepared documents that could be used to execute trades, start a new company, or transform an existing enterprise into a joint-stock company. In January 1992, not long after President Yeltsin decreed that all regions should come up with privatization plans, the PTB presented itself as the most logical and cheapest venue in which to host privatization auctions for the region's biggest companies. It already had in place a network of brokers and a solid reputation, it claimed, declaring itself "ready to share its experience" with the regional government. The PTB offered to host auctions for free, allowing "the resources saved to be used to solve social problems in the city and the region."[20]

Although most people in the Perm region did not interact directly with the PTB, it was nevertheless increasingly presented, in part as a conscious marketing strategy, as the main crucible in which new relationships, markets, and *people* (young brokers, capitalists) were being formed in the first months and years of the post-Soviet period (on this kind of person-formation in the wake of social-

18. P. Koz'min, "Nam est' chto skazat' i pokazat'," *Zvezda*, November 29, 1991.

19. K. Petrov, "Mechty Olega Galashova, kotorye, vidimo, suzhdeno sbyt'sia," *Zvezda*, October 30, 1991.

20. Vladimir Pesternikov, "'Birzhevoi Period' v Istorii PFP-gruppy," *Novyi Kompan'on*, March 1, 2011, and O. Galashov and V. Puchnin, "Gotovy podelit'sia opytom," *Zvezda*, January 29, 1992.

ism, see especially Dunn 2004). Regional newspapers, still widely read at that time, published catalogs of the goods available for purchase on the exchange, along with their prices, as well as a running series of news articles about the exchange itself: profiles of the young brokers, descriptions of the gestures they used on the trading floor, monthly rankings of the most successful brokers and brokerage houses, comparisons between Perm's exchange and others in the Urals and in Russia.[21] The televised evening news began a Tuesday segment that listed items not sold during that day's trading. The Perm Commodity Exchange presented itself and was presented in the media, in short, as the leading edge of the regional transition to capitalism on multiple fronts: in the volume of exchange it facilitated, in the market know-how it taught at its seminars, in the kind of people it was producing, and in its ever-closer links to the regional state apparatus—not least in its efforts to contribute to the resolution of social problems and to be the face of the Perm region in its dealings with the rest of Russia.

In the twenty years between the rise of the PTB and my primary fieldwork, this view of the exchange solidified in repositories of regional collective memory as well. The Perm Regional Studies Museum's treatment of perestroika and the immediate post-Soviet period, for instance, gives the Perm Commodity Exchange pride of place. Whereas one exhibit case focuses on political changes, street demonstrations, and the local dimensions of the unraveling of Soviet power, another exhibit case focuses on the shifting means and modes of exchange. The PTB features prominently here, at what the exhibit calls the beginnings of the "era of capitalism." Next to a collection of various rationing coupons and examples of the dizzying array of currencies and denominations necessitated by rapid inflation are the mallet and gong used by the head auctioneer to announce a concluded sale. Nearby sit examples of brokers' identification cards, certifying membership in the exchange and completion of the required courses of study, including everything from Commodity Exchange Theory to Business Ethics. A published reflection on the tenth anniversary of the first trades at the commodity exchange in 2001, blown up to large size for museum visitors, notes that the PTB "remains in memory as a bright, romantic, and naive period of time, full of hope and expectations."[22]

Fall: The Unmaking of the PTB

All the PTB's image-making, as well as the constant addition of new services and divisions, was essential in large part because the PTB began to fade almost as

21. "Fondovoi Otdel PTB—Eto Eruditsiia, Molodost′, Energiia, Poriadochnost′," *Zvezda*, October 30, 1991. In 1991 and 1992, there was a monthly two-page spread in *Zvezda* titled "Bulletin of the Perm Exchange."

22. "Eto Sladkoe Slovo—Birzha." *Novyi Kompan′on*, April 3, 2001.

quickly as it had risen to prominence, for the same three interconnected reasons that ultimately brought an end to the exchange movement across the former Soviet Union by 1993. One basic reason for the decline of the PTB is that, in the spring of 1992, Russian federal law finally caught up with the country's many commodity exchanges, which had operated with little oversight and no full legal basis for over a year. A USSR Council of Ministers resolution passed in late March 1991 had for the first time in Soviet history officially permitted market prices to determine the cost of certain goods, several months after the Moscow Commodity Exchange had begun operations and only two weeks before the first trades at the Perm Commodity Exchange. But a full-scale Law on Exchanges and Exchange Trading did not appear until February 1992.[23] The new law struck at the heart of commodity exchange practice—at that time approaching its peak—by imposing significant new taxes on both brokers and exchanges and by banning commodity exchanges' main profit-making activities: the sale of seats to brokers and commissions on trades. However, given the weakness of the federal authorities and the unenforceability of many federal laws at the time, these new laws were not really the primary factor in bringing Russia's exchange movement to an end; the more proximate causes lay in ongoing upheavals in patterns of region-scale exchange.

A second reason that the PTB began to founder was that it lost the battle to keep exchanges running through its trading floor. No legal or institutional mechanism kept brokers from closing deals between clients off the exchange floor, a practice that grew quickly as new relationships among enterprises formed and settled in the Soviet aftermath. Indeed, Timothy Frye cites research suggesting that, across Russian commodity exchanges, there were approximately twenty to forty off-exchange trades for every one on-exchange trade in 1992 (1995, 46). The PTB's primary service of linking sellers and buyers, in other words, was most needed just as the familiar constellation of Soviet ministries, Communist Party structures, central planning offices, and unofficial fixers disappeared. As new connections jelled over 1990–92, its services were less needed. Many of the PTB's innovations—ongoing seminars and classes, new trading divisions, services from auditing to contract arbitration, simultaneous trading in multiple locations and contracts with exchanges outside the region, and even a club for new businesspeople, should be understood as efforts to hold onto trading business that could easily save on commission fees by simply cutting a deal off the trading floor. Not unlike business schools throughout the capitalist world, that is, the PTB sought to present itself as supreme arbiter of the kinds of knowledge necessary to

23. Russian Federation Law 2383-I, February 20, 1992, "O Tovarnykh birzhakh i birzhevoi torgovle."

market-based exchange—a claim made necessary in significant part by the obvious fact that much exchange could take place perfectly well without it and its associated costs. Although the PTB lasted longer than many of its fellow commodity exchanges in Russia, by 1993, these strategies were no longer keeping up with proliferating retail operations and direct, nonexchange-traded deals for many kinds of goods, many of them on barter terms as money became scarcer and scarcer.

A third reason for the fall of the PTB and its fellow exchanges, and perhaps the most significant for the overall story of the making of the post-Soviet Perm region, was that most prices in Russia were officially liberalized on January 1, 1992. Ever since the PTB's first trades on April 11, 1991, much of the profit made by brokers, clients, and the exchange itself came through taking advantage of the difference between the Soviet-era state prices at which enterprises had acquired (and, often, could still acquire) goods and the much higher prices that could be obtained at PTB open auctions. Given the fixed and low nature of Soviet state prices, many trades were pure profit for sellers, and it was, in fact, quite easy for the brokers and founders of the PTB to make the first postsocialist millions in the Perm region by putting up for sale items that had, until recently, been Soviet state property. The removal of price controls at the beginning of 1992, which coincided with the diminishment of Soviet-era hoarded goods and supplies that were so eagerly sold at the PTB in its first months, meant that a significant portion of the commodity exchange's business dried up. The smaller profit margins available in an environment of floating prices were simply far less attractive and far less worth the cost of brokerage services and commissions than they had been in 1991.

This final reason for the fall of the PTB bears underscoring. Western observers of Russia's commodity exchange movement have argued that the early post-Soviet state was either unable to enforce exchange contracts and therefore "too weak" (Frye 1995) or still very influential in the setting of certain prices, and therefore "too strong" (Wegren 1994, 223). Such perspectives owe a great deal to normative ideals of how markets should work and how Russian commodity exchanges of the early 1990s did or did not measure up to those projected ideals. Beginning from a less normative perspective, we might note that commodity exchanges have long been a potent nexus for a myriad of different kinds of encounters, collusions, and contests between crystallizations of political power and patterns of economic exchange (see, for instance, Weber [1894] 2000; Zaloom 2006). In the case of the Perm Commodity Exchange, several aspects of these encounters are notable: the exchange's specifically regional scope; the larger context of disintegrating Soviet supply networks and proliferating barter relationships; the high degree of involvement of the former Soviet political and economic elite in the very founding and organization of the exchange; and the high proportion of pure profit sales made before price liberalization, sales that quickly converted Soviet enterprise hoards and state property into new fortunes—for many sellers and especially for the leadership of the PTB itself. A final, and crucial, aspect of the Perm Commodity Exchange's short life is the arrival of regional oil and oil products on its trading floor.

Second Wind: The Arrival of Oil at the PTB

Oil and oil products were not common items of trade on the PTB in its heyday. The auctioneer I spoke with at length about his time at the exchange did not recall their playing a significant role, and published statistics from the time confirm his memory that exchange transactions consisted mostly of food and household items, on the one hand, and an array of hard-to-find large-ticket items that he himself could barely believe at first, on the other hand: helicopters, railway cars, barges, and other items that enterprises had stashed away for trades in the Soviet second economy. In April 1992, the monthly "News of the Exchange" section of the newspaper *Zvezda* reported that the biggest categories of trade were food products (32.1 percent), wood and paper products (14.1 percent), metals and metal products (14.4 percent), chemical products (8.6 percent), agricultural products (7.3 percent), and basic consumer goods (7.2 percent).[24] The major reason for oil's absence on the PTB was that the Yeltsin administration was intent on keeping the entire energy sector under tighter state control than other sectors, and doing so for a longer time. This was true for both energy distribution mechanisms and prices. Whereas in 1990–92 many other essential goods—including foodstuffs—began to move through the commodity exchange or the deals that took place at its edges, PNOS and Permnefteprodukt (both organizations still acting within State Concern Lukoil) continued to distribute the overwhelming majority of refined oil products in the Perm region. Whereas price differentials of various sorts quickly shaped the markets for other items transacted at the PTB, energy was a major exception to the Yeltsin administration's embrace of rapid price liberalization programs, and so oil and other energy prices remained low and state-controlled well into the mid-1990s (and, depending on how one counts, beyond). There was a mix of reasons for this lag in oil price liberalization, including the significance of energy sources for all sectors of production and transportation; an entrenched and powerful energy lobby in Mos-

24. "Syry ne zasizheny—tseny snizheny!" *Zvezda*, April 10, 1992. Wegren 1994 treats the importance of commodity exchanges for agricultural transformations.

cow; and a widespread perception that the combination of Russian winter and energy price deregulation would be catastrophic (see Åslund 1995, 156–61).

Nevertheless, limited increases in energy prices—far below full-scale liberalization—took place in September 1992 and again in February 1993 (Sagers 1993, 350; Åslund 1995, 158), in both cases in a fairly haphazard and uneven fashion, with some regional state administrations, including the Perm region's, prevailing on their local distributors not to go along with price increases. The resulting instability and unpredictability came along just at the moment that the three factors outlined above were drying up business and profits at the PTB, and they thus provided an opportunity for Kuziaev and his companions to reimagine their commodity exchange. A revitalized version of the PTB, they began to project, would trade in a much more limited number of primary commodities, including timber, oil, and metals, functioning on the model of a more traditional capitalist commodity exchange like the Chicago Board of Trade or the London Metal Exchange rather than a *barakholka*, as observers often called it—a trading floor for anything and everything under the sun. By the early fall of 1992, the PTB became a regional leader in the trade of oil products by using its extensive networks and specialists to execute sales on behalf of PNOS.[25] This alliance represented a considerable expansion of PNOS's regional connections and influence, for the PTB staff had come to know and work with all the Perm region's major buyers and sellers over the previous couple of years. By February 1993, as oil and gasoline prices began to rise more precipitously, the PTB began a new service that PNOS lacked the expertise to perform itself: executing the region's first options trades for oil products.[26] Options, in contrast to the spot and short-term forward contracts in which the PTB had previously specialized, enabled large purchasers, along with PNOS itself, to attempt to minimize the risks of an uncertain market by locking in prices in advance, for a fee.

The new regional market for oil options was, to be sure, yet another strategy designed to prop up the PTB as its business threatened to dissipate into more diffuse, less concentrated exchange circuits. But it was also a crucial moment in the reshaping and concentration of regional exchange patterns: for the first time, oil and oil products were circulating at the Perm Commodity Exchange, at the very center of emergent regional exchange networks. They were the PTB's "second wind," as one of my interviewees put it, the newest and latest means by which the Perm region's elite were working to solidify their place at the top of new kinds of regional hierarchies. A handful of new corporations lay at the center of these

25. A. Nichiperovich and V. Uzhegov, "Birzha v reklame ne nuzhdaetsia," *Zvezda*, October 10, 1992.
26. "Birzhevye vedomosti: Birzha umerla. Da zdravstvuet birzha," *Zvezda*, April 8, 1993.

exchanges: the PTB, PNOS, and Lukoil. In these and subsequent incarnations, these corporations, and the exchanges of oil, money, and other goods that they came to specialize in, came to anchor much larger regional patterns of exchange. In fact, the delayed appearance of oil on the PTB made it a convenient bridge over which Andrei Kuziaev and his team could cross into another epoch, another corporation, and another reconfiguration of regional exchange.

The Perm Financial-Productive Group

By early 1993, only around two dozen of the five hundred registered commodity exchanges in Russia were still in operation. Perm's commodity exchange was among them, but, with the writing on the wall as early as 1992, its leadership team had already begun founding a number of new companies, including the Universal Trading House, specializing in consumer goods; the Urals Corporation for Mutual Credits, which facilitated nonmonetized exchanges; the Perm Real Estate Agency; and others—all of them designed to outlast the PTB when, it had become clear, the exchange would close. By far the most significant of these new corporations turned out to be Oil Products Marketing (Neftsintezmarket), a joint venture between the PTB and PNOS that was established in order to house the PTB's increasing participation in the regional circulation of oil products and the just-beginning market for options contracts. Oil Products Marketing brought the reputation and region-wide exchange networks of the PTB to the oil sector. Its expertise in all things related to postsocialist markets even extended to the protection services that were obligatory at the time (Varese 2001; Volkov 2002), and a portion of Oil Products Marketing's sales force was recruited from the same population as the proliferating mafias of the early 1990s: hulking veterans of the Afghan war, bodybuilders, and other sportsmen cast off by the closing of the Soviet Union's Olympic training facilities. "Just try to not pay those guys for your oil products!" laughed one of my interlocutors at the memory.

Oil Products Marketing linked oil refining and distribution together via a new route, one that ran not through the old Soviet system of official ministries plus informal exchanges and not—yet—through a vertically integrated oil company like Lukoil, but through a dedicated region-based company embedded in the structures of the PTB. Indeed, it is worth noting that it represented a step *away* from vertical integration at the regional level, for PNOS itself ceded a good deal of control over distribution to a company outside the oil sector entirely. As the PTB gradually wound down operations in 1993, Oil Products Marketing took up residence in another sprawling, integrated regional company: the Perm Financial-Productive Group.

A Regional Financial-Industrial Group

In March 1993, Andrei Kuziaev and his team from the PTB started still another venture, this one a large holding company designed to replace the commodity exchange: the Perm Financial-Industrial Group (the Permskaia Finansovo-Promyshlennaia Gruppa, or PFPG). The PFPG was the first major financial-industrial holding in the Perm region, and it contained within itself Oil Products Marketing and many of the other companies that had grown up within the structure of the Perm Commodity Exchange. At the time, financial-industrial groups (FIGs) were emerging all over Russia to take advantage of the coming privatizations of major natural resource companies; the largest and most influential FIGs ultimately served, in many cases, as incubators for what would become Russia's oligarchs.

Like a significant number of these FIGs, the PFPG did not specialize in any one sector or product; its activities were, rather, spread across multiple sectors, generally not including a full supply chain in any one of them. This brand of organization enabled the rapid, nonmarket movement of goods and services among constituent companies, often without monetary transactions, and was especially useful for maintaining flexibility in the constantly shifting economic circumstances of the 1990s. Integration across key sectors also concentrated enormous power and influence in the company's leadership. ("What *didn't* the PFPG control back then?" reflected one of my interlocutors when I asked about the company.)

As was the case with the founding of the Perm Commodity Exchange in the early 1990s, however, Kuziaev's team and their supporters in the Perm regional state administration had moved more quickly than federal regulations. In December 1993, several months after the PFPG registered itself as a FIG in Moscow, President Yeltsin issued a decree regulating the formation of financial-industrial groups. The decree included provisions that seemed designed to limit the number of groups that could officially register themselves with this title and enjoy privileged access to the coming privatization auctions that came with registration. Part of the idea behind offering concessions of various sorts to officially registered FIGs was to use the privatization process as leverage over the kinds of companies—ideally more rather than less stable, for instance—that would end up controlling formerly state-owned assets. (FIGs were required, for example, to include a bank within their structure.)[27] After some negotiations with the federal authorities, it became apparent that, despite the backing of the Perm region's political leadership, Kuziaev and his colleagues from the PTB would be shut out of the rush to form FIGs.

---

27. On the negotiations that led to this decree, see Starodubrovskaya 1995.

Gennadii Igumenov, at that point one of the two first deputy governors of the region (the other was Kuziaev's old mentor Evgenii Sapiro) recalls this moment in his 2008 memoir:

> We started to think what to do next. And concluded that the point is not in the name, but in the activity, which should be directed at solving regional problems and developing quickly. And so Kuziaev and I agreed that we should just change a single word in the name and, in place of industrial [*promyshlennaia*], name it a productive [*proizvodstvennaia*] group. This gave us the possibility of registering it here in the region. And this is what we did. (Igumenov 2008, 296)

Andrei Agishev, a former vice president at the PFPG, recalled the same discussions following the federal law on FIGs:

> There were a lot of words and not many thoughts in that [law]. We read that document and decided to change the name from "industrial" to "productive" so as not to fall into the quickly proliferating number of structures that the federal state was intending to manage.[28]

Negotiations in 1993–94 were doubtless more complicated than Igumenov and Agishev recount, but these two key players' emphasis and phrasing highlights the extent to which the PFPG, and with it the deepening ties between Perm's regional elite and the circulation of PNOS's refined oil products, were excluded from the federal-level processes on which most analysis of FIGs has centered. Subsequent scholarship on the rise of the FIGs, for instance, has generally divided them into two camps: first, those led by so-called red directors—socialist managers trying to hang on to their enterprises and wring additional subsidies out of the state—and, second, those led by more dynamic, national banks aiming to invest in industrial restructuring in various sectors (Gorbatova 1995; Johnson 2000). The PFPG, however, arose out of the ashes of a commodity exchange and was built around a set of companies, especially Oil Products Marketing, that specialized in regional exchange and financial operations. It included neither a bank nor a marquee enterprise within its own structure—at least at the time of its incorporation as a holding company. It therefore fits neither model of a FIG particularly well, although it would certainly fit in the more omnibus category of regional "conglomerates" preferred, for instance, by Susanne Wengle (2015).[29]

---

28. "Iubilei Biznesa Prikam'ia," *Novyi Kompan'on*, February 27, 2001.

29. Later in the 1990s, the precise requirements of the federal law on FIGs became less important and the variety of FIGs expanded. Moreover, as the tax credits and other benefits originally expected from this status never materialized, officially registered FIGs and de facto FIGs such as the PFPG

The PFPG's deregistration in Moscow and subsequent renaming and reregistration in Perm was, however, crucial for the creation of regional networks: instead of linking the region more tightly into larger, Russia-wide circuits of capital and connection on the eve of major privatizations, it focused them inward, toward enhanced region-level ties (see also Barnes 2006, 120–28 on "second-tier" regional FIGs).

There were, to be sure, other FIGs active in the Perm region at the time. Menatep Bank, part of the Yukos FIG controlled by Mikhail Khodorkovsky, acquired the Berezniki titanium-magnesium plant AVISMA in 1993. Alpha FIG acquired the Perm Pulp and Paper Mill in 2000, part of its effort to combine ownership of several paper plants across Russia. The PFPG was not even the only FIG in the Perm region that specialized in oil products—the much smaller and less well connected DAN, headed by Pavel Anokhin, was founded somewhat later, in 1995; it controversially supplied the Perm region with oil products imported from Chechnya (Fedotova 2006, 217–19). None of these organizations, however, was nearly as significant for the entire Perm region as the PFPG. From its suite of offices in a two-story house not far from the Kama River embankment, the PFPG's small office staff coordinated an ever-increasing proportion of the oil products and food supplies moving through the Perm region.

Petrobarter and the Regional Imagination

Even if it was not registered in Moscow as a FIG, the PFPG certainly engaged in many of the operations that were characteristic of official regional FIGs and other sorts of conglomerates, many of which were "designed to help industries and regions achieve stability in unstable conditions by creating vertically integrated and closed production cycles, centralizing contract enforcement, facilitating barter arrangements, and spreading risk over a number of enterprises" (Johnson 2000, 160). Given that it emerged from the Perm Commodity Exchange, the PFPG was especially focused on circulation and exchange, which meant, more and more in the early 1990s, barter. In fact, we might summarize this dimension of the PFPG's initial operations by saying that it took the limited petrobarter transactions in which PNOS had engaged in the very early 1990s, infused them into the region-wide networks of exchange that the Perm Commodity Exchange had built up, and thereby established the circulation of oil products as central to the new forms and patterns of exchange across the region.

---

became, in practice, almost indistinguishable from each other. On other types of regional FIGs in the period I discuss here, see Prokop 1995, and on the participation of trading companies (of which the PTB was one variety) in regional FIGs, see Freinkman 1995, 54–56. Makarychev 2000 and Botkin, Kozlov, and Collins 1997 discuss FIGs elsewhere in the Urals.

These petrobarter operations took many shapes, most of which included multiple subsidiaries within the larger PFPG holding structure. Some recalled the targeted international petrobarter that had brought Toyotas to the Perm region in 1990. In one exchange recounted to me several times, for instance, the PFPG, through Oil Products Marketing, bartered PNOS oil products directly for an entire barge of Cuban sugar at a time when sugar had not been seen in the Perm region for nearly a year (see also Alonso and Galliano 1999). It is some indication of the high value placed on the PFPG's international petrobarter transactions at the time that, rather than hire laborers, the PFPG office staff itself unloaded the barge under armed guard and supervised its transfer to the Universal Trading House—the PFPG subsidiary that specialized in supplying stores and markets with consumer goods.

Other exchanges were considerably more complicated. One of my interlocutors, a former employee of the PFPG, gave an example of the kinds of exchanges that, he said, characterized their work in those early years:

> There were lots of long chains of exchange. So we provided fuel oil to the logging enterprises somewhere up north, and, roughly, got [uncut] timber from them. We processed it into packaging timber and sent it off to the Krasnodar region [some 1,500 miles away, in the Caucasus]. And there we took not money but apples. We brought the apples here, sold them, and got money for them. That's the kind of operation we did—to the fifth or sixth link.

Or, as economist and initiator of the Perm Commodity Exchange Evgenii Sapiro remembered the same exchange circuit, with an ironic commentary on Marxian economic theory thrown in:

> Andrei Kuziaev and his team proved that the foundational formula money–commodity–money is unidirectional and primitive. [The PFPG] practiced a quite ordinary formula: money–crude oil–oil products–roundwood–processed timber–packaging timber–apples (from Krasnodar, I recall)–money.[30]

PFPG's petrobarter operations were, moreover, closely entangled with the interests and projects of the regional state government, which continued to cope with pervasive complaints and borderline unrest arising from the ongoing shortages of both food and fuel across the Perm region throughout the 1990s. The timber part of these exchanges, for instance, took place with logging operations in the far north of the Perm region, but these were not simply corporate transac-

30. Evgenii Sapiro, "Ot PTB k PFPG," *Novyi Kompan'on*, March 1, 2011.

tions. The regional state administration had contracted with the PFPG to carry out the region's northern delivery (*severnyi zavoz*)—the annual supply of remote northern districts with food and fuel by river routes before winter set in. The operation was ideal for the PFPG as it had set itself up: a regional state contract, bulk delivery, no competitors, and the acquisition of goods—in this case timber—for which there was demand outside the region.

Indeed, the PFPG soon took on the role of official state contractor for all acquisition, refining, and distribution of oil and oil products for regional state administration needs, most notably support for the agricultural sector, where the withdrawal of socialist-era state subsidies meant that tractors and combines often sat idle for lack of funds to purchase gasoline. As Gennadii Igumenov explained to me in intricate detail, the PFPG's multiple constituent units, in collaboration with state agencies, would facilitate the exchange of gasoline at the time of spring sowing for crops at the time of fall harvesting, and then use a portion of those crops to form a barter fund that enabled trade with other Russian regions. In some cases, these barter chains brought for the regional administration not products to be distributed but resources to complete crucial infrastructural projects, such as the repair of the floodgates at the Kama Hydroelectric Station (Igumenov 2008, 297). A similar chain, with the inclusion of the construction industry along the way, led to the building of a new apartment building that housed faculty members from Perm State University.

These extensive petrobarter chains—far more extensive and significant than the Toyotas and down jackets deals of 1990–91—were another crucial chapter in the emergent intersection of oil, state, and elite in the Perm region. One Perm newspaper described the PFPG's petrobarter operations as the ultimate insiders' network: "[Oil Products Marketing] worked out and carried out dizzying chains of mutual exchanges and credits, the result of which were schemes that it was impossible to understand. Unless you participated in them."[31] This description handily captures both the complexity of petrobarter deals at the time and their control by a small concentration of companies and their supporters in the regional state apparatus. It also hints at the social and material relations that undergirded these deals—relations that, rooted in barter rather than monetized exchange, were dramatically different from the ways in which oil, exchange patterns, and politics have entwined in other contexts around the world. Whereas the fluctuating monetary prices at which oil is exchanged, both at the pump and on international markets, have often been associated with instability, turbulence, and incoherence (see esp. Guyer 2015), the exchange of oil through petrobarter in the Perm region marked stability, a degree of predictability, and a ref-

31. Valerii Mazanov, "PFP-Gruppa: tak soshlis' zvezdy," *Novyi Kompan'on*, February 14, 2006.

uge from wildly gyrating prices and money supplies in Russia more broadly. This was so in part because, although the socialist oil complex was morphing rapidly along multiple axes, there was not yet an entrenched relationship between oil and money.

A number of influential studies of the circulation of oil and oil profits in oil-producing states on the capitalist periphery focus on the magical properties of money gained from oil proceeds, such as oil money's ability to grow with little apparent effort, and then trace the importance of this process for the formation of particular kinds of states and political and cultural imaginations. Andrew Apter's analysis of oil boom and bust in Nigeria, for instance, rests on the claim that oil "standard[ized] the relative values in terms of which other commodities were bought and sold and thus approach[ed] the general equivalent of money itself" (2005, 35). Apter's ethnography beautifully captures the way in which the magical growth properties of oil money as general equivalent entered the popular Nigerian imagination, from an advertisement that billed Saxon photocopiers as "Money Makers" because they could "copy anything" at the push of a button to Mr. Emmanuel Omatshola's Magic Barrel, an oil company public relations stunt in which commodities ranging from diesel fuel to perfume to plastic bags were conjured out of a barrel of crude oil (Apter 2005, 43, 24). Writing about Venezuela, Fernando Coronil argues that,

> circulating through the body politic as money, oil ceased to be identified as a material substance and became a synonym for money. . . . Just as oil came to be seen abstractly as money, the state became a general representative of a political community of shared ownership of the nation's natural body. (1997, 390)

For Coronil, oil as generalized equivalent was central to the imagination of the state; indeed, he opens *The Magical State* with the words of Venezuelan playwright José Ignacio Cabrujas: "Where did we get our public institutions and our notions of 'state' from? From a hat, from a routine trick of prestidigitation. . . . The state is a magnanimous sorcerer" (Coronil 1997, 1). As Michael Watts (1994) puts it succinctly in a study of Nigeria informed by both Marx's and Simmel's accounts of money, the issues here are abstraction and generalization: money enables certain imaginations through its capacity to make unlike items exchangeable and, as part of this process, serves both to integrate and disintegrate social groups.

Although the abstracting and generalizing tendencies of money are quite widespread, Apter, Coronil, and Watts claim that they take on some special characteristics in postcolonial oil-exporting states within the global capitalist system. In these states, an elite class with privileged access to the subsoil realizes profits on the international market (in dollars) and then reinserts those dollars into local currency circuits at home through sectors that it controls. Money as oil rent and its rapid circulation has, in this view, crucial implications for the resulting incarnations of state power and its cultural entailments, whether they be statist, nationalist, pan-African, developmentalist, modernizing, or Occidentalist. Furthermore, these movements are quite distinct from those of industrial, factory-based capitalism, because they rely almost entirely on increased monetary circulation and attendant dreams of rapid progress rather than on transformed relations of production.

Money behaved in very odd—even magical—ways in the postsocialist world as well. I have already noted that bouts of demonetization and remonetization alternated with little warning and were among the reasons for the formation of early post-Soviet suzerainties and the turn to barter. The massive expansion and centralization of petrobarter at the PFPG beginning in 1993 also coincided with all kinds of new contemplations of and anxieties about the money form (e.g., Lemon 1998; Ries 2002) and the rise and spectacular collapse of pyramid schemes. Writing of pyramid schemes in postsocialist Romania, Verdery argues that one of their effects was "producing an abstract sphere in which money circulates and multiplies without clear agency" (1996, 183). Russia's most famous and largest pyramid, the Moscow-based MMM, ballooned in late 1993, reported 1,000 percent returns on investment by early 1994, and was shut down by the Russian government that same June. There was no shortage of homegrown pyramid scheme activity in the Perm region at the same time; indeed, part of Verdery's argument was that a certain set of emergent regional coalitions—usually those shut out of establishment postsocialist structures like the PTB, the PFPG, or lucrative privatization deals—saw running pyramid schemes as their best available accumulation strategy.

The increasing importance and centrality of oil to exchange in the Perm region, however, emerged in good part *in contrast to* the kind of wild, magical, unfamiliar, and dangerous money epitomized in pyramid schemes of those same years. To the extent that transforming oil products into jackets, Toyotas, apples, or timber entailed a certain degree of magic, its tricks and transformations known only to an initiated few, it was magic without the general equivalent of money described for postcolonial petrostates like Nigeria or Venezuela, without the abstract sphere and spectacular booms and busts of postsocialist pyramid schemes, and without the overall turbulence so often associated with shifting oil prices around the world. Flowing along the lines of petrobarter facilitated by the PFPG, 1990s Permian oil and its derivatives were often framed as concrete, reliable, local, tangible, nongeneralizable—precisely a contrast to the unpredictable and massively abstracted nature of rubles and money in the banking, salary, streetcorner exchange, and pyramid scheme sectors. Oil bartered into sugar and apples, not to mention crops sown and harvested in the region thanks to collaboration between the PFPG and the regional state administration, were repeatedly deployed as evidence that oil in the Perm region was *not* simply being transformed into money. In a lengthy newspaper article directed to those who criticized the PFPG for its rapidly accumulating wealth, for instance, the holding's leadership argued that local businesspeople like themselves were much preferable to Moscow structures and foreign ownership, for "local circles" of business owners would be most likely to concern themselves with the development of the regional economy.[32] The Perm region's postsocialist petrobarter thus indexed locality, regional imaginaries, and the ability to stand up against international and national centers of power by fleeing from the most exchangeable and abstract currency—money—into local regimes of value that had some tangible materiality, some connection to the Perm region.

The associations among barter, oil, and the PFPG were so pervasive that they inflected all kinds of imaginations of regional social relations, and continued to do so nearly two decades after their zenith. To give but one example: in 2012, I spoke to a journalist who covered local industry in the 1990s, and she recalled that the PFPG was the first company in the Perm region to open a public relations office and to actively court regional journalists. In contrast to the many financial and industrial leaders who were anxious to keep their activities as quiet as possible, Andrei Kuziaev and his colleagues were talkative, hosting press conferences and producing some of the first press releases in the region. This journalist described the scene:

> In 1994, the PFPG was the only company that organized press conferences for journalists, the first company to roll out that technology. At least once a quarter Kuziaev [spoke] about Oil Products Marketing, about the oil products distributed to the region. [Once, the Universal Trading House] had just opened some new stores, and [he called a press conference]. The journalists all ate and ate, because Kuziaev set out a table like you wouldn't believe. All the journalists came and ate and then wrote articles for free. They all happily reciprocated. It was its own kind of barter.[33]

• • •

In the increasingly practiced hands of the PFPG, then, petrobarter grew from a limited survival tactic for PNOS at the end of the Soviet Union to become both

32. Natal'ia Kopylova, "Svoia rubashka blizhe k telu," *Zvezda*, March 24, 1995. My analysis here complements that of Yoshiko Herrera (2005), who argued that the ways in which different Russian regions imagined their economies was closely tied to political movements, including movements for increased regional autonomy.

33. For further discussion of the afterlives of 1990s petrobarter, see Rogers 2014a.

a primary business model and strategy of accumulation for a new regional elite—often at the expense of the federal center and its monetary currency—and central to that elite's efforts to legitimate itself, stave off popular critique, and reimagine the Perm region in the process. If oil dollars obtained on international markets have been central to the institutions and imaginations of postcolonial petrostates, then, in this postsocialist case, the barter of oil was central to the *unimagining* of the federal state in the 1990s and the contemporaneous rise of a specifically regional oil elite. In light of the ways in which scholars have understood states and oil corporations to be entangled in the rest of the world, it is worth underscoring that the configuration I have been tracing here crystallized long before the monetization of oil markets, the regular collection of taxes in money, the full privatization of regional oil production, or the assembly of a regional vertically integrated oil company. Instead, the consolidation of political power and the circulation of oil took root in barter transactions along and among the omnipresent tentacles of the PFPG.

Monetary exchange, though, clearly played some role—barter is never a closed system (Humphrey 1985; Humphrey and Hugh-Jones 1992). One might ask, for instance, what came of the money that the PFPG made on the sale of apples and sugar through the network of stores run by its Universal Trading House subsidiary. It is hard to say, even for those most well informed about those chaotic years. Certainly one place the money went was to still another subsidiary of the PFPG, the Securities League (Lig Tsennykh Bumag), which specialized in purchasing shares in privatizing companies throughout the Perm region. A particular focus was shares in Permneft, the struggling regional oil production association, and this accumulation of shares turned out to be as consequential for the region as the creation of Oil Products Marketing.

3

THE LUKOILIZATION OF PRODUCTION

Space, Capital, and Surrogate Currencies

"What's good for Lukoil is good for the region." So spoke Evgenii Sapiro, architect of the Perm Commodity Exchange, in 1996, when he was serving as speaker of the Legislative Assembly of the Perm region.[1] Vagit Alekperov's Lukoil was, at the time, in the process of completing the acquisition of Permneft, the Perm region's oil production association, and there were considerable anxieties about whether this development was in the region's best interest. Sapiro's transformation of General Motors president Charles Wilson's famous phrase, "I always thought that what was good for the country was good for General Motors and vice versa"—uttered at his confirmation hearing to be President Eisenhower's secretary of defense 1953—is notable for two reasons. First and most obviously, Sapiro substituted region for country, and, as he sought to reassure critics, underscored the significance of corporation–region relationships in 1990s Russia.

Second, General Motors and Lukoil occupied very different positions in the national and global oil complexes at the times when Wilson and Sapiro spoke. General Motors was coming out of World War II as one of the biggest military contractors in the United States and entering a postwar period in which the V8 automobile engine, suburbanization, and highway construction would become central to American lifeways (Huber 2013; see also Lutz 2014); these lifeways came to depend on increasingly high amounts of imported oil (largely from the Middle East) and helped shape the Cold War–era international order (Mitchell 2011, 41–42). In the 1990s Perm region, the corporate–state symbiosis that Sapiro en-

1. S. Fedotova and A. Neroslov, "Vremia L—1996 god," *Permskaia Neft'*, August 12, 2011.

couraged was rooted in a different dynamic: the gradual assembly of Lukoil as a globally competitive vertically integrated oil company at a time of massive unpredictability and federal state weakness in Russia. Oil was on the way to becoming no less central to lifeways in the Perm region, but it was doing so along quite different paths than the postwar United States of Wilson's famous comment.

Production Crisis

Permneft began the post-Soviet period in considerably worse shape than the refinery Permneftorgsintez (PNOS). In a 1990 interview with the head of Permneft's geology division, a newspaper correspondent pointed out that PNOS's executives were riding around the city in new Toyotas. "And we drive uaziks," replied the geologist—a reference to the Soviet jeeps and vans turned out by the Ulianovsk Automobile Plant. The contrast was doubly hierarchical: flashy and exotic imported Toyotas versus standard-issue Soviet automobiles, and the paved Perm streets along which PNOS's executives drove to work versus the rutted roads that ran to the region's oil production facilities.[2] It was, in keeping with the heritage of the socialist oil complex at the regional level, refining rather than production that brought the earliest signs of glitz and glamour to the Perm region's oil sector.

There were several reasons for Permneft's malaise in the early 1990s (see also Kurbatova 2006, 520–22). First, much of the specialized material used in the region's drilling operations, including drilling pipes, had been produced in Azerbaijan and elsewhere in the former Soviet bloc.[3] Soviet central planning had brought these materials to the Perm region with reasonable regularity, but central planning had ceased to exist and these now-international suppliers had little independent interest in supplying Permneft with the materials it needed to repair existing wells, let alone to drill new ones. For a year or two, Permneft was able to coast on previously hoarded supplies, but they were rapidly depleted. Second, as I noted in my discussion of the Perm Commodity Exchange, price controls for oil remained in place officially until the beginning of 1992 and in practice in the Perm region for some time after that. These prices were artificially low by design, intended to keep the Russian economy from even further contraction by not raising transportation costs too rapidly, but they had a devastating impact on Permneft, which was not protected from rapidly rising costs. With many state

2. V. Deriagin, "Issiakaet neftianoi istochnik," *Zvezda*, May 22, 1990.

3. Azerbaijan was, for instance, home to the only Soviet factory that produced "Christmas trees"—the valve assemblies that sit atop oil wells and enable production to be regulated or shut off. With Azerbaijani independence and conflicts spreading in the Caucasus, drilling new wells in Russia in the 1990s required alternate and creative sourcing (Harley Balzer, personal communication).

and industrial customers not paying PNOS for refined oil products, the refinery was itself often unable to pay even artificially low prices for crude oil, further exacerbating the crisis. Salaries at Permneft were soon delayed, repairs put off or wells taken out of production, and new exploration curtailed dramatically. Moreover, Permneft's available workarounds were few and far between. With prices fixed, profit-making arbitrage through the Perm Commodity Exchange was not an option and, unlike a great many other items, crude oil was not—as was the case in the Soviet period—very desirable or easy to barter because it always needed to be processed through a refinery before its use value could be realized. By itself, Permneft had nowhere near the barter and redistribution possibilities—and hence the Toyotas—that PNOS enjoyed in the very early 1990s.[4]

A third set of problems concerned basic issues of organization and control brought on by the end of the Soviet system. Everything from Soviet-era legal regulations about rights to the subsoil to informal networks that greased the oil industry's relationship with surrounding communities was suddenly up for renegotiation and improvisation. In the era of parcelized sovereignty, lower-level units of all sorts were increasingly empowered—including NGDUs, which often worked with local political networks to their own advantage rather than that of their parent association, Permneft. Knowing very well that oil could not be pumped from just anywhere, for instance, collective and state farms in production districts joined with local political leaders to demand additional concessions for access to lands they saw as under their primary control. The Krasnovishersk district, in the far north of the Perm region, went the furthest in this direction, declaring that natural resources beneath its soil were the exclusive property of the district. As a price for Permneft's continued access for extraction, the district proposed several new apartment buildings, a school, a kindergarten, the reconstruction of the district's heating infrastructure, and 10 percent of whatever Permneft received for the oil (Vikkel′, Fedotova, and Iuzifovich 2009, 87). In Chernushka in the south, the regional administration followed a similar logic and refused to continue supplying the Permneft workers' cafeteria with meat products, claiming that their primary responsibility was to schools and kindergartens and that the oil company should fend for itself.[5] For the leaders of Permneft in the early 1990s, it was therefore all they could do to hold their constituent units

4. Only one NGDU in the Permneft production association, Krasnokamsk, was able to send crude oil abroad through the Druzhba pipeline (Abaturova 2003, 130). Such exports were often permitted by the federal government on the condition that proceeds would return to the Perm region for specific, pre-agreed uses (Sapiro 2009, 78–79). Although it is likely that regional criminal organizations made a good deal of money moving supplies of crude oil to refineries near and far, it is hard to gauge the effect of these transactions on Permneft.

5. V. Deriagin, "Issiakaet neftianoi istochnik," *Zvezda*, May 22, 1990.

together into a single production association and stay on good terms with the localities in which they produced oil.

Finally, the part of the Volga-Urals oil basin on which the Perm region sat was low on the desirability list of those who were scrambling to gain control over former Soviet oilfields in the early 1990s. Production had been declining since the mid-1970s, and it was generally assumed that there were no significant new oilfields to be discovered in the region. Both emerging Russian companies and their major Western counterparts thus focused their attention elsewhere: the Caspian; West and East Siberia; and Sakhalin and the Arctic Sea. In the Volga-Urals oil basin, the chief focus of attention was not on the Perm region but on Tatarstan and its supergiant Romashkino field.[6] Permneft did enter into a limited number of joint production agreements with second- or third-tier international oil companies. PermTOTIneft, for instance, was jointly owned by Permneft and the Ecuadorian company Totisa del Ecuador, and worked on a small collection of wells in the Osa district. Permteks, jointly owned by Permneft and the U.S.-based oil company SOCO, set to work in the north of the Perm region, with the SOCO side promising sorely needed materials and infrastructure for drilling and developing the deposit.[7] But these deals did little for short-term improvements in Permneft's prospects. Hamstrung by falling production, fixed low prices, a very low quota for exporting crude directly abroad, constant challenges to its unified organization, and the comparative difficulty in making barter transactions, Permneft sank deeper and deeper into crisis in the early 1990s. By 1993, 38 of 41 divisions were losing money.[8] Production fell to 9.3 million tons a year in 1994—barely a third of the late-1970s peak.

Although none of this was far out of step with the larger picture of Russian oil production, which fell by nearly 50 percent between the years 1987 and 1993 (Sagers and Grace 1993, 855), Permneft's production crisis set the stage for some specifically regional dynamics and transformations. This chapter charts Permneft's path out of crisis and into big Lukoil over the course of 1991 to 2004, with a particular focus on privatization, incorporation, surrogate currencies, and the uneasy alliance between finance capital and industrial capital as they unfolded in—and increasingly drove the remaking of—the post-Soviet Perm region. These elements, together with those discussed in chapter 2, although somewhat later in their peak significance, were central to the regional oil complex as it was reconfigured in the fast-moving 1990s.

---

6. Permneft does not even receive a passing mention in Sagers and Grace's 1993 overview of the first generation of deals in the Russian oil sector.

7. I. Egorov, "S Amerikantsami—na sever," *Zvezda*, November 15, 1993.

8. S. Fedotova and V. Neroslov "Vremia L—1993 god," *Permskaia Neft'*, April 4, 2011.

Privatization as Incorporation

In contrast to PNOS's relatively speedy and easy entry into big Lukoil in 1990, Permneft found itself at the center of a high-stakes tug-of-war among three parties: Vagit Alekperov's big Lukoil; other emergent state-corporate coalitions in Moscow; and the Perm regional network that had grown out of the petrobarter-facilitated alliances among PNOS, the Perm Financial-Productive Group (PFPG), and the regional state administration. When the dust cleared by the mid-1990s, these regional networks emerged as most influential, repeatedly shaping the course of events and winning significant concessions from both the federal state and big Lukoil. Moscow-centric stories of oil-sector privatization and vertical integration in Russia do not typically account for the influence and power of such regional networks more than in passing; here, I both demonstrate regional networks' influence and trace their origins to the years before privatization.

I also see privatization as a much more complicated and multifaceted process than a battle for control over oilfields and shares in companies. A number of ethnographic accounts of the unmaking and remaking of socialist property have made clear that the issues at play in privatization are far larger and more variegated than simply the creation of property rights, individual owners or corporate shareholders, or the legal and institutional frameworks orchestrating them. Privatization was, in addition to these things, an embedded and encompassing social and cultural process (see, e.g., Hann 1998; Verdery 2003; and Dunn 2004). As I tell the story here, it included fundamental transformations in notions of what it meant to own or control objects or domains and major reformulations of the ways in which people inhabited spaces—a district or a town or, perhaps, the entire Perm region. It included the creation of new entities called corporations and new relationships between, among, and within them. Especially when we consider the regional oil industry, I will demonstrate, privatization was inextricably tied to the circulation of oil and oil products and to ongoing upheavals in the very means and patterns of regional exchange. In the early 1990s Perm region, this meant that privatization was inextricably tied to the Perm Commodity Exchange (PTB) and the PFPG.

Whither Permneft?

Given its multidimensional crisis, Permneft and its constituent units were eager participants in the Perm Commodity Exchange's turn to oil and oil products as its other trading activities wound down in 1992–93. The extensive barter networks facilitated by Oil Products Marketing—networks that were much deeper and wider than Permneft's, given the structure of the PTB and the oil production association's comparatively low prestige and lack of connections in the region—finally began to bring Permneft new supplies of equipment that would permit the repair of old production infrastructure. More important, Oil Products Marketing's success in expanding the consumption of PNOS's oil products through both regional state contracts and short- and long-distance barter chains meant that PNOS was finally able to reduce its accumulated debts to Permneft. In 1993 alone, complex barter deals facilitated by Oil Products Marketing erased millions of rubles of indebtedness between refinery and production association.[9] Even before privatization, that is, new kinds of pathways and linkages were beginning to emerge for the movement of oil from subsoil through refinery and to distribution points. As these new pathways coalesced, however, their participants also had one eye on the blueprints for the privatization of the oil sector that were taking shape in Moscow. Along with the oil products, money, apples, timber, and Toyotas moving through the Perm region, that is, there would soon be new items of significance: privatization checks and shares in privatizing corporations.

President Boris Yeltsin issued a dedicated presidential decree, number 1403, on November 17, 1992, to regulate the privatization of the Russian oil industry. The decree officially established Russia's major new vertically integrated oil companies—Lukoil, Yukos, and Surgutneftegaz—and allocated a number of subsidiaries to each.[10] Lukoil received the three West Siberian production fields that had been part of Vagit Alekperov's State Concern Lukoil since 1990, along with two major refineries (including PNOS), and a group of regional oil distribution companies, including Permnefteprodukt. These subsidiaries would be privatized from within their respective holding companies, with shares allocated to local employees, management, and the holding company itself. Decree 1403 also created Rosneft, a state corporation charged with holding a packet of state-owned shares (equivalent to 30 percent) in each of the new vertically integrated companies for a period of at least three years. Rosneft was also tasked with trusteeship over and management of the gradual privatization for the remaining 259 organizations in the oil sector—that is, those that did not yet belong to the three big new companies—which represented about half of Russian oil production at the time. Permneft fell into this large collection of orphans. It was privatized independently in December 1993, with shares going to employees, pensioners, management, and, in exchange for privatization vouchers, to the general public and investors.[11] At that point, Rosneft held back the standard 30 percent of shares to protect the Russian state's

9. E. Plotnikov, "Pervoprokhodtsy," *Zvezda*, March 3, 2003.

10. On Decree 1403, see Sagers 1993, 350–51; Sim 2008, 22–24; and Gustafson 2012, 76–78.

11. Permneft privatized according to "Option 1," meaning that employees and management received 40 percent of shares, with additional shares being offered to the public in a later round of distribution.

interest in the oil sector. But it was always clear that Permneft would not last long on its own, for much of the following three years in the Russian oil sector was a mad scramble for the orphans created by Decree 1403, with Lukoil, Yukos, and Surgutneftegaz seeking still more subsidiaries and new vertically integrated companies being carved out of Rosneft's holdings every few months, among them Slavneft, Sidanko, and Onako.

This jockeying played out not just in Moscow, where it has been most extensively reported on (Hoffman 2002; Goldman 2008), but along new and existing channels that extended deep, and consequentially, into oil-producing regions. In Moscow, Lukoil president Vagit Alekperov made no secret of the fact that he was hoping to bring Permneft under the Lukoil umbrella. His argument was that, if the goal was to create vertically integrated companies that would compete on a world stage, then it was only logical to unite production and refining in the Perm region into the same corporate structure.[12] Alekperov did not prevail easily, though. In negotiations between the Ministry of Fuel and Energy and the Kremlin, it was decided that Permneft would be allocated to Slavneft, a new, state-owned joint Russian-Belarusian oil company, with production facilities in Siberia and the Krasnoiarsk region. Slavneft had a powerful champion in Anatolii Fomin, who was at the time both first deputy minister in the Ministry of Fuel and Energy and chair of the board at Slavneft. Knowing that Lukoil-PNOS would not welcome its oil for refining, Slavneft was proposing to build a $300 million pipeline that would route Permneft's crude around PNOS and to Slavneft's own refineries.

With 70 percent of the shares of Permneft outside of Rosneft's control, however, Moscow could only apply pressure to Permneft and the Perm region. This pressure could be significant: Fomin, Rosneft, and their allies were successful in prevailing upon Aleksandr Cherkasov, director of Permneft at the time, to support an entry into Slavneft over Lukoil. They were far less successful, however, in persuading the political leadership of the Perm region. In fact, there was considerable concern in the Perm region about the fate of Permneft, inasmuch as the regional distribution of food and fuel—and therefore much of the regional government's legitimacy—depended so heavily on the extensive barter networks anchored by PNOS, PFPG's Oil Products Marketing, and the still-independent Permneft. There was a similarly low level of enthusiasm for a new pipeline that would cut into PNOS's intake of crude oil for refining and sale. Veniamin Sukharev, still general director of PNOS and representative of Lukoil in the entire region, recalled that he approached Andrei Kuziaev of the PFPG and Boris Kuznetsov,

12. V. Kostarev, "Na opekunstvo 'Permneft'' pretenduet tri kompanii," *Zvezda*, September 6, 1995. Susanne Wengle (2015, 99–133) deftly shows that this argument from "industrial geography" was common and frequently persuasive in the scramble to privatize not just corporations, but networks of corporations.

then governor of the Perm region, about the possibility of steering Permneft in the direction of Lukoil rather than Slavneft in early 1994 (Vikkel′, Fedotova, and Iuzifovich 2009, 89), but it is nearly certain that informal discussions were under way well before that. The governor approved of the effort to resist Moscow's plan for Permneft, designating his deputy Gennadii Igumenov to help coordinate regional efforts.

In addition to lobbying Moscow, a major task of this regional coalition was to gain control of as much stock in Permneft as possible. The PFPG, through its Securities League subsidiary, had held a central place on the regional securities market beginning in its days as the PTB, and it had already accumulated a large number of privatization vouchers from ordinary Russian citizens.[13] With this head start, the Securities League began accumulating public shares of Permneft as soon as they began circulating in December 1993. But there were other shares of Permneft to be had, too: those that had been distributed directly to Permneft employees—nearly fifty thousand of them spread out across the length and breadth of the region. One of my acquaintances in the region, who was active in Perm's fund market at that time, recalled:

> It was clear to everyone that, sooner or later, and probably sooner, shares in Permneft would be converted into shares in Lukoil, and it was probably going to be a good conversion rate. And so in 1994 and 1995, everyone who worked on the funds market rode around villages, around the districts, around the collective farms, and bought up those shares. Because people who got those shares for free, well, first they didn't understand what they were, and, second, they were ready to sell them for a few kopeks, so long as they were "live" kopeks [i.e., cash]. Some brokers even came especially from Moscow . . . even Lukoil knew that the shares were being bought up, and hired their own buyers. And so there were wars in all those [oil] towns, in those Kungurs, those Chernushkas. There would be one wagon buying shares at one price, and nearby another buying at another price. It was forty below [Celsius] . . . and there was a real war going on.

As a result of these efforts, the PFPG ultimately acquired a 25 percent stake in Permneft's voting shares.[14] At the first full Permneft shareholders meeting in the

---

13. Russian citizens received their privatization checks beginning on October 1, 1992. Part of the overall scheme of transition designed in Moscow with the advice of international advisers and agencies, these privatization checks were intended to give every Russian citizen a stake in the privatization process by permitting him or her to own shares in a company that was scheduled for conversion to private ownership over the coming months and years.

14. Not all these efforts were completely transparent, something that was noted at the time; see, for instance, N. Kopylova, "Vladeiut tem, ne znaia chem: riadovye aktsionery predpriatii neftianogo

fall of 1994, this voting bloc was the basis for assembling a coalition large enough to push out director Cherkasov, who was following his orders from Moscow to put Permneft on track to enter Slavneft as soon as the 30 percent of shares still controlled by Rosneft could be transferred in the fall of 1995. Cherkasov was replaced by Anatolii Tul′nikov, a career oilman from the Kungur district of the Perm region who had most recently headed energy affairs in the Perm regional administration and was a reliable member of the coalition that was intent on pushing back against the Slavneft option. Under Tul′nikov, Permneft quickly began a series of restructuring moves to increase its financial viability; not coincidentally, several of these moves drew it closer to big Lukoil, including through collaboration on drilling and export that involved the exchange of both personnel and limited finances.[15]

The replacement of Cherkasov did not please Slavneft's champions in the Ministry of Fuel and Energy, who continued to pressure the Perm regional administration well into 1995. Ultimately, however, they were unsuccessful. In the fall of 1995, Rosneft and its allies in Moscow conceded, and Rosneft's 30 percent stake in Permneft was transferred to big Lukoil. The personal connections and petrobarter-based exchange relationships that emerged out of the Lukoil-PNOS alliance beginning in 1990 had created a powerful regional coalition that successfully stood up to plans for oil sector privatization hatched at the federal level. Although the influence of coalitions hailing from oil-producing regions is an aspect of the story of post-Soviet oil rarely told in much detail outside of Russia, its plot should not be surprising in the context of an exceptionally weak federal state and at the height of 1990s-era parcelized sovereignty. The next set of struggles over Permneft revealed that big Lukoil did not always have the upper hand either.

One Region, Two Lukoils

Lukoil quickly encountered a new problem in its planned absorption of Permneft, in the shape of the very region-based allies with whom it had worked to beat out Slavneft. The regional coalition uniting PNOS, the PFPG, the local state administration, and Permneft was not willing simply to turn over the oil pro-

---

kompleksa Prikam′ia," *Zvezda*, October 12, 1994. I would not be very surprised to learn that the leading participants in these deals used one or another extra- or semilegal means to accumulate their stake or to fix the voting outcome at the relevant shareholder meetings. However, even among those in a position to know and at nearly two decades removed, no fantastic stories—such as the archetypical one in which Vladimir Bogdanov, director of Surgutneftegaz in West Siberia, shut out competing bidders by arranging to have the local airport closed on the day of the auction—circulate in the Perm region. Although it might add useful details, evidence of such activities would be unlikely to change my overall analysis.

15. "Permneft′ stala iablokom razdora," *Kommersant*, April 4, 1995.

duction association to big Lukoil without substantial concessions and guarantees. As Gennadii Igumenov, deputy governor at the time, put it in his memoir, regional political leadership was afraid that big Lukoil would quickly turn its back on the region:

> Wouldn't a part of the taxpaying base move to Moscow? How would decisions about the social sphere and ecology arising in the region be made? Wouldn't the Permian oil deposits' low attractiveness for investment compared to other deposits [induce] Lukoil to reduce exploration, drilling, and production [in the Perm region] due to the redistribution of funds within a vertically integrated company? (Igumenov 2008, 297)

What, that is, would happen to the benefits of cross-sectoral integration that the regional state administration was enjoying by participating directly in the passage of oil from subsoil to refinery and on into distribution along the networks managed by Oil Products Marketing and the PFPG? There were financial interests, as well: the price of Permneft's shares had been driven by speculation that they would fetch a high price when ultimately swapped for shares in big Lukoil; with the struggle with Slavneft now over, there was fear that Lukoil would offer a lower, market-based price, pegged to the actual prospects of oil production in the Perm region rather than competition with Slavneft. The PFPG was not interested in losing money on its hard-won investment.

These are matters that would not ordinarily have concerned big Lukoil, which was at the time rapidly accumulating subsidiaries from Rosneft's portfolio, including Nizhnevolzhksneft, Astrakhan'neft, and Kaliningradmorneftgaz in 1995 alone.[16] It had acquired just over 50 percent of the voting shares in Permneft by combining Rosneft's 30 percent stake with the shares its own agents had bought from Permneft employees in the region in the 1994 free-for-all. But big Lukoil needed a two-thirds majority in order to approve the conversion rate between its shares and shares of Permneft—the final stage of takeover. The PFPG's 25 percent stake, when combined with other packets held by regional interests, thus formed a blocking stake in Permneft, and, for the first time in its short history, big Lukoil was forced to sit down at the bargaining table with a regional coalition rather than acquire a new subsidiary outright (Fedotova 2009, 210). After a great deal of negotiation among big Lukoil, the PFPG, and the Perm regional administration, a compromise was reached, one that was approved at a general shareholders meeting in November 1995. In a first-of-its-kind arrangement in Russia, two separate subsidiaries of big Lukoil were created out of the former

16. Pappe gives one of the best overviews of big Lukoil's privatizations, mergers, and acquisitions in this period (2000, 100–115).

Permneft. The first was to be called OOO Lukoil-Permneft. It was composed of the lion's share of Permneft's production districts and management structure, including most of the south of the region, and it was a fully controlled subsidiary within big Lukoil. The second company was to be called ZAO Lukoil-Perm, and it was allocated several oil deposits on the eastern side of the Kama River and in the north of the Perm region. ZAO Lukoil-Perm, in contrast to its southern sibling, was *not* a wholly owned subsidiary of big Lukoil. Ownership of shares was, instead, evenly split between big Lukoil and the PFPG. Andrei Kuziaev, of the PTB and then PFPG, was appointed president of ZAO Lukoil-Perm, and much of his team moved with him to this new company.[17] As part of this deal, the PFPG turned over its shares in OOO Lukoil-Permneft to big Lukoil only after ZAO Lukoil-Perm had gained control of oil production operations in the north and east of the Perm region.

To formalize this negotiated settlement, the Perm regional administration prepared an "Agreement" with big Lukoil. In addition to general declarations of mutual support and mutual interests, this 1995 agreement specified how ZAO Lukoil-Perm would use its access to the oil deposits it had been allocated. ZAO Lukoil-Perm would pump oil specifically for the needs of the Perm regional state administration—at an initial level set at 2.5 million tons a year—and submit that oil for processing at PNOS. The PFPG, through Oil Products Marketing and its other structures, would continue to use this oil in the barter chains that kept both food and fuel moving through the Perm region. Indeed, one of the points of the agreement stated directly that by expanding into the Perm region, Lukoil would be participating in circuits of food as much as oil and oil products: "The sides consider it important to organize the supply of available food products from other regions through exchanges of oil products, for the benefit of the population of the Perm region."[18] Vertical integration at the national level, that is, continued to unfold through cross-sectoral integration at the regional level. The agreement further stipulated that any oil ZAO Lukoil-Perm managed to extract above those 2.5 million tons could be sold for profit, including through big Lukoil's international export channels. The Perm region also won pledges from big Lukoil to continue to modernize facilities at PNOS and to invest in regional oil production infrastructure. In exchange, big Lukoil obtained assurances that regional officials would ensure all the access its subsidiaries needed for exploration and production—the full weight of the Perm region would be brought to bear to ensure, for instance, that there would be no more outlandish requests from privatized state and collective farms in exchange for access to production sites.

---

17. The transfer of Permneft for big Lukoil shares finally took place in 1997, with a 5:1 conversion ratio; see Iu. Manzhosin, "Borot'sia s LUKoilom—idti protif techeniia," *Zvezda*, May 15, 1996.

18. S. Fedotova and A. Neroslov, "Vremia L—1995 god," *Permskaia Neft'*, May 19, 2011.

Beginning at that November 1995 shareholders meeting and for a period of nearly ten years thereafter, the story of oil production in the Perm region was a story of two Lukoils, with one of them explicitly formed and managed by the regional elite that had emerged out of the petrobarter chains of the early 1990s. Nevertheless, for many observers in the Perm region, big Lukoil's acquisition of Permneft meant that, despite the assurances of the regional state administration, control over oil resources was slipping farther out of the hands of individual shareholders and the Perm region itself. In the concluding section of a long article summing up the competition among Moscow- and Perm-based interests for control over Permneft, the central Perm newspaper *Kapital Weekly* lamented, "and now who will determine where Permian oil will flow?"[19] A column in *Zvezda* in the same month turned to *Hamlet* to make the point:

> "All the world's a stage, and we are all actors on it." . . . [Shakespeare's] classic claims this because it saw in every person a personality (*lichnost'*) above all else. Alas, in our time a person in most cases does not resonate so proudly, and "all of us" means nothing more than extras who decide nothing. Or at least that's the conclusion to be drawn when one looks at the theater of politicians, lawmakers, and lords of capital.[20]

It is worth pausing over both the phrase "Permian oil" and the *Hamlet* reference. The first asserts that regional oil itself had some Permian characteristics that were threatened by big Lukoil's appearance on the scene, whereas the second draws attention to the accumulation of once-individually owned shares into the PFPG and Moscow-based big Lukoil—an accumulative process that sidelined individual investors in the Perm region as so many extras. At this point, these framings of Permneft's fate were an element of simmering critique of an emerging corporation, a version of the widespread and highly visible critiques of oligarchs, "new Russians," and social stratification in general as threats to existing communities of various shapes and sizes. Like the ecological problems associated with PNOS in the perestroika period, these claims were impossible for those making deals at the commanding heights of the economy to ignore. Indeed, it was eminently clear that one of the early challenges for both of the new Lukoil subsidiaries would be countering the widespread impression that Permian resources and aspects of regional particularity had just been expropriated to Moscow. Indeed, the Perm-based coalition that battled first Moscow politicians and Rosneft executives and then big Lukoil was already responding to this brand of anticorporate critique when it set the terms of its agreement with big Lukoil in the fall of 1995. Whatever

19. I. Georgiev, "Lukavstvo ot Lukoila?" *Kapital Weekly*, November 22, 1995, 1–2.
20. Ibid.

side one took, however, it was clear that much of what was at stake in the movement of Permneft into Lukoil was the fate of a region and its people—a status oil had never had in the Soviet system.

This, then, was the context in which Evgenii Sapiro claimed that, "What's good for Lukoil is good for the region." He was making an argument rather than stating the obvious, and he was arguing back against those who viewed the coming of Lukoil as a destructive force in the region. Gennadii Igumenov made a similar case in retrospect in his 2008 autobiography, suggesting that the consolidation of state and oil interests at a regional level, as elaborated in the 1995 agreement, produced an enviable outcome in which "the Perm region became an experimental platform for the introduction of new technologies, new equipment, the improvement of ecology, the development of social programs, and the increasing effectiveness of the whole cycle of production, refining, transportation, and sale of oil and oil products" (2008, 300). On this view, the Perm region was set to become the leading edge of the making of new vertically integrated oil corporations in Russia—in the fullest meaning of the term "corporation," one that extended far beyond the privatization of ownership.

Oil and Surrogate Currencies

At the same time that these privatization battles were playing out in the oil sector, Russia was sinking deeper and deeper into what became known as the mid-1990s nonpayments crisis: inter-enterprise debts and tax arrears began to pile higher and higher, and money again retreated as a means of exchange. By 1995–96, the shortage of food products and household items had more or less ended, at least as compared with the late 1980s and early 1990s years of acute shortages and rationing tickets. The demonetization that began in earnest in 1995 unfolded largely on the level of state agencies and corporations rather than households, and it was caused chiefly by the Russian government's attempts to keep inflation in check and the ensuing failure to consolidate control over monetary exchange (Woodruff 1999). In this environment, Russia's regions and newly privatized firms began looking for a new set of ways to cope with skyrocketing inter-enterprise debts and tax arrears that no one had the cash to pay. Barter and offset trade were common but generally insufficient, given the scope and scale of the mutually accumulated debt. The preferred strategy that emerged across Russian regions was the circulation of *veksels* (from the German *weschsel*). Usually rendered in English as promissory notes or bills of exchange, veksels were financial instruments issued in specific denominations that entitled the bearer to a quantity of goods or money. Although they could technically be redeemed for the amount of cash or goods on their face, and sometimes were, the true purpose of veksels was to provide a short-term, limited-circulation surrogate currency for paying down debts. The state's rubles, that is, still served as unit of account (by law, veksels had to have a face value in rubles), but veksels were increasingly the medium of exchange among corporations and state agencies.

Regional veksel systems, as Woodruff shows so brilliantly in his *Money Unmade* (1999), are an important way in which we can glimpse the devolution of sovereignty from the federal Russian government to smaller units in the 1990s. Woodruff demonstrates that the chief political battles of the mid-1990s were over the terms of what counted as a payment; until the financial crisis of 1998 had passed, Russia's regions were winning this battle over the federal center, naming their own terms of payment—such as veksels—and, in the process, enhancing their claims to sovereignty at the expense of the federal center. By attending to the specifics of the veksel system that emerged in the Perm region, we can extend some of Woodruff's analysis by seeing the devolution of sovereignty not just as a temporary lack of federal-level sovereignty, but as an important generator of region-level networks and imaginaries. Veksels, that is, were the next important moment in oil's march from periphery to center in the Perm region political economy, a march that was inextricable from the larger remaking of regional identities, networks, corporations, and state agencies. Like the petrobarter that preceded it, veksels allow us an instructive glimpse into modes and patterns of oil exchange that do not take place through national monetary currencies. They make visible sets of relations that are often—including in the global petrodollar regime that has existed since the early 1970s (see Spiro 1999)—obscured by the abstractions of the money form.

Fuel Veksels

In mid-1995, the Perm regional administration invited proposals from the local business establishment to design and run a veksel system for the region. Proposals for each of the main variants of veksel that were emerging in other Russian regions—bank-issued, corporate-issued, and government-issued—were put forward over the next months. Bank-issued veksels were common in many Russian regions, but the frequency with which banks were folding in the Perm region at the time did not inspire much confidence. Moreover, there was already a large, influential, and well-connected alternative to bank-issued veksels. Representatives of the alliance between the PFPG and the region's oil enterprises were also proposing a veksel system, and their proposal for fuel veksels (*toplivnye vekselia*) was chosen.

Fuel veksels were also deemed to be more liquid than bank veksels because the constant demand for oil products would ensure their circulation (Igumenov 2008, 201).[21]

Beginning on September 1, 1995, Oil Products Marketing, the PFPG subsidiary specializing in oil products sales and other transactions, issued veksels—numbered, paper, promissory notes—that could be redeemed on demand for a specified quantity of oil products.[22] The amount of veksels issued was set to track with the amount of oil products turned out by PNOS, so that redemption demand could never exceed the regional supply of oil products; the first issuance of veksels, for instance, covered 250,000 tons of oil products each month, between a quarter and a third of the monthly output of PNOS at the time.[23] They were issued in denominations of not less than one million rubles, meaning that they were generally circulating among enterprises and state agencies, although some small-scale businesses, including independent farmers, could use them for large purchases of fuel. In order to take into account PNOS's fluctuating output, series of veksels were set to expire in 180 days; before expiration, veksels would need to be either redeemed for oil products or traded in for veksels in a subsequent issuance. By federal law, veksels had to be redeemable for cash as well as products, but this was discouraged in practice by the stipulation that the cash payout would take place not less than five years in the future, whereas oil products could be claimed immediately upon the presentation of the veksel.[24]

The initial veksel circuits were limited to the region's oil sector enterprises, the PFPG, PermEnergo (the regional electricity company), and the regional state administration, all of which were signatories to a mutual agreement to accept Oil Products Marketing's veksels in lieu of cash. So, for example, the region's oil sector enterprises were permitted to exchange veksels among themselves, and to use them instead of money to settle their tax debts to the region. Permenergo was permitted to take veksels as payment and then use those veksels to pay its own tax debts to the region (Igumenov 2008, 302). The regional state administration, in turn, often redeemed veksels for PNOS's oil products to supply the fuel-hungry agricultural sector. (Indeed, the 1996 decree that set out the rules for the Perm regional veksel system begins by referencing the need for reliable gasoline supplies

---

21. For many examples of barter and veksel networks in different configurations and different regions, see Seabright 2000. In the summer of 1996, the Perm regional administration began issuing its own veksels, which were designed to replace fuel veksels for purposes of regional tax collection. They were to be called "budget veksels." See V. Kostarev, "Inye veksely, inye nravy," *Zvezda*, July 31, 1996, and O. Oputin, "Veksel′ umer: Da zdravstvuet novyi?" *Zvezda*, July 24, 1996.

22. Other Permian companies did try to issue veksels, but they never achieved either the trust or circulation of those issued by Oil Products Marketing and, later, ZAO Lukoil-Perm. See "Initsiativy Permskikh Predpriiatii," *Kommersant*, August 12, 1995.

23. A. Bezdeneva, "Parallel′nye den′gi?" *Zvezda*, December 14, 1995.

24. S. Fedotova and A. Neroslov, "Vremia L—1995 god," *Permskaia Neft′*, May 19, 2011.

during sowing and harvesting as a reason for establishing the fuel veksel system in the first place.) In this way, PNOS exchanged 80 percent of its output of oil products without money in 1995.[25] In their first months of usage, the heaviest traffic in these veksels helped to stabilize the balance sheets of PNOS and Permneft ahead of Permneft's acquisition by Lukoil—just one of the ways in which shareholding battles were closely tied to the regional means of exchange at the time. Considering how central the movement of oil products through the region had become to the entire Perm region, it is no surprise that the regional administration pushed so hard to place Permneft within Lukoil's holding company rather than Slavneft's, with its plan to build a pipeline routing Permian oil around PNOS.

At a meeting of many of the region's political and financial elite at a Perm hotel in the fall of 1995—just as Permneft's entry into big Lukoil was being finalized—it was officially agreed that the Perm region would incorporate both new regional Lukoil subsidiaries into its veksel strategy. Some later called the gathering the Perm region's "Council at Fili," (Vikkel′, Fedotova, and Iuzifovich 2009, 97) a reference to the 1812 gathering of Russian generals outside Moscow following the Battle of Borodino, at which Field Marshall Kutuzov decided that Russian forces would retreat through Moscow rather than confront Napoleon's armies from a disadvantageous position. The analogy is apt, for although the onslaught was financial rather than military, the Perm-based politicians and financiers had also decided to withdraw—into their own, oil-backed surrogate currency. The use of veksels was folded into the November 1995 agreement signed between the Perm region and big Lukoil and, in January 1996, ZAO Lukoil-Perm took over veksel emissions from PFPG's Oil Products Marketing (although the accounting was still carried out by a subsidiary of the PFPG).[26]

Although at first limited to a small selection of large corporations and the regional state apparatus—"you can't pay a school teacher or a doctor in a veksel," one participant in these chains emphasized to me—the use of veksels quickly expanded. In late 1995, district-level administrations in the Perm region's oil-producing districts were permitted to enter veksel circuits, accepting veksels from OOO Lukoil-Permneft in lieu of taxes. Later, in July 1996, the city of Perm authorized its agencies to accept and make payments in fuel veksels. By the middle of 1996, 26.7 percent of the region's budget was composed of veksel transfers and an estimated 4.1 trillion rubles of regional debt had been paid down with regionally circulating veksels.[27] A secondary market for veksels developed quickly, with all manner of large and small entities participating in direct trades and even in market exchange where different quantities and series of veksels could be transacted

25. Ibid.
26. Bezdeneva, "Parallel'nye den'gi?"
27. S. Aristova, "Vekseliami Berete?" *Kapital Weekly*, October 7, 1996.

in bulk.[28] Through this secondary market, smaller enterprises would use their veksels to obtain fuel from the distribution network of Lukoil-Permnefteprodukt, whereas larger enterprises usually went directly to Lukoil-PNOS. In a quite different way from the petrobarter deals of the very early 1990s, then, oil remained central to basic patterns of circulation in the region.

Both state agencies and corporations in the Perm region generally presented fuel veksels as a collaboration between industry and state that enabled and served the entire Perm region. ZAO Lukoil-Perm, for instance, continued to maintain its veksel plan for several years, pitching the supply of oil products for state usage through the veksel program as part of the benefits that districts and cities could expect from collaboration with the company.[29] Both in his official memoir and in an interview with me, former governor Gennadii Igumenov likewise claimed that the circulation of veksels was key to the survival of the region at the time. All this may be true, but it is also the case that veksels, again like petrobarter, solidified and amplified the significance and centrality of the oil industry and the PFPG in several ways. First, the issued veksels were non-interest-bearing and long-term, which led some observers to view them as a hidden discount to the oil industry in conditions of inflation.[30] Second, one effect of the veksel system, and one that was envisioned quite intentionally by its architects, was to expand Lukoil's sales in the Perm region. Although oil refined at PNOS was often headed to regional distribution centers, there were exceptions, and the refinery had several competitors for the sale of gasoline trucked into the region from Tatarstan or Bashkortostan in the south or even farther away. (The Perm-based company DAN, for instance, competed with Lukoil at gas stations with oil products brought to the region from refineries in the Caucasus.)[31] By drawing large purchasers of oil products to PNOS or Lukoil-Permnefteprodukt distribution centers where they could redeem their veksels, Lukoil expanded its role in the downstream segments of the vertically integrated company it was gradually piecing together (see also Ledeneva and Seabright 2000, 97).

Third, and especially significant for the emergence of state-corporate relationships, the special tax relationship between Lukoil's regional subsidiaries and the PFPG, on the one hand, and the regional state administration, on the other hand, was a terrific advantage to the oil industry: the oil industry, more than any other

28. Veksels circulating in the secondary market, where they could be bundled and resold, were not unlike the collateral debt obligations that became famous on Wall Street a decade and a half later; indeed, the international financial press noted at the time that Moscow traders were becoming very sophisticated in the purchase and sale of debt.

29. Tatiana Vlasenko, "Bol'she Nefti—Bol'she Deneg," *Novyi Kompan'on*, July 13, 1999.

30. V. Kostarev, "Chto v veksele tebe moem," *Zvezda*, July 26, 1996.

31. On DAN's challenge to Lukoil on the distribution and consumption fronts, see Sapiro 2009, 424; Fedotova 2006, 213–20.

sector, could pay its taxes in veksels and hang on to cash reserves for use in investment, salaries, or for other purposes. What portion of taxes was paid in "live" money and what portion in veksels was a matter of periodic negotiation between state structures and oil sector enterprises. Initially, the veksels were emitted by Oil Products Marketing and ZAO Lukoil-Perm, but were accepted as tax payments from Permneft and PNOS as well; indeed, they were the primary means by which these companies paid taxes, prompting cries of favoritism from other companies who still had to pay in cash and from citizens agitating for more live money to be paid into the regional budget. Veksels allowed the oil industry, by virtue of its central position in the regional economy, to keep more of that live money for itself. Inequality of spheres of circulation was thus built into veksel circuits and served to draw Lukoil's regional subsidiaries and the Perm regional government even closer together. Gennadii Igumenov, governor of the Perm region at the peak of the veksel years, notes in his autobiography that, "Without a doubt, the realization of the [veksel] project within the framework of the already-signed Agreement allowed the growth of Lukoil's prestige and authority in our region" (Igumenov 2008, 302).

The Materiality of Mutual Indebtedness

Thus far, my analysis of veksels follows Woodruff's in many respects, particularly in its emphasis on the ways in which veksels as surrogate currencies continued to undermine already-weak federal sovereignty and empowered networks of regional elites—both vis-à-vis the federal center and against other regional networks that were shut out of the opportunities afforded by participating in veksel circuits. But we can say more about veksels in the Perm region as well. The weekly newspaper *Kapital* traced the path of one fuel veksel for its readers in mid-1996, near the peak of their influence:

> As an example, let us look at the history of just one fuel veksel. It [was issued in the sum of] 10 million rubles. First, the veksel was received by the regional administration from OOO Lukoil-Permneft as tax payment into the regional budget. Then it was transferred to the administration of the Okhansk district, and from there passed into the control of [a certain] independent peasant farmer by the name of Bessonov [probably in exchange for milk or meat products for Okhansk schools or kindergartens]. Farmer Bessonov used the veksel to obtain oil products, and in this way the veksel made its way to Lukoil-Permnefteprodukt, and, after that, to Lukoil-PNOS. Then, farther, the veksel again went to the finance department of the regional administration, and set off on a new

> circuit: the Regional Committee for Social Protection–a psychoneurological dispensary–the company Gidroelektromontazh–the association Tiazhmashopttorg, which, finally, sold it to [PFPG's] Perm Fund Company. And a single veksel goes through hand-to-hand transfers in this way up to forty times. It is not surprising that, with the help of fuel veksels, [the Perm region] has been successful in paying down 4.2 trillion rubles in debts, and that the number of participants in veksel circuits has risen to 1,860 enterprises.

The article went on to cite Andrei Agishev, a PFPG vice president:

> Behind every fuel veksel stands some share of a Permian oil product. At the base of the veksel program is a principle: the total sum of veksels in circulation at any given moment does not exceed the cost of the monthly output of trade production from ZAO Lukoil-Perm [through Lukoil-PNOS], which is around 250 billion rubles. In mid-1996, around 180 billion rubles' worth of oil-backed veksels were circulating in the regional economy.[32]

Elsewhere, Agishev described the utility of veksels precisely in terms of mutual indebtedness: "Today, when everyone, down to the last man (*pogolovno*) is in debt to everyone else, using veksels makes accounting procedures a great deal easier."[33]

Not only did veksels make accounting easier, they did so in a stunning way: by making materially legible the circuits of debt and reciprocity out of which the Perm region was being reconstituted in the mid-1990s. Unlike state currency, that is, veksels embedded tangible traces of the circles of indebtedness they moved through, and they called forth their own region-based forms of sociality. Each and every veksel transaction, for instance, had to be recorded in three ways in order to be officially legal. The first was in acts of purchase–sale, in which the two parties to each exchange recorded the transaction, complete with passport identifications, signatures, and official stamps, in their own accounting books. The second was in the general logs kept by the Perm Fund Company (a PFPG subsidiary), which recorded every transaction. Indeed, tracking veksels through the region was a small industry in and of itself, requiring a constant stream of visits among accounting office personnel and phone calls to the Perm Fund Company to verify authenticity, register transactions, and so on.[34] Before holidays, lines

32. V. Kostarev, "Kak Fermer Bessonov indossantom stal, ili istoriia iz zhizni toplivnogo vekselia," *Kapital Weekly*, July 17, 1996.

33. Bezdeneva, "Parallel'nye den'gi?"

34. Postanovlenie Administratsii g. Permi ot 15.07.1996 N 1323, "O Primenenii toplivnykh vekselei 'ZAO Lukoil-Perm'' pri ispolnenii gorodskogo biudzheta; Postanovlenie Gubernatora Permskoi

would form outside the PFPG offices to ensure that transactions were registered in a timely fashion. Acquaintances recounted to me the ways in which veksels were rolled up and carried around: given that veksels were so carefully tracked, theft was much less of a concern than it was for cash, so office staff at the PFPG's Perm Fund Company did not hesitate to travel around Perm's tram and trolley network with millions of rubles' worth of veksels tucked into their bags for delivery to clients.[35]

The third and perhaps most interesting way was *on the veksel itself*. As each veksel circulated, transactions were logged in dedicated lines on the back of the note, and additional pages could be attached as needed. Personal signatures were required (as opposed to stamps or other mechanical means). In many cases, veksels grew to meters in length, a topic of frequent comment. When I asked former governor Igumenov whether the Perm region's veksels were traded in Moscow, his response quickly moved past the issues of regional sovereignty that I was probing for and to the veksels themselves:

> No . . . It was just our internal affair. And a veksel . . . I saved one for a long time, and only just gave it to the museum. It was a paper, a meter and a half long, on which there were the stamps and signatures of the directors and accountants of the businesses that were participating. You hand it over to someone, that one to the next, to the next . . . It was great. It was a terrific program (*aktsiia*) for reducing collective debts at the time.

The very materiality of the veksel, that is, enumerated with extraordinary precision the mutual circles of debt and obligation that made up the Perm region—by listing them, one after the next, in original signatures, for everyone in the chain to see and inscribe themselves into. This precision is, of course, what enabled *Kapital Weekly* to reconstruct the circuit through which that one fuel veksel moved, something that would be impossible with monetary currency.

Economists Clifford Gaddy and Barry Ickes (2002) popularized the term "virtual economy" to refer to this period in post-Soviet political and economic transformations. Although they are right that prices did not conform to what neoclassical economic theory would predict and prescribe, virtualization is precisely the opposite of what was actually happening in the Perm region's veksel circuits. Although fuel veksels were more abstracted than petrobarter exchanges, they continued to make debts, obligations, and social relationships far more material,

Oblasti ot 04.06.1996, "O Primenenii Vekselei 'ZAO Lukoil-Perm'' i AO TPK 'Neftesintezmarket' pri ispolnenii Oblastnogo Biudzheta."

35. S. Fedotova and A. Neroslov, "Vremia L—1996," *Permskaia Neft'*, August 12, 2011.

visible, and concrete than the abstractions of monetary currency.[36] A far better guide is provided by Gustav Peebles's work (2008) on the semiotics of currencies. As national currencies emerged, Peebles argues, they redirected signs of international value (gold or silver) to signs of national value (the "national horde"). With circulating veksels in the Perm region, we see a different version of this process, with national currency (rubles) locked up in debts and a surrogate currency backed by the regional oil industry serving in its place as a key means of exchange. If, in Peebles's account, national paper currency served as an "inverted panopticon" (2008, 234) in which citizens look back to the state as guarantor of their daily exchange activity, then fuel veksels accomplished something similar in a different and regional-scale way. Veksels pointed neither to the national horde, nor to the Soviet factory elite, nor to the Communist Party, nor even to Perm's regional government. The centerpiece and guarantor of the Perm region, the entity toward which all eyes turned, was written on the face of the veksels: ZAO Lukoil-Perm. In later years, ZAO Lukoil-Perm prided itself on having some of the best marketing and publicity specialists in the region. But it is hard to imagine a more comprehensive way for the company to establish itself at the center of the Perm region immediately after the acquisition of Permneft than by issuing and guaranteeing the region's primary currency.[37]

• • •

As the nonpayments crisis that was particularly acute in 1995 and 1996 slowly wound down, the Perm regional administration decreased the amount of tax payments from the regional Lukoil subsidiaries that it would accept in veksels. In 1996 OOO Lukoil-Permneft had paid nearly all its debts to the region in veksels; in 1997, the regional administration aimed for 50 percent in veksels and 50 percent in rubles. It hoped to decrease veksel payments still further in 1998. In the end, veksels did not so much disappear as peter out. Under pressure from the International Monetary Fund, the Russian Federation passed a law prohibiting regional governments from issuing government-backed, so-called budget veksels in 1997 (although many regions, including the Perm region, continued to accept a number

36. See also Maurer 2005. For more on the ways in which monetary exchanges—and neoclassical economics—obscure relationships of social indebtedness, see Graeber 2011.

37. A brief note on regional elected bodies is necessary. Henry Hale (2006, 163–66) shows that alliances between corporations and regional FIGs were often especially active in electoral politics in the late 1990s, fielding and aggressively supporting slates of candidates for regional dumas and expecting them to pursue the company's interests while in office. This was certainly true in the Perm region—indeed, one of Hale's primary examples involves Lukoil subsidiaries in the Perm region—but the presence of Lukoil-backed elected officials in the Perm Regional Duma was only one corner of region-level corporate–state interactions in the 1990s. In comparison to the veksel circuits of the mid-1990s and the social and cultural project movement of the 2000s, political offices held by oil sector executives were among the less visible and consequential intersections of oil, politics, and economics.

of bank and corporate veksels for some time thereafter).[38] The federal Treasury ceased accepting veksels of all sorts in lieu of taxes in 1999, and the stabilization that followed Russia's ruble devaluation of 1998 reduced the previously pressing need for surrogate currencies. However, large businesses continued to issue veksels, and by the end of the 1990s, they had become a standard financial tool for major Russian companies, particularly in the energy sector, where they continued to exist alongside stocks and bonds in the standard Russian corporate toolkit of financial instruments. Veksels continued to have the advantage—and disadvantage—of being lightly regulated, and were thus mostly tools for banks, large corporations, and professional financial operations. Circulating more narrowly and no longer serving in place of taxes, veksels thus gradually receded from public view and consciousness. But they had left their mark. In the era of veksels, eyes across the Perm region were turned inward, to specifically regional circuits of exchange (as against national circuits signified by money). They were also, increasingly, turned downward, to the depths from which the oil flowed, and to the corporation whose regional subsidiaries were solidifying control over that flow and its intersections with all others: Lukoil.

OOO Lukoil-Permneft: Industrial Capital in the Southern Perm Region

The early entry of PNOS into big Lukoil, the growing centrality of oil to the PTB and the PFPG, the privatization of Permneft, and the circulation of fuel veksels were all caught up in a broader dynamic that played out across the post-Soviet oil industry: the relationship between *neftianiki* and *finansisty*. Neftianiki were the oil generals, the red directors of the oil sector, often engineers who had come up through the Soviet ranks to run factories and production associations, and therefore to occupy key places in the regional and federal party-state bureaucracy (Veniamin Sukharev of PNOS is a prime example in the Perm region). Neftianiki were specialists in production, refining, and the engineering side of oilfield management. Finansisty were the new financiers, younger men who had come to be involved in the oil industry only through the process of privatization and largely by virtue of the fact that they controlled banks or financial-industrial groups. This distinction has analogues across many post-Soviet industrial sectors and has often been noted inside and outside Russia. To give but one example, a rolling

38. The question of whose veksels could be accepted in lieu of cash for tax purposes continued to be a matter of regional contention until at least 1999 and the run up to the gubernatorial elections in 2000 (Fedotova 2006, 213).

set of conflicts and compromises between oil generals and financiers—both within individual companies and in the oil sector as a whole—structures much of the grand narrative of Gustafson's *Wheel of Fortune* (2012). Gustafson sees the difference between neftianiki and finansisty as crucially a matter of management style and mentality, and shows how different skills, instincts, and leadership styles intersected fatefully with shifting political conditions. (Mikhail Khodorkovsky's brash and outspoken financier ways, Gustafson argues, were central to his company Yukos's downfall.)[39]

I prefer to see neftianiki and finansisty less as personality types and more as crystallizations in the post-Soviet oil sector of different kinds of networks and relationships—those connected to industrial capital and finance capital, respectively—that are brought together in capitalist corporations of all sorts. From this general perspective, a crucial task for the ethnography of corporations is to explore and theorize the diversity of ways in which these kinds of capital actually articulate.[40] In the Perm region's oil complex, perhaps the most striking aspect of this process was its dramatic spatialization. Recall that the negotiated agreement between the regional state administration, big Lukoil, the PFPG, and Permneft created two separate Lukoil production subsidiaries in the Perm region. Although the same big Lukoil vice president—Ravil Maganov, a neftianik whose career had taken off at the Langepas oilfield in West Siberia (Chernikov 2003)—chaired both boards of directors, the two regional subsidiaries were otherwise quite different. In the south was OOO Lukoil-Permneft, which inherited most of the Perm region's aging oil deposits and production infrastructure, as well as most of the regional oil industry's administrative staff. It was, that is, the primary domain of the Perm region's neftianiki, and its struggles in the second half of the 1990s were those of transforming production, labor relations, and the oil industry's old company towns in the face of declining production (even following the years of acute crisis in the early 1990s, some forecasts projected that OOO Lukoil-Permneft's oil production would drop by 25 percent between 1998 and 2006).[41] In the north was ZAO Lukoil-Perm, with its ownership split fifty–fifty between big Lukoil and the PFPG. ZAO Lukoil-Perm's central office staff was composed mostly of veterans of the petrobarter and fuel veksel operations run out of the

39. For other takes on the neftianiki/finansisty dynamic, see Dienes 2004, Gaddy 2004, and Poussenkova 2004. Due to its origins in the Soviet oil industry, Lukoil is often considered to be a classic domain of neftianiki; the region-based analysis here is intended to considerably complicate that image. For a study of factions of capital and their competition in the oil industry, including in the post-Soviet case, see Labban 2008.

40. The anthropology of flexible labor and production is very large; in the postsocialist world, see especially Dunn 2004. Karen Ho (2009) provides a crucial boost to this production-focused literature by tracing expert networks associated with finance capital outward from Wall Street; in the early 1990s, postsocialist companies were one important destination of that expertise.

41. Valerii Kostarev, "Neftianaia reka issiakaet?" *Kapital Weekly*, July 7, 1998.

PTB and PFPG—the region's new finansisty, led by Andrei Kuziaev and his deputies. Although it also produced oil, of course, first and foremost to fulfill regional state contracts, ZAO Lukoil-Perm was increasingly involved in intricate financing deals that began to tie the Perm region's oil industry to many of big Lukoil's other projects around the world and to the international oil complex. "We ended up with two Lukoils: one *very* Soviet, one *very* bourgeois," was the way one of my interlocutors summed up this situation for me.

• • •

In my discussion of Soviet-era oil sector company towns in chapter 1, I noted neftianik Nikolai Kobakov's description of Chernushka as a "natural economy." Returning to Chernushka in the post-Soviet period usefully illustrates some of the transformations taking place across OOO Lukoil-Permneft's domain in the south of the Perm region after its incorporation into big Lukoil. At the close of the Soviet period, Permneft's Chernushka NGDU produced 45 percent of the Perm region's oil and employed around seven thousand people. Chernushka's oil, like much of the oil in the south of the Perm region, was viscous and sulfurous, and it proved especially hard to find refineries willing to take it in the early 1990s, a situation that led to the temporary stopping of production in and around Chernushka.[42] Prospects for future expansion were likewise bleak: over 80 percent of Chernushka's reserves were classified as "difficult to retrieve" (*trudnoizvlekaemye*). As was the case in other oil-producing towns of the southern Perm region, an enormous social sphere had grown up around the Chernushka NGDU over the course of the late Soviet period, including stores, clubs, schools, kindergartens, and over 280,000 square meters of apartment space (Bondarenko 2003, 122). In 1992–94, still before the coming of Lukoil to the region's production fields, and in keeping with larger trends, the Chernushka NGDU transferred many of its stores and schools to the balance sheet of the city, although it was often still the local corporation that was most influential in the operation of the social sphere (see Healey, Leskin, and Svetsov 1999; Rogers 2006). At that point, Permneft continued to hold on to the apartment blocks that housed oil workers and their families. Indeed, selective municipalization notwithstanding, the mutual embeddedness of Permneft's Chernushka NGDU and city of Chernushka was only enhanced in the very early 1990s, as the devolution of sovereignties to lower and lower units solidified the local power of Chernushka-based networks of politicians and neftianiki and enabled them to make increasing demands on both Permneft and regional state agencies.

42. I. Gurin, "Porogi i otmeli neftianoi reki," *Zvezda*, April 16, 1996.

This was the situation—declining production, dim prospects for new discoveries, and powerful local networks—that confronted the management of Permneft as their company became OOO Lukoil-Permneft 1995–96. The sharpest problems remained demonetization and mutual indebtedness, and so, between 1995 and 1997, OOO Lukoil-Permneft embarked on a wide-ranging plan to bring its financial house in order. Its massive debts were largely eliminated by 1999, in large part through the company's participation in the veksel system managed by its sibling ZAO Lukoil-Perm in the north, and to a lesser extent through improved possibilities for international export enabled by big Lukoil (around 21 percent of OOO Lukoil-Permneft's production, much of it from Chernushka, went for export in 1997).[43] Chernushka's was one of the three NGDUs that entered OOO Lukoil-Permneft at the point of the Lukoil takeover in 1995, and its earliest years as part of a Lukoil subsidiary were understood by most as a significant improvement over the early 1990s—and over the prevailing situation in non-oil-producing areas of the southern Perm region. Both oil and salaries, that is, had begun flowing again.

Following financial stabilization, however, OOO Lukoil-Permneft embarked on a major workforce restructuring program designed to break up the "natural economy" model of an oil town and, in the company's own framing, to put oil production at the center of its operations. A centerpiece of this restructuring was an effort to weaken the power of the old NGDUs, like the one based in Chernushka. In 1999, the company began to reorganize into a two-layer system of management that involved a corporate center in Perm and eleven nongeographically organized sections, each responsible for aspects of the oil supply chain as it ran from the subsoil to the point of transfer to pipelines and each reporting directly to Perm's corporate center.[44] This restructuring eliminated a significant portion of Permneft's middle-management bureaucracy—which was based in geographically defined NGDUs and had, in the Soviet and early post-Soviet eras, mediated between local oil operations and Permneft's corporate center. No longer, that is, would each NGDU be a " 'small feudal estate' where equipment, resources, and people accumulated over the decades."[45] Well over 50 percent of the office jobs that had once been based in Chernushka were eliminated, along with the NGDU itself as a unit of administration (Bondarenko 2003, 136). A new office

43. Some efforts at restructuring were already taking place in the 1994–95 window, among them the closing of Krasnokamskneft, the least productive of the five NGDUs that were part of Permneft at the time. This was a significant blow to the city of Krasnokamsk because liquidating the NGDU—and continuing production under another organization, based in Perm itself—would reduce Krasnokamsk's tax income significantly (Abaturova 2003, 138).

44. See G. Volchek, "Nas sviazivaiut tysiachi nitei," *Neftianik* 3 (80), February 2002.

45. S. Fedotova and A. Neroslov, "Vremia L—1998 god," *Permskaia Neft'*, September 26, 2011.

in Chernushka was established, but it was an extension of central OOO Lukoil-Permneft leadership in Perm rather than a more autonomous local unit deeply embedded in Chernushka affairs.

At a lower level, this reform also meant the reorganization of what were once the constituent divisions and brigades in each NGDU. Some were redistributed into the eleven new corporation-wide units, with positions that had been duplicated in each NGDU eliminated. A greater portion, however, was spun off entirely, as OOO Lukoil-Permneft's restructuring plan also called for retaining within the company's structure only operations that were directly concerned with oil production itself. This meant that a wide range of divisions specializing in other parts of the oil production process—from transport to well repair to all manner of technical and social support—were reborn as independent companies. The idea here, quite common across the global oil industry in the 1980s and 1990s (see, e.g., Shever 2012) was that OOO Lukoil-Permneft would then contract independently with these companies for specific services in aid of its primary production goals. Some small, separate companies were permitted to compete with each other for contract work on multiple oil deposits in the region. Others became units of oil services companies spun off from other NGDUs in the Perm region. Some of them even moved under the larger Lukoil umbrella by affiliating with oil services branches of big Lukoil such as ZAO-Lukoil-Burenie-Perm, a drilling company managed directly from Moscow.[46] In general, though, there were not nearly enough jobs in these new oil services companies to handle the outflow of workers associated with restructuring in OOO Lukoil-Permneft. By 2003, the Chernushka district was down to around one thousand employees of OOO Lukoil-Permneft, with an additional four thousand employed in some aspect of oil services through an array of different companies. Nikolai Kobakov, the neftianik in charge of this reorganization, summed up the company's plan:

> From the old, unwieldy Permneft we created a flexible, specialized enterprise, very sensitive to changes in the surrounding environment. We carried out deep changes in the system of management. From the management of structural units (NGDUs) we moved to the management of facilities, that is, oil deposits. (Bondarenko 2003, 138)

Three things are notable about Kobakov's description of this restructuring plan. First, it was a decisive move away from the geographically organized company

46. Workers from Chernushka were among those who hired themselves out for shift work above the Arctic Circle that was being run out of ZAO Lukoil-Perm. I. Gurin, "Vlast′ na mestakh: Otchety," *Zvezda*, February 26, 2005.

town model, especially in its determination to break the entrenched power of district-level NGDUs and their relationship with local politics and replace it with cross-company, task-oriented, nonduplicative divisions. In other words, OOO Lukoil-Permneft was aiming to become increasingly disembedded from the local population—not only by shucking off elements of the social sphere—a process that had begun in the early 1990s and continued apace, but by replacing locally powerful networks with a combination of corporate-wide divisions reporting to central OOO Lukoil-Permneft and independent service companies that contracted to work all over the place, many of them even outside the Perm region.

Second, the plan set as its goal what Kobakov called the "management of oil deposits"—a quite new way of conceptualizing the regional oil industry. The Soviet oil industry, recall from chapter 1, produced oil in conditions of socialist central planning and shortage, conditions in which the management of an oilfield translated into attempting to maximize its output at nearly all costs. What Kobakov meant by becoming "sensitive to the surrounding environment" was that factors such as global and local oil prices, reserve forecasts, and other market signals would now drive production decisions. This was a strategy directed at maximizing profits by managing shortage—that is, adjusting supply levels. Managing shortage to keep oil prices as high as possible without causing demand to fall has, of course, long been a central pillar of the global capitalist oil industry's accumulation strategy, but it was, at that point, quite new and unfamiliar to OOO Lukoil-Permneft and its workforce. A history of Chernushka oil, published in 2003, ended with a question posed by an imagined reader: "It used to be, industry veterans say, that we fought to meet the plan. The country demanded: 'Oil, give us oil!' Is it really not that way anymore? Do oil workers really not fight to increase the production of oil?" The book's answer, detouring through economic theory and the latest industry conferences in Moscow, was that, no, oil workers did not always strive for the highest possible production levels anymore. The contemporary oil industry, it confirmed, does what is "economically promising," which might mean that, at times, it would be better to keep the oil in the ground (Bondarenko 2003, 157–58).

Third, if Kobakov's framing of restructuring in the language of workforce flexibility and specialization seems like a textbook model for enterprise transformation in post-Fordist times, this is because it was drawn from precisely that textbook, and this language reverberated across the southern Perm region in the late 1990s and early 2000s. In 2001, for instance, the company issued a "Memorandum for the Development of Lukoil-Permneft in the Twenty-First Century" to all its workers, explaining the company's goal of being the most dynamic subsidiary in big Lukoil and impressing on its workers three key terms that would be central to the company's development: innovation ("search!"), systematicity ("think!"), and dynamism ("act!").[47] The remainder of the memorandum outlined how these goals would be realized in practice. It should come as no surprise that OOO Lukoil-Permneft's restructuring was carried out with the help of the management consulting firm PAKK, based in Moscow and with ties to the international business community. (PAKK worked with a number of Perm-based companies in the 1990s.) Indeed, the managers of OOO Lukoil-Permneft, with PAKK's assistance and encouragement, went on to write their own textbook, describing their management changes and establishing their clear links to what they viewed as the latest thinking in international management theory. Titled *Permian Oil: the Art of Being above Circumstances* (OOO Lukoil-Permneft 2003) and listing eight authors—six of them career neftianiki—the book proudly described the deep crisis from which the leadership team of OOO Lukoil-Permneft had extracted itself in the previous eight years and fleshed out in some detail the keywords of the restructuring plan that had enabled them to do so: flexibility, systematicity, and dynamism. *Permian Oil* frequently references the practices of Western oil companies in workforce management, comparing its own earlier "Development Memorandum" to a similar document that, it notes with pride, came out two years later: "The ChevronTexaco Way." The book concludes that, "Permneft and ChevronTexaco are moving in a single direction, the most progressive and responsive to the contemporary circumstances of the oil industry in the world and the level of technological development" (OOO Lukoil-Permneft 2003, 212).

Whether or not OOO Lukoil-Permneft was on its way to becoming as flexible and specialized an oil company as its leadership was fond of claiming was a matter of one's perspective—and considerable debate in the Perm region. Clearly, the massive layoffs in Chernushka and other oil towns marked a stark departure from the Soviet era and from the early 1990s, as did the elimination of older, embedded NGDUs and the retreat from Soviet-style company towns. Clearly, as well, scarcity was beginning to take on capitalist dimensions—the management of oil supply in order to maximize profit—rather than all-out production in a socialist shortage economy. However, I came across few who would claim that OOO Lukoil-Permneft was anything close to as adaptive and flexible as its sibling to the north, ZAO Lukoil-Perm. Indeed, restructuring notwithstanding and by comparison to ZAO Lukoil-Perm, the oil towns of the southern Perm region still looked to one of my acquaintances as if they were feudal estates run by what he called "Soviet barons"—just barons who had learned to speak the language of flexibility and oil services contracting.

47. These terms appeared everywhere in OOO Lukoil-Permneft's literature and public relations material at the time. See, for instance, V. Kostarev, "Lukoil-Permneft′ ishet, dumaet, i deistvuet," *Zvezda*, November 11, 1997.

ZAO Lukoil-Perm: Finance Capital in the Northern Perm Region

ZAO Lukoil-Perm was created in early 1996, with two production units transferred over from OOO Lukoil-Permneft: the Polazna NGDU and the remains of the Krasnokamsk NGDU (which, as the oldest and least productive of the Perm region's five NGDUs, had been disbanded in the fall of 1995, at the same shareholders meeting at which Permneft entered big Lukoil).[48] Crucially, the transfer that created ZAO Lukoil-Perm was entirely asset-based, with OOO Lukoil-Permneft retaining all the debts that had accrued to Permneft at the point of transfer. Lukoil's new half-owned subsidiary in the northern Perm region thus began with a clean balance sheet, whereas OOO Lukoil-Permneft was, at the time, working its way through 900 billion rubles of debt.[49] Although ZAO Lukoil-Perm and OOO Lukoil-Permneft collaborated on many issues—including production and transport in certain areas—there were some fundamental differences between the two companies as they began separate operations as in 1996, differences that worked mostly to the advantage of ZAO Lukoil-Perm. In the first place, OOO Lukoil-Permneft in the south was a wholly owned subsidiary of big Lukoil, and its production and financial goals were in all cases negotiated through Moscow, meaning that—new claims of agility and flexibility notwithstanding—the possibilities for local innovation were comparatively limited. By contrast, several insiders I spoke with noted that the 50 percent local control and regional connections enjoyed by ZAO Lukoil-Perm were key components of its success because they allowed the company to make its own decisions relatively quickly. Second, oil from the northern Perm region was also in many cases lighter and sweeter than that from the south, making refining a less intensive and less costly proposition; refineries often competed with each other for oil coming in from the northern Perm region.

ZAO Lukoil-Perm implemented some of the same kinds of management reforms that took place in Chernushka and the other oil towns of the south, but its much smaller workforce made these both less complicated and less contro-

48. There was, in fact, an earlier company named ZAO Lukoil-Perm. It was formed in 1992 as an engineering company dedicated to producing equipment for oil production within Russia following the breaking of supply chains from Azerbaijan with the dissolution of the Soviet Union. It was headed by Nikolai Kobakov, a Permneft veteran, and included collaboration with Perm's Motovilikha and Iskra factories, which were aiming to transition out of declining defense sector production. In the general crisis and chaos that came in the mid-1990s, however, it proved impossible to make or sell new equipment, and ZAO Lukoil-Perm was essentially nonfunctioning by 1994. Its legal name, however, remained registered, and the company was revived in the negotiations in which big Lukoil acquired Permneft in 1995. See G. Volchek, "Nas sviazivaiut tysiachi nitei" *Neftianik* 3 (80), February 2002.

49. S. Fedotova and A. Neroslov, "Vremia L—1996." *Permskaia Neft'*, August 12, 2011.

versial. The major focus of ZAO Lukoil-Perm from its earliest days had less to do with transforming relationships of production than with the legal, financial, and shareholding structures that make up a modern oil corporation. Whereas OOO Lukoil-Permneft was the new home of the old Soviet neftianiki, the "skeleton" (Vikkel', Fedotova, and Iuzivofich 2009, 99) of ZAO Lukoil-Perm's central office staff and leadership was made up of veterans of the PFPG (especially Oil Products Marketing), together with the networks that extended into the regional state administration. ZAO Lukoil-Perm's glossy "The First Five-Year Plan" booklet, issued in 2001, reads, in fact, like a PTB and PFPG yearbook. Andrei Kuziaev was the president. Vsevolod Bel'tiukov, one of Kuziaev's right-hand men since the early days of the Perm Commodity Exchange, was his first deputy for corporate affairs. Bel'tiukov was joined by Viktor Lobanov, a respected neftianik originally from Osa, as deputy director for drilling operations, and Aleksandr Bulgakov, who until 1996 had been the deputy head of the Perm region's tax office, as deputy in charge of finance. The head of the accounting office was Tatiana Zagruskaia, former head of accounting for the PFPG's Oil Products Marketing, and legal affairs were handled by Evgenii Voskoboinikov, a former Soviet Gossnab official who had moved to Perm in 1990 and taken a position at the Perm Commodity Exchange, and Igor Fomin, who also hailed from the PTB and PFPG. Pavel Tiulenev, who moved from PNOS to Oil Products Marketing, stayed on at ZAO Lukoil-Perm as director of refining and distribution. With a few exceptions dedicated to production, these were the finansisty of the Perm region's oil complex (see ZAO Lukoil-Perm 2001).

I have already noted this team's central role in designing and implementing the Perm region's fuel veksel program. They also embarked on other finance projects. With support from both the Perm region and big Lukoil, for instance, ZAO Lukoil-Perm was able to obtain major loans from European banks to invest in new technologies that boosted oil production in its old oilfields and intensified the search for new deposits in the comparatively unexplored northern reaches of the Perm region. Lukoil-Perm's production rapidly exceeded the 2.5 million barrels that, according to the terms of the original agreement between big Lukoil and the Perm region, went into the regional state's operations and barter fund; around 30 percent of the company's oil went for export in 2000 (ZAO Lukoil-Perm 2001, 92). In 1997, as the veksel program that had occupied ZAO Lukoil-Perm since January 1996 wound down, the company created a special new division dedicated to the specific problems presented by working with subsidiaries, jointly owned companies, licenses, and mergers and acquisitions. Vladimir Mikhnevich, a neftianik originally from Krasnokamsk, headed this effort; his support staff had experience both at the PFPG and in the intricate negotiations that created OOO Lukoil-Permneft and ZAO Lukoil-Perm (ZAO Lukoil-Perm 2002, 43–45). This division was soon at the forefront of the company's operations, obtaining financing for capital-intensive, high-technology exploration projects that would be repaid with oil sales only decades later and executing a number of complicated mergers and acquisitions that further solidified big Lukoil's presence in and around the Perm region.

All of these were jobs led by the region's finansisty. If OOO Lukoil-Permneft in the south was increasingly speaking the language of flexible labor regimes, contracting companies, and the management of oilfields rather than social spheres, then ZAO Lukoil-Perm in the north was speaking the language of stock swaps, corporate valuation, and complex loans. If OOO Lukoil-Permneft management cited "The ChevronTexaco Way" in their own book, Andrei Kuziaev of ZAO Lukoil-Perm frequently mentioned that reading books on the importance of shareholder value had been transformative for him (e.g., Vikkel′, Fedotova, and Iuzifovich 2009, 99). If OOO Lukoil-Permneft was working with the PAKK consulting firm in Moscow, then ZAO Lukoil-Perm was working with consultants from global PriceWaterhouseCooper. Through such distinctions and differences, industrial capital and finance capital, and neftianiki and finansisty, were highly spatialized in the Perm region's oil complex over the decade from 1995 to 2004.

Expanding Production in the North

ZAO Lukoil-Perm began operations in 1996 with control over fifteen oil deposits in the northern Perm region, enough to provide for the regional state administration's needs at the time, and it moved rapidly to expand its holdings in a variety of ways. Recall that, in the early 1990s, the general trends toward parcelized sovereignty and decentralization of production had created a number of small oil companies in the region, some of them internationally collaborating joint-stock companies. Shortly after its founding, ZAO Lukoil-Perm began a concerted effort to gain control over these companies, completing stock deals to acquire, in whole or in part, KamaNeft, Russkaia Toplivnaia Kompaniia, PermTOTIneft, and Permteks. Permteks continued to be a joint operation between ZAO Lukoil-Perm and the U.S. company SOCO International, and, with support from ZAO Lukoil-Perm, was able to receive a $10 million line of credit from the European Bank for Reconstruction and Development.

The case of Krasnovishersk's oil is particularly instructive. The Krasnovishersk district, in the far northeast of the Perm region, was at the forefront of the dissolution of the production side of the oil industry in the early 1990s, declaring its exclusive ownership of the subsoil beneath the district. Although the oil workers of Krasnovishersk had failed to achieve full independence in the early 1990s, they had obtained very favorable terms on a twenty-year lease on all Permneft's equipment in the district, with no payments due until the second decade. (Before its entry into big Lukoil, Permneft was focusing its operations on the south and did not plan further exploration and development in the north for many years.) By 1998, the district was working on plans to build its own small refinery, which would have allowed it to process a portion of the oil pumped from beneath it. However, this autonomy was not well received at ZAO Lukoil-Perm, which hoped to include further exploration in the Krasnovishersk district in its rapidly developing plans for the northern Perm region. ZAO Lukoil-Perm was ultimately successful in its effort to incorporate the company Visheraneftegaz, becoming its sole shareholder in 1999 and offering, as part of the deal, to continue mitigating the effects of the underground nuclear explosions conducted in the area in the 1980s and to structure its development support for the region in ways that would respond to local expectations for the support of the social sphere.[50]

Not every possibility for reincorporation and exploration was so easily accomplished as that in the Krasnovishersk district. Conflicts over access to new deposits swirled most notably around the mining city of Berezniki. Sixty-five percent of Russia's reserves of potash lay near Berezniki in the VKMKS—the Upper Kama Potassium Salts deposit. These deposits were situated in the first 600 meters or so of the subsoil, and had been under extensive exploitation for decades. It had long been known that oil deposits—perhaps up to 30 percent of the Perm region's untapped oil reserves—lay under the potash deposits, at depths of 2,000–3,000 meters. Disputes had been running since the Soviet era about how to access this oil without endangering the fragile potash deposits, which could easily be damaged by water or oil—or so, at least, insisted the leadership of Uralkalii and Silvinit, the two companies engaged in mining operations in the VKMKS and among the world's largest potash producers. Already in 1996, the use of new kinds of exploratory technologies was pointing to new areas in the north of the Perm region that might be promising for oil wells, and new drilling and maintenance technologies were under development to keep the layers of potash in the upper subsoil unharmed. But the potash industry still refused to allow extensive oil operations in its zone of influence. ZAO Lukoil-Perm responded by pointing to a range of new studies from both mining engineers and industry scientists claiming that the oil could be safely pumped, especially in light of new technologies: horizontal drilling, reinforced wells, and drilling fluids tailored not to react negatively with potassium compounds. It proposed, moreover, to drill for oil only in places where

50. O. Oputin, "'Akuly' neftebiznesa pokazyvaiut zuby," *Kapital Weekly*, April 4, 1998. Although ZAO Lukoil-Perm's own operations were primarily in the north, the fact that its finansisty were charged with bringing joint and smaller-scale oil operations back into the overall Lukoil fold meant that, over the years, the company came to control oil production in several southern areas of the Perm region as well, especially through new deals with the half-Ecuadorian PermTOTIneft and Russkaia Toplivnaia Kompaniia.

potash mining was not in process or planned in the foreseeable future. These proposals, too, were initially rejected.

Behind these claims and counterclaims lay something else. The concentrated power of the potash mining operations in the north were perhaps the only networks that could, at least for a while, stand up to the alliance of ZAO Lukoil-Perm, the PFPG, and the regional state apparatus. The high prices obtained for potash sold as fertilizer on international markets in the years since the end of socialism meant that the mining cities of Berezniki and Solikamsk were very wealthy and in little need of support from Perm's regional state apparatus. Promises to increase the flow of state expenditures in the city and district were easily rebuffed—they meant little in the context of already-concentrated wealth. Quiet political pressure was no less persuasive. District-level politicians representing areas of the VKMKS answered reliably to the interests of the mining companies, and it was often difficult even for the Perm regional administration to get its way cleanly in this part of the region. As negotiations over drilling in the VKMKS dragged on into the summer of 2000, both Andrei Kuziaev of ZAO Lukoil-Perm and Governor Gennadii Igumenov resorted to publicly claiming that Uralkalii and Silvinit's objections were inhibiting the development of the entire Perm region. More than a decade of tight collaboration on multiple fronts had made this argument—that the fate of the region and of Lukoil were inextricably bound together—seem entirely logical. But it was a measure of the strength of the political and financial networks densely gathered around the potash mining companies that this argument even needed to be deployed.[51]

The issues were eventually resolved, not entirely to either side's satisfaction. ZAO Lukoil-Perm gained access to a number of new oilfields and new drilling locations in existing oilfields. The large "Siberian" deposit in the Usol′e district, which had been developed in small ways since 1985, was slated to go from 77 to 279 wells in the following years, and it was soon the most productive in the entire Perm region, with most wells still in the fountain stage—having enough internal pressure to produce without expensive technologies for increasing production in old wells.[52] But the proceeds did not go directly to ZAO Lukoil-Perm. The final deal that allowed limited oil drilling in the VKMKS called for the formation of yet another joint company that funneled proceeds from oil sales partly to ZAO Lukoil-Perm and partly to a set of district interests associated with the potash industry (see Vikkel′, Fedotova, and Iuzifovich 2009, 166–67).

---

51. T. Vlasenko, "Neft′ i sol′ zemli prikamskoi," *Novyi Kompan′on*, July 4, 2000. On the role of this dispute in the Perm region's gubernatorial elections in 2000, see Fedotova 2006, 213–16.

52. V. Zuev, "Samoe krutoe otkrytie," *Neftianik*, June 12, 2000.

Beyond the Perm Region

The finansisty of ZAO Lukoil-Perm also became increasingly central to big Lukoil's Russia-wide and global ambitions. In the chaotic early and mid-1990s, big Lukoil had acquired, in one way or another, minor stakes in a number of small oil companies around Russia. Following its successes in the Perm region, ZAO Lukoil-Perm was assigned the task of using its clean balance sheet, its financial expertise, and its friendly relationships with the Perm regional administration to become big Lukoil's "consolidation center" (Vikkel′, Fedotova, and Iuzifovich 2009, 103) for much of Russia.

In fairly short order, this plan came to include eight such companies, in the nearby Khanty-Mansi autonomous district, the Komi republic, and the Volgograd region. Drilling began, for instance, in the Komi-Permiak autonomous district in 1998 through a ZAO Lukoil-Perm subsidiary company Maikorskoe. (In fact, oil workers from the former Krasnokamsk NGDU played a major role in the development of the Maikor oilfield, one of several cases in which the oldest and declining oilfields in the Perm region provided the labor for the opening of new fields and new deposits outside the region [Abaturova 2003, 150–53].) SOCO international, the U.S.-based company that had been working with ZAO Lukoil-Perm on deposits in the north of the Perm region through the collaborative Permteks, sold its block of shares to ZAO Lukoil-Perm in 2001, making the company a wholly owned subsidiary of ZAO Lukoil-Perm.[53] Parma-Oil, another collaborative ZAO Lukoil-Perm venture, specialized in exploration and drilling operations over the northern border of the Perm region, in the Timan-Pechora basin of the Komi republic. It operated according to a new agreement worked out among Lukoil, the Perm region, and the Komi republic.[54]

In 2001, taking into account all these expansions and acquisitions, ZAO Lukoil-Perm's annual production figures reached 6.5 million tons—nearly three times what the company started with in 1996. The newspaper *Permskii Obozrevatel′* estimated that Lukoil had contributed 35 billion rubles in taxes to the Perm region in the 1996–2001 period, a third of that to the regional budget and the rest to district and city budgets. The company had also invested about the same amount in its own operations—45 percent of the total investment in the Perm region in that five-year period. As a result of these mergers and acquisitions, by 2000, nearly a fifth of the 100,000 shares in big Lukoil were held in the Perm region—a figure large enough that big Lukoil held its annual shareholders meeting in Perm.

53. T. Vlasenko, "Neftianaia Konsolidatsiia Pod Egidoi 'Lukoila,'" *Novyi Kompan′on*, August 28, 2001.

54. G. Volchek, "Znakomtes′: 'Parma-Oil,'" *Neftianik* 10 (67), November 2001. For a fuller accounting of these 2001 acquisitions, see also Markelova 2004, 79–80.

Soon, these operations extended beyond Russia as well. Big Lukoil had established Lukoil Overseas Holding in 1997 to manage its existing interests in the former Soviet Union, especially in Kazakhstan and Azerbaijan, and its emerging global ambitions in Iraq, Egypt, and beyond. (Lukoil was, at the time, the first Russian oil company to have a subsidiary dedicated to overseas operations.) In early 2000, in a major reshuffling of big Lukoil's subsidiaries undertaken at the federal level, Lukoil Overseas Holding moved to Perm, and Alekperov appointed Andrei Kuziaev as president. Kuziaev and his staff—drawn largely from ZAO Lukoil-Perm—began to represent big Lukoil at elite gatherings of international investors, and were soon traveling the world to negotiate deals for big Lukoil. As Perm's newspapers frequently pointed out, Kuziaev and his team had come a long way from their first trades on the Perm Commodity Exchange.

It is worth emphasizing that ZAO Lukoil-Perm was the key catalyst in the worldwide assembly of big Lukoil as a vertically integrated oil company. In its early years, Lukoil Overseas Holding developed not only with staffing and expertise drawn from the Perm region but also with region-based investment: from 2001 to 2006, the fledgling Lukoil Overseas Holding received all its investment capital not from big Lukoil in Moscow, but from ZAO Lukoil-Perm (Vikkel′, Fedotova, and Iuzifovich 2009, 106). Big Lukoil's expansion into Iraq, Egypt, and beyond, then, was enabled quite directly by the regional dual-subsidiary arrangement that I have described. Although this arrangement originated in a failed raid by big Lukoil rather than anyone's strategic planning, it held the neftianiki at OOO Lukoil-Permneft and the finansisty at ZAO Lukoil-Perm at arm's length for nearly a decade, allowing ZAO Lukoil-Perm to have the benefits of start-up capital from the PFPG and ongoing close relationships with the regional state apparatus while at the same time insulating itself from the declining production fields and labor-heavy company towns that fell to OOO Lukoil-Permneft. Even readers who have thus far been skeptical of a region-focused account of Russian oil should now see, I hope, that any detailed account of how Lukoil became Russia's leading oil company on the global stage runs through relationships, networks, and possibilities that originated not only in Moscow but also in a rapidly transforming regional oil complex.

Reunification: A Regional Vertical

Although talk of reunification of the Perm region's two Lukoil subsidiaries had begun to circulate in 2002, plans were formalized in 2003, and on January 1, 2004, OOO Lukoil-Permneft and ZAO Lukoil-Perm were reunited into a single oil production company named Lukoil-Perm. Production sites within Russia but outside the Perm region—such as those acquired by ZAO Lukoil-Perm in the Komi Republic and West Siberia during the company's time as big Lukoil's consolidation center—were relinquished to the control of Lukoil subsidiaries in those regions. Nikolai Kobakov, who was at that time head of OOO Lukoil-Permneft, was named the new head of the unified Lukoil-Perm. Also as part of the restructuring, the PFPG sold its 50 percent stake in ZAO Lukoil-Perm to big Lukoil—the stake that had been so significant for the Perm region in the veksel years and beyond—for nearly $400 million, a sale completed by the middle of July 2003. The reunited Lukoil-Perm continued to lead big Lukoil's internationalization plans, and from the time of reunification in 2004 until 2006, Lukoil-Perm was, in fact, a subsidiary of Lukoil Overseas Holding (which continued to be based in Perm). Perm-based neftianiki and finansisty, led by Andrei Kuziaev at Lukoil Overseas Holding, that is, continued to provide the expertise guiding Lukoil's operations outside of Russia (Vikkel′, Fedotova, and Iuzifovich 2009, 107.)[55] In 2006, Lukoil Overseas Holding—and with it Andrei Kuziaev and much of his team—moved to Moscow, and the unified Lukoil-Perm assumed the institutional location it retained until the time of my primary fieldwork for this project in 2009–12: a single, region-based oil-producing subsidiary of big Lukoil, working alongside and in concert with other subsidiaries in the region, including the refinery Lukoil-PNOS and the distribution network Lukoil-Permnefteprodukt. It became common, at least in the company's own publications and self-presentation, to refer to Lukoil's operations in the Perm region as a "regional vertical."

As a final step in this reorganization in 2005, Vagit Alekperov named Aleksandr Leifrid, a career oilman from Kogalym in West Siberia, director of the unified Lukoil-Perm, later adding the designation of "representative of the president of Lukoil in the Perm region" and giving him oversight over regional refining and distribution as well. Leifrid's appointment initially came as a surprise and a jolt in an industry that had long relied on specifically regional networks and, in the post-Soviet period, was composed almost exclusively of people who had known each other and worked together for many years. Leifrid brought with him an entire executive team—"they rented ten of the biggest apartments in Perm," one friend commented—and they quickly replaced many of the regional insiders who had so closely knit together oil companies and state agencies in the first postsocialist decade. According to some observers, this was precisely the goal: from the perspective of big Lukoil, relationships between Perm regional subsidiaries and the regional state administration had, over the 1990s and early 2000s, grown too cozy and were beginning to interfere in corporate projects and profits. As much as regional networks spanning state and corporation had served the purposes of big

55. G. Volchek, "Ob″edinenie—eto normal′no," *Zvezda*, April 1, 2003.

Lukoil in those years, an outsider like Leifrid was now needed to take firmer control of operations. It is impossible to know whether or not this was the true motivation for Leifrid's appointment. What is clear, however, is that he was no less interested than his predecessors in striking deals with the regional state administration, and enough of the existing management team (including Kuziaev's longtime deputy Vsevolod Bel'tiukov) stayed on to facilitate this transition. Indeed, by the time of reunification and Leifrid's arrival in Perm, both OOO Lukoil-Permneft and ZAO Lukoil-Perm had also become heavily engaged in a number of other significant projects in the region (many of which appeared to have little to do with oil itself) and the decentralized and chaotic 1990s had shifted to the monetization and state building of the 2000s. This set of conjunctures, including the place of corporate reunification within them, is the subject of part II.

• • •

Corporate vertical integration, in Russia as elsewhere, is best understood as an ongoing process, not an event (see esp. Welker 2014). Even within longstanding vertically integrated companies, that is, subsidiaries are acquired, sold, and reshuffled; labor forces are structured and restructured; financial arrangements linking units are pursued and abandoned; relationships with states and state agencies morph and shift. The reunification of the Perm region's two Lukoil subsidiaries in the mid-2000s brought to an end a period of exceptionally rapid incorporation and vertical integration. My analysis of this period has demonstrated that, in the conditions of weak central authority that prevailed in the 1990s, these processes were far more than the Moscow-based battle over the privatization of production fields that is often portrayed. They extended to refining and circulation, to specifically regional configurations of finance capital and industrial capital in the oil sector, and, through petrobarter and fuel veksels, to the very means of everyday exchange that stitched together people and institutions across the region. In all of these ways, oil and oil companies were becoming ever more closely connected to the ways in which the Perm region itself could be imagined.

Recall that in the late Soviet period none of the ten phone numbers on the desk of the head of the Regional Communist Party Executive Committee connected him to Permneft, a good indication of the comparatively low prestige of the oil production industry at the time. When Governor Gennadii Igumenov lost his reelection bid in 2000, his relationship with the oil industry had grown so tight that, rather than retire from public life, he moved across the street to take up a new position as head of the staff of the board of directors for ZAO Lukoil-Perm. More than a decade later, after I took a tour of the Lukoil-Perm museum, my guide pressed Igumenov's autobiography (written and published with the company's help) into my hands (Igumenov 2008). It tells the story of Igumenov's life, from his early days in the Perm region's once-prestigious coal industry, where he worked and first entered political life, through his time in the Perm regional administration, including his term as governor. It concludes with his move to Lukoil-Perm. Igumenov's was far from the only career trajectory that began elsewhere in the Soviet period and ended up in the regional oil industry; indeed, making that point was a major goal of both the museum and the autobiography.

Part II

THE BOOM YEARS

4
STATE/CORPORATION
The Social and Cultural Project Movement

In 2004, the Children's Arts School in the city of Berezniki received a grant from Lukoil-Perm for a multifaceted educational program that the organizers called Unity (*Edinenie*). From May to October, schoolchildren and teachers, along with a number of the city's library, museum, and House of Culture employees, embarked on a wide-ranging effort to reacquaint Berezniki with the cultural traditions of the northern Perm region. They constructed an electronic "virtual museum" of the area in which Berezniki was located; traveled to the neighboring Cherdyn district in search of bits of folklore; held a handicrafts fair that featured everything from ceramics to rugs to belts; and sought to incorporate pieces of traditional culture into the school's primary teaching streams: choreography, visual arts, and music. All in all, nearly five hundred people participated in the public events associated with Unity over the project's six-month run.

After the grant wrapped up, Unity's organizers reported to Lukoil-Perm not only these basic facts and figures but also another set of transformations they believed central to the efficacy of the project. Citing a survey and set of interviews with schoolchildren and teachers that they had completed, the organizers wrote:

> The children note that, thanks to this project, they were able to study their folk traditions. Participating in the project helped them become kinder, more honest, more responsible, better mannered, and more cultured. . . . The teachers note that, thanks to the project, new personal qualities and possibilities opened up for the schoolchildren: they became more sure of themselves in concert performances, more independent and organized.

Unity, the organizers continued, successfully addressed important issues such as developing children's aesthetic sensibilities; shaping their personalities; forming civic qualities (*grazhdanskie kachestva*); and fostering independence and responsibility. They reported that 95 percent of those surveyed knew that the project was sponsored by Lukoil-Perm.[1]

Given the often embellished style of grant reports—a style shared by reports on the (over)fulfillment of socialist plans—what the students, teachers, and residents of Berezniki really thought about the range of activities sponsored by Lukoil-Perm in 2004 is something of an open question. But it is clear that those writing this report knew what the Lukoil-Perm office in charge of funding grants for social and cultural projects wanted to read about: projects focused on the revival of traditional culture in a variety of ways; the formation of new kinds of post-Soviet subjects—independent, well-mannered, responsible, civic-minded; and the furtherance of the company's branding, marketing, and corporate social responsibility efforts. The authors of this report would seem to have been quite persuasive, for the Berezniki Children's Arts School went on to participate in generous grants from Lukoil-Perm in subsequent years for the projects I Love this Land and A Time of Changes. In all cases, these grants incorporated cofinancing from the regional state budget, either through direct allocations or through the regional state administration's own grant competitions.

Berezniki's Unity was by no means exceptional in the Perm region of the 2000s. The report excerpted here appears in an archival collection—about which more below—that houses materials related to many hundreds of such social and cultural projects spanning the length and breadth of the Perm region, nearly all of them sponsored by Lukoil subsidiaries, region-level state agencies, or some combination thereof. These projects ranged from grants to purchase new library books for children to new health-care initiatives, from ecological awareness campaigns to museum exhibits, from the celebration of folk culture to the reconstruction of churches and mosques, from children's summer camps to adult fitness programs. By the mid-2000s, Lukoil-Perm had, in short, joined regional state agencies in those classic state projects of molding communities, making citizens, searching for cultural pasts, and transforming subjects. Across these domains, a new conceptual vocabulary was spreading rapidly, one that featured talk of grant proposals, selection commissions, audits, sponsorship, corporate social responsibility, and dozens of other terms and procedures that were initially unfamiliar to all concerned.

The chapters of part II show that social and cultural projects such as Unity were central to a new kind of state-corporate field that began to coalesce in the Perm

---

1. "DShM im. L.A. Starkova, Virtual'nyi Muzei Istoriia Verkhnekam'ia," PermGANI f. 1206, op. 1, d. 388.

region following the Russian financial crisis of 1998 and enduring, in various permutations, for over a decade. In these years, social and cultural development joined the orchestration of industrial and financial capital as matters of central concern for both Lukoil-Perm and the regional state administration. Although I will also point to many significant continuities, I agree with the conventional wisdom that Russia's August 1998 default on its debt and devaluation of the ruble was a crucial break between two epochs of the post-Soviet era: the chaos and decentralization of the 1990s and the oil boom and recentralization of federal state power of the early 2000s.[2] Of the many ensuing conflicts and reorientations over the next decade, the protracted struggle between Russian federal state organs and the oil company Yukos—culminating in the arrest, trial, and lengthy imprisonment of Yukos head Mikhail Khodorkovsky—has received far and away the most attention.[3] There is good reason for this; after all, the Putin administration's offensive against Khodorkovsky and Yukos was meant precisely to broadcast the reemergence of the federal state and bring the powerful oligarchs to heel. The Yukos Affair certainly resonated in the Perm region, even though the company had no significant operations there. (One of my earliest interlocutors in this project, a Lukoil-Perm employee, told me in the mid-2000s that the whole company was waiting to see what federal tax inspectors, who had at that point been sitting in the Lukoil-Perm central offices for nearly two months, would or would not conclude about the company's finances, and that the result would make a big difference in whether I would be able to do the fieldwork we were discussing.)

In the battle between the Russian state and Yukos, the state is usually assumed to have won, as evidenced by Khodorkovsky's imprisonment and the subsequent absorption of Yukos into the state-controlled Rosneft. However, this picture, along with all the public and scholarly attention trained on the Yukos Affair at the expense of other topics, obscures a much broader range of transformations that reshaped the corporate-state field in the oil sector in the same decade. I show that if the Yukos Affair sent a signal that big business was to stay out of federal politics, then this was not the case at the regional or district level at all; indeed, President Putin's simultaneous demands that Russia's major companies turn their attention to development and helping to solve social problems at home—rather than pad-

2. As Gustafson (2012, 184) rightly emphasizes, the combination of rising global oil prices and the boost to the value of Russia's exports brought about by devaluation meant both enormous profits for oil companies and rapidly rising receipts for the Russian state treasury.

3. On the Yukos Affair, see, for instance, Balzer 2005; Thompson 2005; Goldman 2008, 105–20; Gustafson 2012, 272–319; and Sakwa 2014. Vadim Volkov (2008) provides an instructive and detailed comparison of the Yukos Affair and the early twentieth-century antitrust cases against Standard Oil in the United States, viewing both as cases of expanding state power in the context of "early capitalism."

ding their bank accounts in Cyprus and Geneva—only enhanced the collaboration of corporation and state at lower levels of organization and administration (see also Orttung 2004).

In an influential paper on postsocialist state building, Venelin Ganev argued that scholars should locate their understandings of postsocialist states within the large literature on modern state building in political science and historical sociology. "What are the analytical ramifications," he asked, "of the fact that the state-building process in post-communism takes place simultaneously with the disintegration of a state-owned economy?" (2005, 435). In these chapters, I modify Ganev's question to fit Russia in the 2000s: What are the analytic ramifications of the fact that Russian state building in the 2000s coincided with both a massive oil boom *and* a dramatic global rise in the power of corporations as direct participants in matters of governing human social and cultural lives?

The first of these coincidences—between state building and oil boom—points us less to early modern Europe (Ganev's chief foil, via Charles Tilly) than to the archetypical petrostates of the twentieth century, especially as theorized in Terry Lynn Karl's *Paradox of Plenty* (1997). The mid-twentieth century's major oil-exporting states, from Venezuela to Nigeria to Saudi Arabia, were developmentalist states that sought to capture oil rents and use them for modernizing projects. Although international oil companies might tend to their labor forces in small production enclaves and participate in one or another aspect of industrial development beyond their own operations (such as the construction and operation of petrochemical plants and other heavy industry projects in the case of Saudi Aramco—see Hertog 2008, 650), the state was decidedly at the helm when it came to collecting tax revenues and selectively channeling them into broader social and cultural development and modernization (or in other directions, such as military buildup). Indeed, inasmuch as many of these states had only recently achieved postcolonial independence and the oil companies operating on their territories were international oil companies, the central role of the state was very much the point. Karl's argument was that the "perverse outcomes" (1997, xv) of economic inequality and corrupt politics commonly found in these petrostates were traceable to the fact that torrents of oil money arrived at the same time as initial efforts to build a modern state were under way, powerfully shaping them. Countries where modern state formation was more advanced at the time of the oil boom—such as Norway—did not, she argued, fall victim to the resource curse because their sturdier state institutions and bureaucracies were able to channel oil money in more productive and less distorting ways. This model of a rent-seeking and rent-distributing petrostate continues to be the basic framework of debate for rentier state and/or resource curse models as applied to Russia (Fish 2005, 114–38; Goldman 2008; Ross 2012).

Although I have registered my objections to the resource curse approach on overall analytical grounds (see the Introduction), there is no question that the coincidence of Russian state building in the 2000s and an oil boom cry out for investigation. It is also certainly true that, like its mid-twentieth-century predecessors, the reconsolidating Russian federal state of the 2000s envisioned a powerful and long-lasting role for itself in Russian development through the exploitation of natural resources, as a quick glance at Vladimir Putin's prepresidential dissertation on state-led modernization projects through natural resource rents indicates (Balzer 2006). Nevertheless, hundreds of on-the-ground projects like Berezniki's Unity, with their central and direct involvement by Lukoil-Perm, should give us pause. They point, I will argue, to a second coincidence that deserves our attention: that of postsocialist state building and the increasing involvement of corporations in the direct governance of human social and cultural life in recent decades. In the current configuration of global capitalism, natural resource rents enter development ambitions and projects not only via state taxes, royalties, or licensing fees—as was typically the case in the classic petrostates—but also through complex hybrids and layers of state and corporate entanglement that extend far beyond production enclaves. In this global era of corporate social responsibility programs (a topic I introduce in more detail in the next chapter), the accumulation of resource rents at the Russian federal center and high-stakes battles between the Kremlin and Yukos are only part of a much larger picture. When we take into account the processes unfolding in an oil region as well—consistent with my overall contention that states are built along a center–region axis—our understanding of the Russian oil complex becomes both more complicated than the simple victory of the Russian state over the oligarchs and more closely tied to very contemporary global processes of state and corporate entanglement than is typically acknowledged. Although my analysis will take a quite different route, I am, nevertheless, in full agreement with Pauline Jones Luong and Erika Weinthal's insistence, in their pathbreaking and scrupulously argued *Oil Is Not a Curse* (2010, esp. 322–36), that post-Soviet cases serve to illuminate just how contingent were the configurations that prevailed in the classic oil-exporting petrostates of the 1960s–90s (and, therefore, just how contingent are the theoretical models of resource curse that emerged to explain them).[4]

• • •

A word on sources is instructive at this point. Collection number 1206, "Social and Cultural Projects Realized in the Perm Region," housed at the Perm region's

4. On state-corporate ownership structures in the Russian oil sector in the 2000s, see also Rutland 2008, 1057–59.

State Archive of Contemporary History (PermGANI), has furnished a significant portion of the information on which I base these chapters. The story of this collection's assembly and organization is itself an important clue in unraveling the relationships that made up the regional state-corporate field in this period. In 2005, the director of PermGANI, an energetic and well-respected historian determined to catalog new political processes in the post-Soviet period, approached Lukoil-Perm to ask whether the company would contribute a portion of its files related to social and cultural projects to the archive. The company agreed, and soon after turned over many cartons of its internal documentation, including everything from grant applications to final project reports, and from drafts of publicity statements to highly polished annual reports. Separately, and in a more standard transfer of documents originating in the regional state offices and ministries—although one that was by no means guaranteed in those years—the director also acquired the papers of the regional state administration's sponsored social and cultural projects.

The collection was not yet fully organized and open to researchers at the time of my primary fieldwork, but I was able to view those files that had been prepared for public viewing and to consult extensively with archive staff about the remainder of the collection. Even without the collection fully cataloged, it was clear that the archive's specialists had made the determination that these separate acquisitions were part of the same historical moment and process. The boxes originating in state and corporate offices were stored together, and often contained duplicative material and cross correspondence. The same state officials and Lukoil-Perm managers, and the same agencies applying for funding, appeared regularly in both sets of documents. Fittingly, the archive's official description of documents (*opisi*) moves seamlessly between state administration and Lukoil-Perm projects as aspects of what they term a single "trend" (*napravlenie*) and others I spoke with simply called the project "movement" (*dvizhenie*).[5] Independent of my own inquiries, that is, the assembly, organization, and structure of this archival collection points usefully to some of the ways in which corporation and state became deeply intertwined in the administration of social and cultural projects.

I also base much of my discussion in these two chapters on interviews with some of the key participants at the time. Here, too, we can see some indication of the shape of the regional state-corporate field that emerged in the early 2000s. With only rare exceptions, everyone I discuss in these two chapters had herself or himself moved between jobs at Lukoil-Perm and the regional state adminis-

5. Introduction to PermGANI f. 1206, op.1. The term "project culture" was also popular, but I stay with "project movement" in order to decrease the number of meanings of the term "culture" in play at the same time.

tration at some point. With its origins in the 1990s, this cycling accelerated in earnest in the early 2000s, and showed no signs of slowing at the time of my primary fieldwork.[6] When I pointed this out to several of my interlocutors, few found it noteworthy or exceptional. "Of course," one told me, "those people are the best managers (*praviteli*)." Even if their particular networks were temporarily out of favor in either the regional oil industry or the regional state apparatus, he went on, it was generally understood that there would almost always be a place for them elsewhere in the state-corporate field. One category of exception to this cycling of personnel was cases in which local elites successfully inserted themselves into Moscow structures—such as Andrei Kuziaev and his team at Lukoil Overseas Holding. Very occasionally, officials left state or corporate life altogether. Such departures, however, were rare in the first decade of the 2000s, when the Russian state bureaucracy was expanding by leaps and bounds and increasing federal regulation made room for more and more levels and divisions of management in corporate structures, especially if those managers were already familiar with navigating the state bureaucracy. With the exceptions of a few career state employees and cradle-to-grave oilmen, then, the state-corporate field I am describing was, in key part, constituted by a revolving door of employment in regional state administration and regional oil company.

These materials and sources lay bare the composition of committees and commissions; the trajectory of personal and political networks; the definition and attempted resolution of problems in the social and cultural spheres; the manner in which state and corporate funds were combined, carefully separated, and allocated; and the fates of all these efforts as they shifted over the course of the decade. Such are the mechanisms by which the regional state administration and Lukoil subsidiaries came to serve as joint metacoordinators of nearly all domains of social and cultural life in the Perm region in the 2000s. They permit me to continue this book's account of the ways in which both states and certain corporations—and their intersections at all levels—should be theorized as aspiring to metacoordination among different domains of human social and cultural life. In the 1990s, this metacoordination took place through region-based circuits of exchange such as petrobarter and fuel veksels, both of which stitched together the most diverse sets of actors across the region, making them relatable and charting their mutual constitution and indebtedness. In the 2000s, those circuits disappeared with the return of federal control over the money supply. At the analytical level of federal state building, there would be ample reason to consider

6. The same movement held at the district level, and by the later 2000s it included United Russia party structures as well.

this oil-backed remonetization in the terms often suggested for other oil-exporting states: as a key domain of abstraction and fetishization closely tied to the imagination of a unitary state distinct from society (e.g., Coronil 1997; Apter 2005). At the regional level, however, this remonetization largely took the means of exchange off the table as a joint corporate–state project, and the terrain of state-corporate metacoordination shifted from petrobarter and fuel-backed veksels to social and cultural projects.

It can be somewhat tricky to demonstrate the interpenetration of state and corporation, given how entrenched these terms are and the prevailing assumption—made by many of my interlocutors in the Perm region and, I expect, many readers—that they refer to discrete entities that have definable interests or interact with each other in coherent, unitary ways. My goal is to take seriously these projections of coherence (for they are, precisely in their status *as* projections, crucial aspects of both state and corporation formation), while also demonstrating their blurry edges and extensive interpenetration within a broader field of state-corporate governance. I navigate this representational challenge by focusing in chapter 4 on the state side of the field and in chapter 5 on the corporate side, while gradually building a case for their numerous linkages across both chapters. I hope the result conveys a sense of interpenetration and separation at the same time. This is precisely what deserves our attention in 2000s-era Russia and, indeed, around the world.

The Etatization of Civil Society Building

Two overarching tendencies characterized the social sphere of the Perm region in the 1990s. First, the socialist-era welfare state was in retreat—if not collapse—as a range of state functions and tasks directed at social defense (*sotsial'naia zashchita*) of the population were rolled back, reformed, or eliminated altogether. Moreover, in step with privatization and marketization reforms, and as we saw in the case of Chernushka in the previous chapter, most social and cultural organizations were spun off from the firms that had controlled them in the Soviet period. Kindergartens, clubs, and cultural centers once operated by factories and farms were municipalized, privatized, or, in many instances, shuttered altogether. In practice if not in formal ownership and operation, they often entered into the purview of the highly decentralized suzerainties discussed in earlier chapters, where they joined the queue for patronage-style support (see, e.g., Rogers 2006). Social welfare and cultural institutions were not, that is, a high priority for any level of state administration in the 1990s. But local bosses had no easy refuge from the requests of citizens in trying times: they often ended up folding support for social and cultural institutions, as indeed they did for struggling citizens on a case-by-case basis, into their standard practices of cutting deals of all sorts to make ends meet.

At the same time, a small network of organizations providing new kinds of social services began to emerge, inspired and fostered by major Western nongovernmental organizations (NGOs) seeking to advance Russia's transition away from socialism by "seeding civil society" (Mandel 2002a) or constructing a "third sector" (Hemment 2007) alongside the state and business. In the model promoted by these organizations, an array of voluntary, nongovernmental, noncommercial organizations would replace what was understood to have been the omnipresent Soviet state. These new organizations would attend to the social needs of the population, serve as a bulwark against the return of authoritarianism, and be important engines of capitalist transformation. But, as in other aspects of the transition from socialism, the proliferation of new kinds of organizations did not occur as the models predicted. Whatever lofty goals came under the banners of privatization, democratization, and a slew of other labels to which Western agencies were devoted, Steven Sampson writes, "At a basic level . . . they exist as concrete activities called 'projects.' The transition in Eastern Europe is a world of projects" (1996, 121). These projects must, Sampson was arguing, be tracked historically and ethnographically rather than simply modeled. In the Perm region, the trail leads first to new practices of state building. It eventually opens up into an entire state-corporate field.[7]

From International NGOs to the City of Perm

A number of new organizations and agencies sprung up in Perm in response to the new possibilities offered by international funding agencies in the late 1990s.[8] The city had an especially active chapter of Memorial, the organization dedicated to preserving the memory of victims of Stalin-era repression, and the group was instrumental in setting up Perm-36, a Gulag Museum set in a former prison camp 120 kilometers outside Perm. Through the global sister cities program, a number of residents of Perm made enduring connections with Oxford, England, and, among other things, worked through these contacts to establish a Perm-based chapter of Hospice.[9] Registered as an independent societal organization (*obshchestvennaia organizatsiia*) in 1994, Hospice received funding from the Soros

7. Julie Hemment (2012a; 2012b; 2015) analyzes another major trajectory that civil society–building projects took in the 2000s: into state-sponsored youth groups and festivals.

8. On civil society building in this period, see also Borisova 2005 and Borisova, Reneva et al. 2003.

9. See, for instance, Varvara Kal'pidi's 2010 film *Perm-Oksford: Vremia Deistviia,* which devotes substantial attention to this axis of the development of charitable work in the Perm region.

Foundation in 1996–98 for a project titled Life Before Death, which aimed to recruit new hospice volunteers.[10] Other similar small-scale organizations grew up in Perm at the time, some new and some reincorporated from Soviet organizations into post-Soviet nongovernmental societal organizations: the League of Culture, the Fund for Peace, Badger: A Foundation for the Protection of the Family, and others.[11] These were also the formative years for the Perm Human Rights Center, which grew over the next two decades to be one of the most influential organizations of its kind in Russia, judged both by its success in winning international grants and in shaping regional public and political life from a position outside official state agencies (see also Borisova 2008).

Perm's set of societal organizations was less fully plugged into Western aid and NGO circuits than, for instance, Moscow's or Saint Petersburg's, but the situation in the mid-1990s tracks well with what Julie Hemment (2007) reports for the provincial city of Tver: a small number of organizations had successfully cast themselves as ideal conduits for international aid money aimed at building civil society through project-focused grants. In Perm as in Tver, many—but not all—of these new organizations were led by former Communist Party or Komsomol activists working to reinvent themselves (see also Hemment 2007, 48), and the circuit of NGOs and international aid organizations did not extend much outside the regional center. In more provincial cities and rural areas of the Perm region, the decentralized suzerainties of factories and farms held much more sway.

This situation changed in the late 1990s, when a conjuncture of Perm city politics, economic crisis, and international aid agencies began to catapult social and cultural projects to the forefront of governance, first in the city of Perm, later in the entire Perm region. In 1996, Iurii Trutnev was elected mayor of Perm. Although Trutnev had started out in the Soviet-era oil industry, he had left it behind for a career in Party politics, especially in the Komsomol, by the early 1980s. In 1990, he helped found the enormously successful trading company Eks Limited, which was the conduit for many of the Perm region's imported consumer goods and, by the mid-1990s, had expanded to include a chain of shopping centers and an array of other services, ranging from security firms to real estate to trucking. Colleagues and associates from both his Komsomol days and his Eks Limited days accompanied Trutnev to the Perm mayor's office to take up key positions in his administration. Trutnev appointed Tatiana Ivanovna Margolina, a highly respected regional Komsomol leader from 1972 to 1991, who had been working in education for much of the 1990s, as deputy mayor in charge of social issues in Perm. Margolina's active and influential tenure as one of Trutnev's dep-

10. PermGANI f. 562, "Khospis: Blagotvoritel′noe obshchestvo."

11. PermGANI f. 738, "Barsuk: Fond Zashchita Sem′i"; PermGANI f. 7099, "Fond mira"; PermGANI f. 1232, "Permskaia obshchestvennaia organizatsiia Liga Kul′tury."

uty mayors—rather than the sporadic appearance of societal organizations in the early to mid-1990s—was often identified by my interlocutors as the origin point of the social and cultural projects movement in the Perm region. In interviews with me, however, Margolina pointed to a number of factors aside from her own initiative and the collaboration of others in Trutnev's Perm City government. "It all came together," she told me on one occasion, "because of the patronage and financial collaboration of the Eurasia Foundation."

The Eurasia Foundation was founded in 1993 with funds from the United States Agency for International Development. In 1998, with Trutnev's blessing and with Margolina leading the way, Perm municipal government and the Eurasia Foundation entered into a collaborative agreement to administer a grant competition for societal organizations in Perm. The first grants were awarded in 1998, and the first in a long series of seminars, roundtables, and discussions about the relationships between societal organizations and Perm municipal government began at this time as well. Margolina was familiar with and well respected among these new and reconfigured organizations, and her installation as deputy mayor gave them far more access to city power structures than that to which they had been accustomed. Funds to support these projects and seminars came from the Eurasia Foundation and, to a lesser extent in the first years, Perm municipal government itself. From the perspective of Perm's mayor Iurii Trutnev and his team, this funding was especially useful at the time because it eliminated some pressure on the city budget—pressure that increased exponentially following the crash of summer 1998. "[The second half of the 1990s] were very difficult years," Margolina recalled to me in 2009, "a crisis that is not even comparable to today's."

The Eurasia Foundation had worked on civil society creation efforts in the former Soviet Union for around four years at that point. But the kind of structure that its representatives set up in Perm was a new departure. As Ruth Mandel (2002a) and others have shown, seeding civil society often involved not just administering funds, but also helping to set up the organizations that would apply for and receive those funds. From the perspective of international donor organizations, this process often led to the awkward situation in which state officials or their close collaborators moonlighted as the directors of NGOs, making themselves eligible to receive development funds but hardly exemplifying the funders' vision of grassroots citizen initiative. In the Perm region, beginning in 1998, we find something different: a collaborative agreement that linked the organization and administration of civil society–building grants directly to the Perm City mayor's office. Indeed, as N. Borisova's analysis of Perm's political "quasiautonomy" (2002) from its surrounding region in the late 1990s shows, the high degree of attention to the management of social issues in the city was a crucial part of the electoral coalition that Trutnev assembled during his time as mayor—a coalition that eventually helped him win the governorship of the Perm region.

Although those at the Eurasia Foundation and other participating agencies (such as the British Charities Aid Foundation) viewed this collaboration as a way to spread civil society more efficiently and effectively, it was an important early moment in the movement of grant-based projects from international aid agencies to agencies of the Russian state in the Perm region. Looking back on the agreement between the Eurasia Foundation and the City of Perm from nearly two decades' distance, we see the inflection point at which international grants for social and cultural projects became part of regional state-building, a means of crystallizing control over multiple vectors and domains of power after the decentralization and disorganization of the 1990s. It was not long before Lukoil's regional subsidiaries became major partners in this project as well.

From City to Region (via Federal District)

The first collaborative grants between the Eurasia Foundation and the City of Perm were awarded in 1998. They may or may not have continued under their own steam, but this model for funding the social sphere (from the perspective of Russian state agencies) and seeding civil society (from the perspective of international aid agencies) received a major boost early in Vladimir Putin's first term as president of the Russian Federation. One of Putin's first major efforts to rein in big business and reestablish central control over Russia's powerful regions—announced on May 13, 2000, scarcely two months into his term—was the insertion of a new layer of state bureaucracy between the federal center and Russia's eighty-nine regions.[12] In this reorganization, the Russian Federation was divided into seven federal districts, each administered by a presidential envoy and staff charged with monitoring the implementation of federal laws and initiatives in that district. The Perm region was allocated to the Volga federal district, which was headquartered in Nizhnii Novgorod and included fifteen Russian regions, extending from the Samara region in the southwest to the Perm region in the northwest. The Volga federal district was large and consequential, containing nearly a quarter of Russia's population and a significant portion of the country's industrial and energy complexes. Putin appointed as his envoy Sergei Kirienko, who had once served in the Komsomol and had risen, briefly, to the post of Russian prime minister toward the end of Boris Yeltsin's second term as president. (Kirienko's was not, however, a standard biography for the post of presidential envoy; of the seven men appointed to the first envoy positions by Putin, he was the only one

12. On these reforms, see especially Reddaway and Orttung 2004 and Ross and Campbell 2009.

who had not come up through the power ministries—the security services and Ministry of Internal Affairs [Sharafutdinova and Magomedov 2003, 155].)

It is likely that some combination of Kirienko's Komsomol background and his time as prime minister—both positions in which he had to deal with broad swaths of society and social problems, including international agencies in the latter post—predisposed him to like what he saw in his first official visits to Perm, where he found social and cultural projects closely integrated into Mayor Trutnev's city administration. Grant-based financing of social and cultural projects quickly became one of Kirienko's signature initiatives for the Volga federal district as a whole, and they contributed significantly to his emerging reputation as a so-called technologist interested in remaking the policies and procedures of Russian governance for a new, more centralized time (Sharafutdinova and Magomedov 2003, 156). In November 2000, at the initiative of Kirienko's office, Perm hosted the first ever Social Projects Fair, which showcased the winners of a competition that solicited applications for projects throughout the entire Volga federal district. In that year, 120 projects were funded in four categories: Social Partnership, dedicated to new methods of collaboration among state organs, NGOs, businesses, and media outlets that would solve concrete problems such as homelessness; The Development of Networks and Communities, which emphasized new kinds of connection across time and space, especially the preservation of ecological and cultural heritage; The Cultural Capital of the Year, an opportunity for a city in the Volga federal district to use this title as a catalyst for creative innovation in the cultural realm; and a miscellaneous category called Periscope: A Stock Exchange [*birzha*] of Innovative Ideas.[13]

In the four years that the Volga federal district ran these competitions (2000–2004), all sorts of organizations based in the Perm region received grants based on the judgment and evaluations of Moscow-based experts. The winning organizations tended to hail from urban areas, with most of them based in Perm itself. The 2001 competition, for instance, awarded a total of nearly 4 million rubles to eight projects in the Perm region, including a university-based local history project, run by noted scholar A. A. Abashev, that was dedicated to collecting residents' memories of the city of Perm; a documentary film festival, Flahertiana (named after the famous filmmaker Robert Flaherty); a new grade school program designed to improve educational possibilities for handicapped children; and a charitable fund dedicated to moral and spiritual improvement, to be sited in the industrial district of Perm.[14] In 2003, the Perm region submitted a total of 124 grants (the

13. "Privolzhskii Federal'nyi Okrug: Okruzhnaia Iarmarka Sotisal'nykh i Kul'turnykh Proektov. Perm'," Pamphlet, Perm Regional Library, Regional Studies Division, 2000.

14. "Pobediteli konkursa 'Sotsial'noe Partnerstvo' okruzhnoi Iarmarki sotsial'nykh i kul'turnykh proektov 'Saratov 2001,'" PermGANI f. 1206, op. 1, d. 1.

second highest in the Volga federal district, behind only the Samara region), with just over half of them coming from the city of Perm itself; 47 were from noncommercial, nonstate entities and 70 were from state agencies and departments.[15] The annual social projects fairs, which moved on from Perm in 2000 to Saratov in 2001, Tol'iatti in 2002, and Nizhnii Novgorod in 2003, continued to serve as the primary context in which members of the growing social and cultural projects movement—from experts evaluating grants, to regional state officials supervising them, to the actual grant winners—met and exchanged ideas and techniques.

At this expanding set of conferences and fairs, the leaders of the social and cultural projects movement expended considerable energy developing a new vocabulary for what they understood themselves to be doing, a necessary step, they thought, in the larger projects movement. They devoted special emphasis to encouraging businesses and private individuals to sponsor projects, and a dedicated scholarly conference on the theme of "Trusteeship in Russia: Past and Present" sought to identify historical precedents for the kind of sponsorship relationships that were being created.[16] A circulated "Dictionary of Trusteeship" (*Slovar' popechitel'stvo*) ran from *blagopoluchatel'* and *blagotvoritel'* (beneficiary and benefactor) through a range of English loan words—*volonter*, *grant*, *sponsor*, *fandraizing*—and arriving at *tselevaia sotsial'naia programma* (targeted social program, a coordinated "group of projects, events, and measures directed at solving the population's most important social problems").[17]

Contributors to the pool of money that was doled out to the winning social and cultural projects in the Volga federal district included not only a who's who of Western aid agencies, including Soros, MacArthur, Ford, IREX, the Eurasia Foundation, and others, but also a number of wealthy federal Russian companies, among them Yukos, MDM-Bank, and the Alfa Group, in addition to federal and regional state agencies and businesses.[18] I do not have precise funding figures for the first or second of these Volga federal district competitions, but the third competition, which took place in 2002 and distributed 38.1 million rubles among 98 receiving organizations, envisioned a split among those contributing funds as follows: 25 percent federal-level businesses; 25 percent regions of the Volga

---

15. "Itogi podachi zaiavok na okruzhnoi konkurs sotsial'nykh i kul'turnykh proektov, 'Nizhnii Novgorod-2003,'" PermGANI f. 1206, op. 1, d. 196; "Informatsiia dlia zhurnalistov," PermGANI f. 1206, op. 1, d. 196.

16. "Popechitel'stvo v Rossii—Traditsii i Sovremmenost', 30 Sentiabria 2002 g.," PermGANI f. 1206, op. 1, d. 1. See also the discussion of Lukoil-Perm's embrace of the legacy of the Stroganov family in chapter 6.

17. "Slovar' Popechitelia ot A do Ia," Center for the Development of Trusteeship, Perm. Perm Regional Central Library, Regional Studies Collection. See also Aksartova 2009 on the mechanics of exporting civil society terminology and grant-making practices.

18. "Vpervye v Rossii: Iarmarka Sotsial'nykh Proektov 'Perm'-2000,'" *Pravda*, November 29, 2000.

federal district; 15 percent donating organizations (i.e., Western foundations); and 35 percent region-level businesses.[19] The first annual Social Projects Fair gathered experts and politicians from the regions constituting the Volga federal district as well as those organizations that had been successful in grant competitions for a series of meetings and roundtables. Representatives of major Russian companies were there, too, along with international aid agencies. Much was made of oligarch Vladimir Potanin's recent contributions to education—he had personally offered to pay the stipends of 100 of the most promising Russian college students. A high-level representative of the Open Society Institute lauded the fact that this gathering in Perm was the first of its kind that had taken place at the initiative of state officials, who, he said, usually kept their distance from international aid agencies and the NGOs they were interested in funding.[20]

For all the talk of third sector and civil society development, however, it was clear that this initiative was also a state-building exercise, albeit one not based in the traditional model of tax collection and expenditure. The contributions from major Russian businesses and the high-profile participation of national-level oligarchs were the earliest responses to the Putin administration's demand that Russian businesses begin to collaborate with the Russian state on local development projects. According to Kirienko:

> The district Social and Cultural Projects Fair is the beginning of a major effort in the promotion of project technologies in the practice of state and municipal governance (*upravlenie*). To think and act in project terms (*proekno*) is to think and act strategically, comprehensively, and responsibly . . . and, absolutely, in public dialogue and active collaboration with all communities, institutions, corporations, and sides who are interested in the development of the country. Movement forward is not only innovative technologies and economic growth but also modern education, the preservation of cultural heritage, the protection of human dignity and the rights of peoples to the preservation of their historical fates.[21]

State administration of social and cultural projects was, then, a technique of governance that emerged in part out of the Putin administration's effort to reestablish central control over the regions. At least in the Volga federal district—

19. "Protokol Zasedaniia Popechitel'skogo Soveta Iarmarki Sotsial'nykh i Kul'turnykh Proektov Nizhnii Novgorod 2003," PermGANI f. 1206, op. 1, d. 196.

20. Iarmarka Sotsial'nykh i kulturnykh proektov Privolzhskogo Federal'nogo okruga "Perm'-2000," Institut Ekonomika Goroda, 2000, http://www.urbaneconomics.ru/projects/?mat_id=177 (accessed June 13, 2014).

21. "Privolzhskii Federal'nyi Okrug: Okruzhnaia Iarmarka Sotisalnykh i Kul'turnykh Proektov. Perm'." Pamphlet, Perm Regional Library, Regional Studies Division, 2000.

home to a quarter of the population of Russia—this was a primary way in which "the state" reemerged in the 2000s. It did this by putting state officials at the meeting and coordinating points of nearly all sectors of society, collecting contributions from some (including international agencies) and distributing it to others through a competitive, expert-judged grant-making process. The project movement, as some proponents were beginning to call this emerging set of practices, was uniting multiple domains of life into a single brand of bureaucratic administration. It was at the organizational heart of a new era of state building, and Perm was its showcase city in its showcase federal district.[22] One reason for this showcase status was that Mayor Trutnev was elected governor of the Perm region in December 2000, and he and his team—including his deputy for social projects Tatiana Margolina—moved from the offices of Perm City government down the street to the offices of the Perm regional administration.

The Project Movement as State Building

For the first three years of Perm's municipal social and cultural projects competitions, the Perm regional state administration had taken note but did not follow suit. Newly elected Governor Trutnev, however, declared that on his watch and with the support of the Volga federal district, Perm would become Russia's "Capital of Civil Society," and Tatiana Margolina and her team moved quickly to set up regional competitions for social and cultural projects. By 2003, the height of the social and cultural projects movement, a societal organization or state agency in the Perm region could apply for funding from several layers of state administration: the Volga federal district, the Perm region, and even some districts within the Perm region.[23]

In the regional administration, Trutnev and his team had at their disposal far more tools with which to expand social and cultural projects than they did in the Perm municipal administration. Indeed, they had far more tools than did Sergei Kirienko in his role as presidential envoy to the Volga federal district. (Presidential envoys had comparatively small staffs, relied largely on the power of persuasion and negotiation, and had much less direct influence over the shape of regional laws and budgets than did the still-powerful governors.) One of the Trutnev administration's key actions in its first year was to pass a law "On State Social

22. Elements of the project movement later appeared in other regions as well; see Hemment 2012a for one example from Tver, where a Social Forum took place in December 2004 at the initiative of the region's new governor.

23. See, for example, "Postanovlenie Glavy Mestnogo Samoupravleniia Il'inskogo Raiona Permskoi Oblasti ot 21.15.2003," PermGANI f. 1206, op. 1, d. 204.

Commissions (*sotsial'nyi zakaz*) in the Perm Region" (No. 1141-166, passed September 21, 2000), which established the legal and budgetary procedures through which the regional state apparatus would interact with the social sphere. The new law featured the mechanism of grant applications for social and cultural projects, and created both a council in the governor's office to oversee the process and a commission to evaluate and administer the grants (Perm Regional Administration 2002, 38–44). Law 1141-166 specifically elaborated the importance of raising funds from a variety of sources, including not only the regional budget but also wealthy individuals and businesses in the region. Although international aid agencies like the Eurasia Foundation were not excluded from contributing funds—they were included in a catchall category of donor labeled "other contributions, in accordance with current law"—in practice they made no contributions to these region-level grants.

Along with this new set of legal and administrative tools, Trutnev and his team were responsible for administering a much larger, more complex social and cultural sphere than that to which they had grown accustomed in Perm municipal government, including many small cities and struggling agricultural districts. They also had to come to terms with region-wide corporations, especially Lukoil's subsidiaries. Lukoil-PNOS and the main offices of OOO Lukoil-Permneft and ZAO Lukoil-Perm were located in Perm, and all three regional subsidiaries had been very supportive of the first social and cultural projects in both the city of Perm and the Volga federal district. In recalling the very early years of the social and cultural projects competitions—before her move to the governor's office—in one of her conversations with me, Margolina singled out ZAO Lukoil-Perm:

> From the very beginning of those projects in Perm, I would say [ZAO] Lukoil-Perm was a partner. Without their understanding, their interest in participating, and their funding in that budgetary crisis [of the late 1990s], the idea of [social and cultural projects] could not have been realized. Andrei Kuziaev understood the technology [of social and cultural projects] precisely, as a systematizer [*sistemshchik*], and supported it.

The "tone" that ZAO Lukoil-Perm set through its example, she said,

> enabled the city government to attract other business structures, so that they could participate in their own ways. It was an interesting approach: [competition] categories that were important for the development of the city, and categories in which these interests coincided with the interests of one or another city-based corporation.

After Trutnev's election to the post of governor, the social and cultural projects movement moved into the Lukoil production subsidiaries' most significant, and, in the early 2000s, increasingly lucrative, upstream operations spread throughout the Perm region. The company was already heavily involved in political, economic, and social life in these locales, and the collaboration between corporation and state agencies that had begun in Perm deepened and spread.

Lukoil subsidiaries' participation was, to begin with, built into the new administrative structure set up by Law 1141-166. Margolina, as Governor Trutnev's deputy, chaired the Social and Cultural Projects Commission, charged with overseeing grant competitions for social and cultural projects in the Perm region. She had three of her own appointed deputies on the commission: V. V. Abashev, a literature professor, who headed a highly regarded and culturally focused NGO dedicated to studying the city of Perm and its place in the region; P. I. Blus′, the head of the governor's social projects office (*apparat*), who would directly manage the administration of the grants; and I. V. Marasanova, identified as "head of the division of social technologies in the department of social management at ZAO Lukoil-Perm" (Perm Regional Administration et al. 2002, 64). Beneath these three deputies were some thirty-three members of the commission who would participate in the process of awarding grants after experts had evaluated the proposals. These commission members hailed from every corner of the social and cultural sphere in the Perm region, from the state Department of Culture to nongovernmental musical groups; from the director of the region's AIDS prevention department to the interim head of the ecological protection agency; from representatives of newspapers and small businesses to the directors and deputy directors of several midsized regional businesses (Perm Regional Administration 2002, 64–65). This organizational structure of the commission, with representatives from the regional state administration and ZAO Lukoil-Perm at the top and a full spread of other institutions arrayed beneath them is—like the organization of PermGANI's archive that houses so many of the documents the commission produced—nicely indicative of the state-corporate field that was taking shape in the early 2000s. The first how-to guide for organizations wanting to participate in social and cultural projects competitions—a weighty 128-page tome, thick with laws, application forms, and examples from every step of the process—was likewise coauthored and copublished by the Perm Regional Administration, ZAO Lukoil-Perm's Office of State Relations, and the Urals Center for the Support of Nongovernmental Organizations. It was called, simply, *Social and Cultural Project Competitions: Recommendations for Organization and Execution* (Perm Regional Administration et al. 2002).

Lukoil subsidiaries' participation extended far beyond the makeup of the commission, and quickly came to influence the mechanisms by which funds were appropriated and dispersed. Following Law 1141-166, an annual authorization "On State Social Purchasing [*zakaz*] in the Perm Region" set out the mechanisms of social and cultural project administration. In 2003, for instance, the law directed a variety of state agencies, such as the Committee on Youth Policy, the Committee on Physical Culture, Sport, and Tourism, the Department of Education, the Department of Culture, and the Department of Ecology and Environment, to contribute part of their operating budgets to financing social and cultural projects chosen by the Social and Cultural Projects Commission, rather than dispersing funds directly out of these state agencies' own administrative offices.[24] In the announcement of project winners that year, these agencies appear as *zakazchiki*, or orders/purchasers, of the various social and cultural projects. The first set of grant recipients under the overall social and cultural projects umbrella included groups planning to do everything from publishing a youth newspaper to holding an interreligious dialogue called The Culture of Toleration, and from planting flower gardens outside a local school to running an anti-drug abuse program dubbed High without Narcotics.[25]

One effect of this structure was to remove some direct authority over their budgets from these units of the regional state apparatus and to reconstitute that authority in the Social and Cultural Projects Commission. As Borisova has noted, an important aspect of Trutnev's efforts to reshape the state administration in his governorship, as was the case in his time as mayor (see Borisova 2002), was the establishment of mechanisms for increased coordination and collaboration. The new Social and Cultural Projects Commission was a centerpiece of this effort to recentralize the state bureaucracy itself. Once established near the center of Trutnev's regional state apparatus, its leadership and members worked to expand the project movement in all directions.

Especially for inhabitants of the Perm region's small municipalities and rural districts, grant applications—and even the conception of a project itself—were something new and unfamiliar. One of the primary tasks of the commission in its first years was, therefore, training. There were ubiquitous seminars and workshops in those early years, the lessons of which were subsequently bundled into a how-to guide, *The Road to a Social Project: Practical Advice for Beginners* (Shabanova 2004), a much more user-friendly volume than its 2002 predecessor.[26] The guide

24. See, for instance, the spread of projects described in "O pobediteliakh tret'ego oblastnogo konkursa sotsial'nykh i kul'turnykh proektov," May 15, 2003, PermGANI f. 1206, op. 1, d. 205. These units retained some diluted influence by contributing members of the commission and expert evaluators for judging applications; on this point, see the comprehensive evaluation of results of projects funded through the Department of Education at "Analiticheskaia Zapiska po itogam monitoring sotsial'nykh proektov 2003 g.," PermGANI f. 1206, op. 1, d. 205.

25. "Doklad o realizatsii sotsial'nykh i kul'turnykh proektov v Permskoi oblasti v 2001 g." PermGANI f. 1206, op. 1, d. 24.

26. This volume was accompanied by another widely distributed publication in the same year: *Social Projects in the Perm Region: Methods and Technologies* (Perm Regional Administration and the Center for Social Initiatives 2004) that gave detailed examples of completed projects.

emphasizes the importance of practical knowledge of the entire grant process and takes beginners in exhaustive detail through a number of steps: assembling a team, identifying a project, putting together a plan, setting achievable goals, finding the required cofinancing from other organizations, budgeting and auditing, and reporting.

It begins, logically, with a question that was on the lips of many at the time: What, exactly, is a social project and what does it do? *The Road to a Social Project* defines a social project in its glossary as "a means of social creativity . . . connected with the solution to some sort of problem. [A social project] changes the original situation in a social system" (Shabanova 2004, 9). The booklet specifies precisely how low-level state agencies and non-commercial groups should go about finding a problem:

> The beginning of project activity is connected to the discovery of a difficulty, a discomfort in one's everyday surroundings. By "difficulty," we mean direct discomfort, dissatisfaction, that which does not suit a person in his or her social life. . . . It is very important at this stage not to confuse the "social problem" and its visible consequences. For example, having seen piles of garbage that have not been picked up, do not formulate the problem as a problem of garbage. That is just the consequence. In fact, the problem can be one of the following: (a) a low level of professionalism among representatives of local state agencies; (b) absence of a program of ecological education (*vospitanie*) in school and nonschool education; or (c) a low level of general culture among the population of the particular location, and so on. (Shabanova 2004, 17)

Beginners are then counseled to do further research to identify the problem more precisely, by talking to townspeople, researching in the library, finding out whether other organizations in their locality are also working on the problem, and so on.

In *The Road to a Social Project*, as in all the training sessions conducted by the Social and Cultural Projects Commission, all sorts of issues—from cultural to environmental to youth—were reclassified as variations on a single theme: a problem. They had a single kind of answer: a project. And single regional commission would evaluate them and select winners among them. All social and cultural difficulties, all discomforts—and there were many in those years—were thus fashioned into a single administrative process, a central part of the aspiration to bureaucratic metacoordination of spheres of life that is state building.

Another dimension of metacoordination was also taking place on the other end—that of the evaluators and funders of the projects. The legal basis of the Social and Cultural Projects Commission allowed a wide range of entities to serve in the role of zakazchik—sponsor and funder of a particular category of grant competition. Many of these were state agencies such as the Department of Culture, but they could also be—and were encouraged to be—corporations. ZAO Lukoil-Perm, Lukoil-Permnefteprodukt, and OOO Lukoil-Permneft were among the zakazchiki for successful projects in all the grant competition years, sometimes in competition categories that they themselves had proposed, sometimes by funding specific projects in overarching competition rubrics. In 2003, for instance, ZAO Lukoil-Perm sponsored several major projects through the regional Social and Cultural Projects Commission, among them Ecology and Oil: The Experience of Collaboration, based in the Krasnovishersk Central Library, and The Charitable Corner, sponsored by the Chermoz City Organization of Veterans, a societal organization in the Il′inskii district. ZAO Lukoil-Perm teamed up with Lukoil-Permnefteprodukt to cosponsor two other projects in that same year: The Kama—A River of Friendship—Meetings of Culture, under the general competition rubric of The Kama River Region—Culture of Relationships among Nationalities, and My Fate Is My Region's Fate, in the northern district of Usol′e. Lukoil subsidiaries were, in fact, far and away the largest nonstate funders of social and cultural projects, contributing over 80 percent of the corporate funds donated in 2003 (63 percent of that was from OOO Lukoil-Permneft).[27] Only three other nonstate organizations appear in the list of zakazchiki in 2003: OAO Permturist, a tour agency, and two other businesses, NOVIS and Sibur-khimprom, each of which funded a single project.[28]

To recap: In the state bureaucracy itself, in the definition of projects, and in the coordination of state and corporate sponsorship, the Social and Cultural Projects Commission chaired by Tatiana Margolina was engaged in a massive project of region-level state building. Consider, for instance, part of her job description that Margolina provided to the compilers of the 2004 edition of *Who's Who in the Political Life of the Kama River Region*:

> carrying out a unified state policy in the fields of education, academic and fundamental sciences, culture, arts, cinema, historical-cultural heritage, archives, and interregional cooperation in the sphere of social and cultural partnership . . . [and] implementing coordination with societal and union organizations and businesses (*khoziaistvuiushchie sub″ekty*) in the framework of social partnership. (Podvintsev 2004, 145)

27. "Informatsiia po itogam provedeniia munitsipalnykh konkursov sotsial′nykh i kul′turnykh proektov v 2003," 7, www.perm.ru/files/mynic (accessed May 2, 2013).

28. "Spisok pobeditelei tret′ego oblastnogo konkursa sotsial′nykh i kul′turnykh proektov," PermGANI f. 1206, op. 1, d. 205.

This job description points to precisely the aspiration to metacoordination that, I am arguing, was so central to state building after the decentralized 1990s. Compare Margolina's job description, for instance, to Pierre Bourdieu's definition of a state as the

> culmination of a process of concentration of different species of capital: capital of physical force or instruments of coercion (army, police), economic capital, cultural or (better) informational capital, and symbolic capital. It is this concentration as such which constitutes the state as the holder of a sort of meta-capital granting power over other species of capital and over their holders. (1999, 57)

The concentration of the capital of physical force had only recently, at this point in the mid-2000s, been reclaimed by the Russian state from various mafias and other violence managing agencies (Volkov 2002) that ruled in the 1990s. In the Perm region, the social and cultural projects movement was the means of concentrating and coordinating remaining domains into a unitary "field of power" (Bourdieu 1999, 58).[29]

The ways in which people work through "difficulties" or "discomforts" in their "everyday surroundings"—to return to the primary language of *The Road to a Social Project*—is central to a great deal of scholarship on postsocialist transformations. In many cases, this scholarship works from a particular context—women's organizations in Tver (Hemment 2004, 2007) or an HIV/AIDS treatment and drug rehabilitation clinic in Saint Petersburg (Zigon 2011), to give but two examples that would fall readily under the Perm region's social and cultural projects rubric—and builds on a particular strand of theory to help understand transformations of subjects or communities (feminist anthropology and the anthropology of NGOs in Hemment's case, the anthropology of morality in Zigon's). Thinking through the theoretical lens of aspiring metacoordination, of the sort that Margolina's job description and Bourdieu's definition contemplate, permits us to locate and chart the interaction of hundreds of such instances within a broader field of power managed not just by a state, but at the intersection of state agencies and regional oil companies.

This is not what we would expect from classic petrostate theory, on at least two counts. First, local oil corporations were fully involved at every step, from administration to evaluation to execution, rather than just contributing taxes or other fees to an institutionally separate state apparatus. Second, the entire process was built on the procedure (rather odd in historical and comparative terms)

29. I believe this claim holds whether or not it is couched in Bourdieu's theoretical language of convertible types of capital.

of competitive grant applications, with guidelines and evaluative processes adapted explicitly from Western foundations seeking to build civil society in the postsocialist world. On both counts, I suggest, we must see the state-corporate field that emerged in the Perm region—and emerged in relationship to the federal state through presidential envoy Kirienko's office—as linked to quite recent transformations of state and corporation on a global scale in ways not envisioned by scholarship on the developmentalist petrostates of the twentieth century or, for that matter, by Bourdieu.

An Objective, Disinterested State: Project Grants vs. Local Networks

In Perm, where social and cultural projects had been up and running since the mid- to late 1990s, a critical mass of project specialists had quickly coalesced around Margolina's Social and Cultural Projects Commission and the companies and state agencies that contributed funding and commissioned projects.[30] One of the main tasks Trutnev's team set itself was to bring this project movement to the rest of the region, especially its smaller cities and districts (Perm Regional Administration et al. 2002, 11). In fact, a subset of grants awarded in the Volga federal district competitions went to Perm-based organizations dedicated not to specific cultural, ecological, historical, or health-care projects, but to the spread of the grant-based project movement as a whole: the development and spread of procedures, vocabularies, and patterns of expertise and coordination that would flesh out social and cultural projects as a technology of governance and spread it throughout the Volga federal district.

This subset of "meta" grants is a useful place to look in on another dimension of state building: the way in which the project movement was designed to produce a "state effect": that sense of the unity and coherence that is so central to people's belief that "the state" exists as an organization separate from society or its constituent groups. It was clear to the coalition that made up Trutnev's team that social and cultural projects had emerged out of a specific set of networks and interests that had momentarily aligned. (As Tatiana Margolina put it to me, reflecting on those years nearly a decade later, the project movement was the product of a group of authors, "each with their own mission.") Yet their task was to make that coalition—made up of the Eurasia Foundation, Lukoil subsidiaries,

30. In the Volga federal district's competition, winners were chosen by specialists based in Moscow, who were often affiliated with international aid agencies. As the Perm region began to run its own competitions, it drew evaluators from region-based teams of experts.

and others—appear as an objective, seamless whole empowered to relate fairly and legitimately to the full spread of social and cultural organizations arrayed beneath them in the Perm region. Central to this was another framing borrowed from Western civil society organizations—that grant-based competitions were objective, impersonal, and disinterested ways of allocating resources based on merit rather than personal or patronage connections. In endorsing these new funding mechanisms, for instance, presidential envoy Sergei Kirienko explained that they were,

> a mechanism of state commissions [*gosudarstvennyi zakaz*], in which money is distributed not under the table or on the principle of "I'll give it to those whom I like the most" . . . here we are not only divvying up money, but divvying up responsibility for realizing that order.[31]

The many booklets, trainings, seminars, and other means by which Trutnev's team sought to spread the project movement followed this line of reasoning, insisting on the ways in which their new system was objective, transparent, and removed from any political network. With much fanfare, for instance, they released the selection committee's deliberations and votes to the public. The instruction manual released in 2002 included not only the legal documentation behind the competition design and examples of all the forms to be filled out by applicants but also the minutes of the commission's meetings in which the projects were discussed (Perm Regional Administration 2002, 93–103). Although the minutes appear in summary form, the process by which the commission heard reports from experts, combined them with their own knowledge and evaluation, and voted on the results—including the vote totals in each category—was on display for all to see and consult. Anyone reading would see, the organizers believed, that projects were discussed and debated for their realistic aspirations, relevance to regional goals, and a range of other factors, with the goal of maximum openness—a sharp contrast to the largely invisible transfers that prevailed in a number of other familiar systems, ranging from simple charity to the kinds of informal networks and patronage ties that proliferated both in the Soviet informal economy and in the decentralized suzerainties of the 1990s.

Of the tasks that Trutnev's team took up, this effort to distance the project movement from patronage politics was among the most difficult. In 2001, the Volga federal district awarded a substantial grant (funded in large part by the British Charities Aid Foundation) to the Center for the Development of Charity and Trusteeship in Perm for a project called Unified Resources. The project's

---

31. Iarmarka sotsial'nykh i kul'turnykh proektov privolzhskogo federal'nogo okruga "Perm'-2000," Institut Ekonomika Goroda, 2000.

goal was, according to the organizers' final report to the Charities Aid Foundation, to "facilitate the appearance of people with initiative in small municipal jurisdictions who are ready to master new, social technologies for acquiring extrabudgetary resources."[32] This grant was, that is, dedicated not to any particular social or cultural cause itself, but precisely to establishing the routes of metacoordination among them. The conclusions of their project, they said, "would be of use to all branches of power on both the district and municipal levels."[33] The project focused on three administrative units: the Osa district in the Perm region; the small city of Votkinsk in the Udmurt republic, and the Kotel'nicheskii district in the Kirov region. Collective conferences sponsored under the auspices of the grant included sessions dedicated to the administration, monitoring, and methodology of social and cultural projects—especially at the municipal level in the cities and districts on which they focused their attention.[34]

The project, however, did not go smoothly. The leadership group from Perm ran into a similar problem in two out of their three chosen locations: local political factions and patronage networks that coordinated business, state offices, and the social sphere in other ways, and that were not always receptive to subsuming their work into a larger regional, purportedly objective framework managed by experts. In the city of Votkinsk in Udmurtia, representatives of the Center for the Development of Charity and Trusteeship reported, it was impossible to begin project work in the months before a local election. In these months, local factions and politicians battled for votes on the terrain of the social and cultural sphere, handing out promises and funding to all manner of societal organizations in the hope of securing endorsements and votes—a political strategy known across the postsocialist world as using "administrative resources" (see esp. Allina-Pisano 2010). Local networks were highly mobilized and lines were particularly sharply drawn. In this kind of environment, the new specialists in social and cultural projects reported, any alliance they formed or group they worked with would draw them into one or another side of the election campaign, with the result that, in the organizers' eyes, the entire project would be discredited. The phrasing is important: the problem was not that funds would not be dispersed to a worthy cause, but that they would be politically tainted: the goal of achieving an objective, transparent state bureaucracy would be compromised, the desired state effect weakened by its association with one or another faction.

32. "Zakliuchitel'nyi otchet o realizatsii proekta, 'Ob"edinennye resursy s 1 noiabria 2001 g. po 31 dekabria 2002 g.," PermGANI f. 1206, op. 1, d. 11.

33. Ibid.

34. "Programma zakliuchitel'noi vstrechi uchastnikov proekta 'Ob"edinennye resursy,'" PermGANI f. 1206, op. 1, d. 11.

In the Osa district of the Perm region, the group ran into difficulties that, they reported, eventually necessitated switching their efforts to focus on an entirely different district. In Osa, the team of experts completed their first planned tasks in 2001 and early 2002—including fielding a survey of the local population about what kinds of social problems were most pressing, discerning the general desire to improve them, and then holding a series of informational seminars and roundtables featuring state officials from the Trutnev administration and other members of the project movement—including specialists from OOO Lukoil-Permneft, which operated the oil production facilities around Osa.

But, after these stages, they found themselves unable to work with local officials. Osa was, at the time, embroiled in a deep and rolling set of political controversies that pitted the head of the district, elected in 2000, against a long list of district business mangers and heads of societal organizations. By the account of one of Perm's regional newspapers, the former head of the Osa district's Communist Party Committee, Ivan Kamskikh, had been elected by a coalition of rural voters, pensioners, and former Party allies. Once in office, he had used his position largely to reward friends and punish those who had opposed him in the election. To do this, he used every administrative resource possible, including audits, investigations, and the withholding of his signature from many of the standard administrative documents that made both business and state offices in the Osa district run.[35] Some degree of network and patronage politics was entirely expected by all observers; it seems to have taken place in the Osa district between 2000 and 2003 to an unusual extent, with Kamskikh's opponents repeatedly appealing to the governor of the Perm region for an intervention and, ultimately, launching a bid to recall Kamskikh from office. (That effort finally succeeded in April 2003.) Such controversies and efforts to unseat a sitting head of administration were not unheard of in the Perm region at the time, but they remained rare, and Osa's was particularly sharp. The ins and outs of who had unseated whom and what court cases were pending were closely followed in Perm regional newspapers such as *Zvezda*.[36]

As in Votkinsk in the Udmurt Republic, but to a much greater and more enduring degree, the regional state's project movement emissaries to the Osa district found themselves unable to achieve many of their goals because of entrenched and highly mobilized district-level networks—the standoff between these networks long after the election meant that the kind of overarching alliances, net-

---

35. Zakliuchitel'nyi otchet o realizatsii proekta 'Ob"edinennye resursy' s 1 noiabria 2001 g. po 31 dekabria 2002 g.," PermGANI f. 1206, op. 1, d. 11.

36. See, for instance, Evgenii Plotnikov, "Blizorukost'," *Zvezda*, January 10, 2002; "Esli drug okazalsia vdrug . . . ," *Zvezda*, September 14, 2001; and "Poka petukh ne kliunul," *Zvezda*, November 30, 2001.

works, and objective evaluations that the proponents of the regional project movement sought to build could not easily be made. They were always interpreted as supporting one or another local faction, rather than incorporating all local factions into a larger, region-level, set of collaborations among state, societal organizations, and businesses. The report quite astutely summarized the situation as one in which this new model of uniting the "three sectors" of state agencies, businesses, and noncommercial organizations around the practice of social and cultural project was a *political* task. It involved subordinating other modes of dealing with social problems—chiefly the local use of administrative resources and patronage networks—to this new vision, a task that, these emissaries concluded, could not be expected to occur voluntarily.

They also drew some conclusions and made recommendations. "The idea of social partnership in an administrative territory," they wrote, "can be realized only with the good-willed readiness of executive authorities to be one of those partners. But one probably cannot hope that such good-willed readiness to carry on a dialogue with society will arise by itself."[37] As one strategy to stimulate this interest, they recommended that state agencies at the regional level issue sets of standard, normative documents that could be used to cajole local factions and networks to a single table under the banner of social partnership. In other words, social partnership in building civil society would not simply arise naturally or with a gentle nudge—a view, incidentally, at which many of the seeders of civil society at NGOs based in the West were also arriving.

Despite the project's difficulties and, indeed, its near failure, the effort to spread the project movement to Osa yielded what the report called an "unexpected but interesting result."[38] Although the organizers were unable to work with the head of the local administration and, therefore, with one of the three crucial sectors among which they sought to coordinate, N. N. Evdokimova, an employee at the district Department of Employment, took a very active interest in the project movement. In addition to the seminars held in Osa, she went on to participate in other seminars run under the auspices of the The Unification of Resources project in the Udmurt republic and the Kirov region, and to network extensively with others in the local and regional offices of the Department of Employment. With her help, eight social and cultural projects from the Osa district were funded in 2002—not in the Volga federal district or Perm regional competitions, but in still another class of social and cultural projects competition: a "corporate competition" run by OOO Lukoil-Permneft. The projects received, all together, more than 700 million rubles for social and cultural projects in the Osa District, but they

37. "Zakliuchitel'nyi otchet o realizatsii proekta 'Ob"edinennye resursy' s 1 noiabria 2001 g. po 31 dekabria 2002 g.," PermGANI f. 1206, op. 1, d. 11.

38. Ibid.

came from grants submitted to the local oil company rather than the state. This, in the view of the report's authors, counted as a success: "We think that, having gotten [this] taste of project activity, the authors of these projects are themselves able to stand at the origin point of the project movement in the district, in just the way it was conceived."[39] In the view of this report, this small task in the much larger project of early 2000s-era state building was completed not by the state at all, but by the local oil company engaged in an overlapping effort.

Lukoil-Perm and "The New Leadership"

Trutnev's declaration that Perm would strive to become Russia's Capital of Civil Society thus heralded the emergence of a new kind of powerbroker: the project specialist. Members of this new group—spread through regional state agencies and an array of Perm-based NGOs—were, like Trutnev and Margolina, often former members of the Komsomol. Others were democracy, human rights, or civil society advocates who found the regional state administration far more sympathetic to their activities than had been their experience in the early 1990s. Although these two groups were not necessarily allies in the late Soviet period, by the late 1990s they shared a dissatisfaction with many of the excesses and inequalities of the first postsocialist decade and saw opportunities to engage in a new kind of activism with far more funding and legitimacy than that to which they had been accustomed in the 1990s.

As should be clear, theirs was a state-building project that was highly self-reflective. Its proponents explicitly understood themselves to be developing new, post-Soviet, internationally recognized and valued techniques of governance, and they produced a steady string of publications, reports, plans, and vision statements that summarized their progress. A 2004 report by P. I. Blus′, head of the office of social policy in the Perm region and a key deputy to Tatiana Margolina, provides a concentrated summary of four years of intense effort on this front. The report is titled "The Participation of Business in the Development of Territories," and it begins, like so many of the documents of that era, by attempting to sort through a welter of new and unfamiliar terminology, from "monetary grant" to "social investment." It then explains that there are rising expectations from the sides of state agencies and businesses alike that some sort of mutual participation in the development of territories—beyond the payment of taxes—is both necessary

39. Ibid. The report states that "ZAO Lukoil-Perm" funded these projects, but this is likely a mistake because OOO Lukoil-Permneft was the operative Lukoil subsidiary in the Osa district at the time.

and desirable. It notes some common problems leading to unsatisfactory ways in which this relationship might play out:

> On the one hand, the rise in expectations on the part of regional and district state agencies can lead to the conclusion that a business's social program is required and not the voluntary participation of the company in the development of a territory's social sphere. Such a position leads to increasing pressure on business from the side of state agencies, expressed in the phenomenon of "voluntary-coercive charity," the direct result of which is a lowering of the effectiveness of a business's social programs. On the other hand, business that are financing the social sphere are often dissatisfied with the ways in which organs of local government dispense funds and govern the social sphere. This leads to the desire that business, together with regional and district state officials, should determine the priorities of social policy in the region, [a process in which] business can play an active part.[40]

In this report, Blus′ thus distinguished the administration of social and cultural projects that Trutnev's team envisioned from some other different, and quite common, arrangements: patronage networks, state agencies coercing businesses, the straightforward state-controlled dispersal of tax receipts, and state projects that did not proceed with business-like effectiveness. The models being developed in Perm would, Blus′ projected, even go beyond the kind of financing that had been contributed by national companies such as INTERROS, Yukos, and big Lukoil to the Volga federal district's social and cultural projects competitions, since it was acknowledged by all sides that those contributions had been more or less coerced by the Putin administration. Blus′ envisioned a far more intimate and interlaced kind of relationship between state and corporation:

> The leading Russian corporations are coming to an understanding of the importance of beginning a fundamentally new dialogue with district and regional state agencies and with an initiative-taking network of modern thinking municipal managers. This unified conception is called "the new leadership."[41]

This kind of leadership, the report continued, should take the form of "social investing": long-term, directed, effective support of the social sphere that was fully integrated into companies' business interests, from market development to labor

---

40. P. I. Blus′, "Uchastie biznes v razvitii territorii," 2004, www.perm.ru/files/biznes.doc (accessed April 16, 2013), 4–5.
41. Ibid., 5.

relations, and that would proceed hand in hand with compatible, mutually agreed-upon initiatives run out of state offices.

Only a small portion—perhaps 10 percent—of Russian companies had reached this advanced stage of development in their social efforts, Blus′ noted, but Lukoil-Perm was a prime example. Lukoil-Perm was, the report noted, among the very first companies in Russia to sponsor grants for social and cultural development, beginning with its participation as a zakazchik in the earliest regional social and cultural grants competition. Citing S. N. Buldashov, head of Lukoil-Perm's Connections with Society Division, the report approvingly noted several ways in which Lukoil-Perm had voiced its dissatisfaction with regional and district level social policy, including with its weak mechanisms for financing and planning, its ineffective use of budgetary funds, and its inadequate funding of important initiatives. Lukoil-Perm had offered to help state agencies by sponsoring and participating in a series of seminars on the theme of "Project Culture as a Mechanism of Effectively Dispersing [State] Budgetary Funds," and the state administration had gladly accepted. It was, Blus′ noted with approval,

> a priority in Lukoil-Perm's financing initiative for district-level organs of government to instill innovative mechanisms of managing the social sphere . . . the company is interested in the further development of this project, because it understands that acquiring new technologies can substantially help districts in the rational use of ever more modest state budget resources.[42]

Trutnev's new technology of open competitions for social and cultural projects was, that is, preached at least as vociferously by Lukoil-Perm as it was by state representatives themselves—something that will become still clearer in the next chapter—and this collaboration was increasingly cast by both company and state as new and innovative on a Russia-wide scale.[43]

A high point in the Perm region's social and cultural projects movement came in 2003, when the region's social projects were deemed the best in Russia in a competition judged at the level of the Russian Federation Council. But social and cultural projects—at least in the form developed by Trutnev's team—soon fell from favor, at least in the regional state administration. In March 2004, Iurii Trutnev was appointed Russia's minister of natural resources and moved to Moscow, along with a number of deputies from the business and industry side of his adminis-

---

42. Ibid., 9–10.

43. Not all the original participants in the project movement were enthusiastic about the increasing participation of businesses in the social sphere. For many in the Perm-based human rights community, whose late Soviet-era trajectory was through human rights campaigns and activism, it was clear that businesses should pay their taxes and nothing else (see Borisova 2004, 31).

tration. Sergei Kirienko stepped down as presidential envoy to the Volga federal district in 2005, and most Western foundations were drastically scaling back their civil society–building efforts at this time. Much of Trutnev's social and cultural projects team remained in Perm and involved in one way or another with state administration and/or projects in the social sphere, but the Capital of Civil Society era was over. Ironically, given the way the project movement so often contrasted itself with network- and patronage-based politics in the effort to build an objective, detached state, its form of state building was revealed, in the end, to be the effort of another network and a particular time and place. To replace Trutnev, Vladimir Putin appointed as governor Oleg Chirkunov, who devoted some funding to social and cultural projects of the sort Trutnev had endorsed, but never with the same enthusiasm or centrality to regional governance of his predecessor. Chirkunov was, however, no less devoted to putting Perm on the map, and in 2008 he—and his own team of project managers—began a campaign to have Perm declared a European Capital of Culture. That campaign is the subject of chapter 8.

• • •

The story that I have told thus far will likely discourage, if not horrify, many proponents of civil society building after socialism. A central premise of seeding civil society in the wake of the Soviet Union, after all, was to replace what were assumed to be the all-powerful structures of the Soviet state with engaged groups of independent citizens whose activities in the zone between families and the state would serve as a guarantor of democracy and a check on the emergence of a too-strong state. This is the function that a healthy civil society was alleged to serve in the West, and, in the 1990s, it was the justification that undergirded the attempted export of civil society to the former Soviet bloc (see also Mandel 2012). From this perspective, the story of the project movement in the Perm region that I have related likely reads as part of a story of the abject failure of civil society building, of its gradual co-optation, even hijacking, into the service of a new, Putin-era, project of centralized state building.

But most anthropologists studying the postsocialist world have never accepted the premise that the term "civil society" usefully names a type of actually existing social organization—even in "the West" itself. They have therefore focused on the concept's role as "key symbolic operator . . . rather than organizational realit[y]" (Verdery 1996, 104–5); on elucidating modes of interaction that, despite having the key trappings of civil society as envisioned by outside architects of "transition," have nevertheless gone unrecognized as such, and even derided (see, above all, Creed 2010, 105–30, on the rituals of Bulgarian mumming); and on bringing to light the unexpected outcomes and permutations of civil society–building projects—what Steven Sampson called "the social life of projects" (1996). In the introduction to a collection dedicated to dismantling unitary ideas of civil society—written as many of those unitary ideals were winging their way to the postsocialist world—Chris Hann wrote that "there is something inherently unsatisfactory about the international propagation by Western scholars of an ideal of social organization that seems to bear little relevance to the current realities of their own countries" (1996, 1). If, however, one wanted to look for a brand of social organization that *was* coming to characterize Western countries at around this same time, the increasing salience of corporations in the governance of human lives would be a good candidate.[44] Nowhere, it is safe to say, was this trend clearer than in the global extractive and energy industries.

44. David Stark's work on "heterarchy" and "recombinant property" in postsocialist firms (2001)—especially when read alongside the other chapters in *The Twenty-First Century Firm*—illustrates important ways in which postsocialist transformations are part of global-scale shifts in the nature of corporations. My aim in these chapters is to add additional dimensions to this general strand of analysis.

5

CORPORATION/STATE

Lukoil as General Partner of the Perm Region

If the expansion of markets into the former Soviet bloc was one of the major stories of global capitalism in the 1990s and early 2000s, then surely another was the rise of the corporate social responsibility (CSR) movement. There is reason to trace the expectation that corporations should be attentive to human rights, environmental harms, and other social issues back to the postwar period and, in one fashion or another, for millennia. But the issue of how corporations relate to broader communities and, in the now-common phrasing, "stakeholders" other than their shareholders, came to special salience in the 1990s and early 2000s due to a confluence of factors, among them widespread deregulation and privatization in and after the Reagan-Thatcher years; the spread of transnational environmental, indigenous rights, labor, anticorruption, and allied forms of activism; and a number of high-profile international lawsuits against major global corporations (such as those against Royal Dutch Shell following the 1996 death of Ken Saro-Wiwa and eight other Nigerian human rights activists).

Under the sign of CSR, hundreds of new corporate divisions have been set up and charged with managing corporate relationships with those whose lives are affected by corporate operations. CSR efforts have touched on an exceedingly wide array of domains, from infrastructure to labor, from environment to health care, from culture to religion, and beyond. Indeed, practitioners and analysts have expended a considerable amount of energy attempting to classify and categorize different types of CSR (e.g., Auld, Bernstein, and Cashore, 2008). New international agreements, such as the United Nations Global Compact (Sagafi-Nejad and Dunning 2008) and the Extractive Industries Transparency Initiative (EITI), have set out standards for corporate social responsibility, and an ever-increasing stream of audit documentation—both internal and external—tracks corporations and corporate divisions in relationship to these standards.[1]

My argument in this chapter is most closely related to studies of CSR that consider the shifting relationship between corporations and states in the oil industry and broader extractive sector. Suzana Sawyer and Elana Shever have both used the increasing prominence of oil companies in development initiatives and social projects as a lens through which to theorize transformations of capitalism in its neoliberal phase. Each argues in a somewhat different way that neoliberalism does not represent the withdrawal or the disappearance of the state, but rather its transformation into new forms: a "fiscal manager geared toward facilitating transnational capital" in Sawyer's argument (2004, 9), a powerful resource for imaginaries of kinship and affective claims-making even after privatization in Shever's (2012, 73). There are numerous resonances with both approaches in Russia. Along the Moscow–Perm region axis in the 2000s, however, the state apparatus was not so much being gradually transformed as it was being built from the ground up. The rise of CSR coincided, it follows, with basic state building (see fig. 4). In this context, the multidimensionality and mobility of CSR—the fact it could be directed toward almost any corner of social or cultural life—had a powerful affinity with the processes of state building as aspiring metacoordination described in the previous chapter. One result was an interpenetration of corporation and state that was far more thorough and extensive than we find elsewhere.[2]

---

1. CSR has been variously understood as corporate greenwashing, as a new generation of business plan that provides win–win solutions for business and the environment (or business and labor, or business and culture); and, in some notable recent studies, as a way in which capitalist logics that otherwise might be questioned are imbued with a virtuous morality (Dolan and Rajak 2011; Rajak 2011). As Welker (2014) has shown through an ethnography of CSR that stretched from corporate boardroom through Indonesian mining sites, we should expect that many of these dynamics go on at the same time, even within the same corporation and even, in fact, within the same meeting. To reduce CSR to either fig-leaf-style greenwashing or to the cold rationality of a business case is likely to miss many of the contingencies, multivocalities, and internal tensions that characterize corporations.

2. There remains very little written about Russian CSR practices in the Euro-American scholarly literature, and projects of the sort I discuss here are usually dismissed as leftover Soviet-era obligations to company towns or basically incidental to corporate operations (see, however, Jones Luong and Weinthal [2010] for a wealth of information about oil-sector CSR and Ghodsee [2004] for prescient attention to the increasing interaction of nongovernmental organizations and corporations in postsocialist Eastern Europe). I have therefore learned the most about CSR in Russia from the work of Russia-based scholars, especially S. P. Peregudov and his collaborators (e.g., Peregudov 2003; Peregudov and Semeneko 2008). My analysis below agrees with a number of their points, although I place relatively more emphasis on the heritage of the 1990s and, as the chapters of part III will illustrate in more detail, on the particularities and materialities of oil and the oil industry.

Anticorporate Critique and Response in the Post-Soviet Context

In a highly instructive contribution to the study of corporations, Peter Benson and Stuart Kirsch suggest that a defining feature of contemporary capitalism is the "corporate response to critique" (2010, 459). They discern a three-phase series of responses through which many corporations have moved over recent decades: overt denial of the harms they cause; acknowledgment / symbolic amelioration; and engagement / strategic management. These phases are not intended to provide a fixed typology or universal, unidirectional sequence. Adapting them to the case of a Russian oil company shows how post-Soviet dynamics are very much a part of a global moment in corporate transformations and, at the same time, how trajectories out of socialism—in which, recall, firms and states were tightly intertwined—generated important elements of distinctiveness generally not found elsewhere in the world.

Critiques of the Oil Industry

Recall Lukoil's general standing in the Perm region at the turn of the millennium, as Vladimir Putin, Sergei Kirienko, and Iurii Trutnev took up their positions along the Moscow–Perm federal–regional axis. Lukoil subsidiaries had consolidated control over nearly all of the oil sector in the Perm region, but the negotiated deals of the mid- and late 1990s had produced two structurally separate oil-producing subsidiaries in the region: ZAO Lukoil-Perm in the north, run by Andrei Kuziaev and fellow veterans of the petrobarter and fuel veksel systems, and OOO Lukoil-Permneft in the south, home to the region's heritage oil production and rapidly restructuring its organizational and labor force profile. Lukoil subsidiaries controlled in whole or in part numerous other smaller companies in the oil and oil services business, the refinery Lukoil-PNOS continued to operate in the industrial district of Perm, and Lukoil-Permnefteprodukt accounted for nearly 60 percent of the market for oil and oil products in the Perm region (Ekhlakov and Panov 2002, 141). The regional oil industry, especially ZAO Lukoil-Perm, seemed poised for continued growth and success: profits were rising due to comparatively high global oil prices, the lucrative Siberian deposit in the Usol'e district was boosting production figures, and relationships with the regional state apparatus continued on good terms. On a global scale, big Lukoil was fully engaged in Vagit Alekperov's long-term project of becoming a globally competitive, vertically integrated oil company, one that could compete with the likes of BP and ExxonMobil on their own international turf. (As a largely sym-

FIGURE 4 "Lukoil—For the Good of the Perm Region" at a gas station in Perm.

Photo by author, 2009.

bolic gesture on this last front, Lukoil had acquired more than a thousand service stations in the U.S. Northeast from the former Getty Gas in November 2000.)

However, Lukoil and its subsidiaries, and the entire Russian oil industry, were also beset by critique. The environmental movement that had coalesced in the "Clouds Overhead" controversy of the late 1980s and early 1990s, ultimately compelling substantial changes at PNOS, was a foundational moment of anticorporate critique in the Perm region, forgotten by no one. But the pervasive concern with just getting by in the 1990s (not to mention plunging production and unclear ownership structures) meant that environmental critiques had receded substantially. By the turn of the millennium, in the Perm region as across Russia, the critiques that emerged had to do much more with inequality and the expropriation of common resources than with ecology.[3] The same spikes in global

3. Ecological concerns did continue to percolate, and there was an active environmental/ecological movement in Perm (centered around the group Green Ecumene, led by Roman Iushkov, and its various offshoots) that staged occasional protests outside Lukoil-Perm headquarters, particularly in connection with a big spill in the village of Pavlovo in 1997. Although such environmental critiques were always in

oil prices that had enabled the enrichment of oil companies and drove the Putin administration's campaign to reconsolidate the central state apparatus around natural resource rents had accentuated popular discontent with the massive wealth accumulated by New Russians, and the loans-for-shares deals of the 1990s were increasingly seen as a giveaway of properly national resources to private corporations. These disparities were especially sharp in the north of the Perm region, where ZAO Lukoil-Perm was closely tied to Russian and international circuits of finance capital, but they extended to the south as well, where increasing profitability at OOO Lukoil-Permneft stood in sharp contrast to the ongoing layoffs and other major workforce restructurings in these old oil company towns. Everywhere in the region, significant worry remained about what it meant that Permian oil was now under the control of a Moscow-based corporation. In the 1990s, the very materiality and regionality of petrobarter and then fuel veksel circuits had provided some ways to cast the oil industry as deeply enmeshed in the region—symbolic moves that resemble Benson and Kirsch's second phase of corporate response—but the rapid post-1998 monetization of the Russian economy and influx of oil money strained these possibilities.

A second set of critiques came from various levels of the resurgent Russian state apparatus (see also Jones Luong and Weinthal 2010, 157–65). At the federal level, Putin and his team had made it clear that federal taxes on the sale and export of natural resource wealth were to be the basis for Russia's state-led development for years and perhaps decades to come. The years of systematic tax evasion were over, corporations would need to begin investing in Russia, and, as the aftermath of the Yukos Affair would soon make clear, the renationalization of private companies was not off the table. The tone emanating from the federal center allowed the Perm regional state apparatus to publicly challenge Lukoil subsidiaries as well. Shortly after Iurii Trutnev replaced Gennadii Igumenov as governor of the Perm region in 2000, for instance, the regional government fired a rare series of public warning shots across the bow of Lukoil's regional subsidiaries, threatening to revoke several of the companies' licenses to drill and explore in the Perm region and publicly commenting on the unacceptably high price of oil products.[4] At the lowest organizational level, Lukoil subsidiaries' relationships with their produc-

---

the background, no one I spoke to ranked them among the most significant kinds of criticism to which Lukoil subsidiaries in the region responded (cf. also Stammler 2011, on the predominance of relationships of peaceful coexistence between companies and native populations in Siberia, and M. Balzer 2006, on an array of much tenser relationships gathered around ecology and homeland). On the fate of environmental organizations in the post-Soviet period, see especially Carmin and Fagan 2010; Henry 2010.

4. See "U kompanii Lukoil-Perm nachinaiut otzyvat' litsenzii na neftedobychu," *Novyi Kompan'on*, October 30, 2001, and "Iurii Trutnev: 'Korporativnye tseny na neft' nuzhdaiutsia v razumnoi korrektirovke,'" *Novyi Kompan'on*, April 2, 2002. See also Ekhlakov and Panov 2002, 145–46.

tion districts were often strained as well. The company no longer controlled as much of the social sphere as it once did and district-level politicians were well aware of the massive sums generated by the oil pumped from beneath them. Themselves beset by critiques from residents frustrated with the outcomes of a decade of "transition," these local officials did not hesitate to make claims on the wealthiest businesses operating on their territories. Indeed, they knew quite well that the companies needed ongoing access to lands that they controlled. There was, finally, the matter of international markets and partnerships. Big Lukoil was planning major stock offerings on the New York and London markets in the 2000–2002 period, and, with the global corporate social responsibility movement well established by this time, investors were looking for Lukoil to participate in a range of global transparency initiatives and labor and environmental standards as a condition of market entry. Given that Lukoil Overseas Holding was based in Perm at the time and closely intertwined with ZAO Lukoil-Perm, this concern resonated strongly in the Perm region.

As one Lukoil-Perm employee put it to me, recalling the early 2000s, "The opinion of the oil industry could not have been much worse." If emergent Russian big businesses had been able to ignore critiques with comparative ease in the decade of privatization and primitive accumulation that was the early 1990s, this strategy—Phase 1 in Benson and Kirsch's typology—ceased to be viable in the booming early 2000s. Across Russia, companies had began to move into Phase 2, token acknowledgment and symbolic amelioration, through actions such as increasing their charitable contributions and participating in presidential envoy Sergei Kirienko's fund for social and cultural projects in the Volga federal district. Lukoil's subsidiaries in the Perm region were among the first to move into Phase 3 of responses to critiques of the corporation: strategic engagement with ongoing processes of state building and support for the social sphere.

Connections with Society

Within the Perm region, ZAO-Lukoil-Perm, OOO Lukoil-Permneft, and Lukoil-PNOS reacted somewhat differently to this constellation of critiques. I begin with ZAO Lukoil-Perm because it was closest to the regional state agencies in the 1990s and its early moves to defuse criticism and remake its relationship with the regional state apparatus under Governor Trutnev had, in the end, the most lasting significance for the region as a whole. At Andrei Kuziaev's initiative, ZAO Lukoil-Perm opened a new division in 2000–2001, called "Connections with Society" (*Sviazi s Obshchestvennostiu*), and charged it with reimagining the company's relationships both with society as a whole and with organs of state power.[5] Connections with Society was under the supervision of a vice president and on the same reporting level as the corporate division responsible for relationships with employees and negotiating union contracts. Rather than being an afterthought to corporate town labor relations, as they had earlier been, social and cultural projects directed at the population at large now had their own office. Although other companies in Russia were also establishing offices by this name at the time, ZAO Lukoil-Perm was one of the first to do so in the Perm region, and it was certainly the biggest and most influential (Krasil'nikova 2005).

To set up and run this new division, Andrei Kuziaev hired Sergei Buldashov. Buldashov was an experienced operative in the Soviet period, when he rose through the Communist Party ranks to become second Party secretary at the Motovilika factories—one of Perm's oldest and most distinguished defense-industrial factory complexes. Buldashov had spent the 1990s working at a securities trading firm and dabbling in politics in some of the minor political parties of the era. He recalled to me that Kuziaev told him, upon making the job offer, "if you're coming to do what someone tells you . . . don't bother. Because no one knows what to do. . . . We'll hire you, and then you formulate the ideology and strategy, and you hire [your own] people, and start everything from zero."

Although Kuziaev may not have had a concrete plan for what this new facet of ZAO Lukoil-Perm's relationship with state and society would look like, he certainly had an idea of what kinds of people he needed to build this new division. Buldashov quickly hired as his deputy another former factory-based Party functionary, Irina Marasonova, who had a broadly similar background in securities trading and regional politics in the 1990s. In chapters 2 and 3, I noted the ways in which the 1990s in the Perm region tracked with broader postsocialist trends in their uneasy alliances between former Soviet red directors (in this case neftianiki) and new financiers at the helm of privatized industries. The 2000s saw these groups joined by former Party functionaries: Buldashov's and Marasanova's moves to ZAO Lukoil-Perm are part of the same pattern of former Communist Party officials and operatives returning to powerful positions that we began to see with the hiring of Tatiana Margolina, former Komsomol leader, to the post of deputy mayor (and then vice governor) under Iurii Trutnev.

The movement of these former Party officials in the Trutnev era thus occurred across—and helped to knit together—state and corporate domains. I should

5. "Connections with Society" is often rendered into English as "public relations," but I prefer to retain the more literal translation in order to capture the ways in which, as I describe in more detail below, the task of this division was originally somewhat broader than what is usually indicated by the English term "public relations."

emphasize that this did not by any stretch mean the return of official communist ideology, for Buldashov, Marasanova, and others like them had made their way—often enough successfully—through the rough-and-tumble 1990s and none that I spoke to were eager for the return of the Communist Party or nationalized industry. They were valued for two other things associated with effective Communist Party officials: the ability to make deals and negotiate compromises with a wide range of people and the ability to do ideological work. In the context of the multifaceted critique of ZAO Lukoil-Perm, that is, provisional and personal deals of the sort that predominated in the 1990s could no longer, by themselves, secure the company's legitimacy or operations. What was needed was a more systematic, institutionalized approach, one that combined consummate connecting with the articulation of a new vision and practice of ZAO Lukoil-Perm as a socially responsible corporation.

I often asked my interlocutors about this phenomenon—the reincarnation of former Party figures from the region's factories in the now-central oil industry—because this shift seemed to me emblematic of the Perm region's broader transformation from defense-industrial region into oil region. Few objected to my characterization, but they often added important details and qualifications. Buldashov, in particular, impressed upon me that, from his perspective, his move to ZAO Lukoil-Perm should not be taken as evidence that a decades-old, self-conscious network had moved from Soviet factories to the post-Soviet oil industry. Too much time had passed. He was originally quite surprised by the offer to build a new division at ZAO Lukoil-Perm, he said, and one of the first things he did was begin looking through his old phonebook and wondering what all those people he used to work with were doing now. Would they join his new team? Did they still have the skills and connections he would need? "We didn't all come directly from the factories," one of Buldashov's coworkers agreed, explaining that, "a good chunk of time had passed" since the Soviet period and many former Party colleagues had re-created—or failed to re-create—themselves in other ways. Many now had nothing to do with the oil industry, and some of the younger members of the Connections with Society Division knew little of the Soviet period at all. Still, most of my interlocutors agreed with me that, across the longer-term trajectory of the Perm region, it was notable that key players on the corporate side of the turn to social and cultural projects were disproportionately former factory-based Party functionaries given the new mandate of designing corporate social responsibility programs. These new specialists in CSR quickly found a common cause and language with the former Komsomol leaders who had accompanied Trutnev to the regional administration. Delegates from the Perm region to the Volga federal district's yearly social fairs, for instance, began to include not only regional state officials but also managers from ZAO Lukoil-Perm's Connections with Society Division.[6]

Their experience and common career trajectories notwithstanding, the members of ZAO Lukoil-Perm's newest division were swiftly confronted with new circumstances and problems. Said Buldashov of his move to Connections with Society:

> The first thing we had to think about was: what constitutes relationships with organs of state power? It's not giving bribes, after all! And we came to the conclusion that relationships with local communities are mutually beneficial. Why? Well, at that moment, what kind of tension [*napriazhenie*] was there? Ecological problems, tax problems . . . and so on. So there was, in general, tension in society. . . . I wouldn't say that there were public conflicts, but there was tension about the contrast between oil workers' lives and other people's lives, especially in that period. . . . [People were saying things like] "There's Vasia, he got himself into Lukoil and now he's [got the money] to build himself a new cottage." . . . And we had to think: what to do? Well, there were onetime games [that we could play]. There was the Charity Fund (*fond pomoshchi*), and if you wrote a letter, saying that so-and-so was destitute and needed something. . . . But it [worked] on the principle of who came first, who cried more loudly. And so we gave things out [that way]. And while we were doing that we began to think: what could be our system here? And so we began with formal grants [for social and cultural projects].

The novelty of this thought process is somewhat overstated in his recollection. After all, by the time ZAO Lukoil-Perm's Connections with Society Division was formed in 2000–2001, social and cultural projects were already up and running in both the city of Perm and in the Volga federal district, and representatives of Lukoil subsidiaries had been working with Deputy Mayor Tatiana Margolina to fund a number of projects as early as 1998. Nevertheless, this origin story shows some of the ways in which ZAO Lukoil-Perm's entry into social and cultural projects was conceptualized in terms that overlapped significantly with state-building projects. The best way for Lukoil's regional subsidiaries to respond to critique and preserve their legitimacy, Buldashov was saying, was to enter the field of social and cultural projects, already articulated in the Perm region as the governor's favored path for regional development. Buldashov's

6. "Sostav Delegatsii Permskoi oblasti na Rossiiskom forume 'Nizhnii Novgorod–2003,'" PermGANI f. 1206, op. 1, d. 197.

recollection points to a second overlap with state projects: a concern with systematicity and metacoordination, coupled with a desire to extract itself from patronage- and charity-style relationships. Such aspirations, I have been arguing, were a crucial aspect of Russian state building at the time. The geographic spread of Lukoil's regional subsidiaries, coupled with the multifaceted nature of the critique the company was facing, was, in other words, calling forth a multifaceted, systematic response. The result was the creation of what we should see as an interlaced state-corporate effect, and not simply parallel state and corporate effects.

It is notable, too, that ZAO Lukoil-Perm's turn to CSR through grant competitions did not come through engagement with other international oil companies' practices. Although Buldashov, Marasanova, and their coworkers all told me that they were well aware of the CSR mission statements on the websites of multinational oil companies based in the West, they also affirmed there was no direct dialogue with these companies' corporate divisions responsible for CSR. (This was the case despite the fact that international collaborations and discussions about technical or engineering matters had gone on for many years across Perm's oil complex.) Nor were they following explicit orders from big Lukoil in Moscow on this score. ZAO Lukoil-Perm's emerging CSR operation was, then, largely homegrown, part and parcel of the Trutnev administration's embrace of social and cultural projects and campaign to make Perm Russia's Capital of Civil Society.

From Regional to Corporate Social and Cultural Project Competitions

The previous chapter recounted Irina Marasanova's central role in Perm's regional Social and Cultural Projects Commission and Lukoil subsidiaries' participation as zakazchiki in the earliest rounds of region-based grant competitions. In its formal institutional terms, however, this arrangement did not last long. In 2003, OOO Lukoil-Permneft and ZAO Lukoil-Perm began to run their own corporate competitions for social and cultural projects, parallel to those taking place under the auspices of the regional state administration. The structure of grant competitions and associated paperwork, including an expert commission evaluating projects, was imported almost wholesale from the regional competitions, with the exception that the process was administered entirely out of the offices of OOO Lukoil-Permneft or ZAO Lukoil-Perm, with the companies' directors having final say in all matters. The nomination categories in those years also closely mimicked those that had previously been announced in the regional competitions, ranging over the domains of health, youth, culture, sport, and ecology.[7] As of January 1, 2004, with the unification of the two regional Lukoils, the two project competitions were also combined. ZAO Lukoil-Perm's Connections with Society office, still headed by Sergei Buldashov, assimilated its southern counterpart and extended its activities to include all oil-producing districts in the region.[8] By 2004, Lukoil-Perm's unified corporate social and cultural projects competition had taken the basic structure it would retain, with small adjustments each year or two, for the next decade and beyond. (It was the unified Lukoil-Perm's corporate competition, for instance, that awarded the Unity grant to Berezniki's school discussed in chapter 4.)

Under the banner of corporate social and cultural projects, Lukoil-Perm actually ran three separate grant-based competitions each year, all of them limited to organizations applying from oil-producing districts. The first of these competitions supported projects proposed exclusively by the elected heads of those districts of the Perm region in which Lukoil-Perm had production operations. Although the company's yearly reports on social and cultural projects funding do not provide precise details, it seems as though few, or perhaps none, of these projects were rejected, so long as they complied with the standard application procedures. Every year, that is, the head of each oil-producing district in the Perm region could count on funding from Lukoil-Perm for at least one social or cultural project. The second structural mechanism of corporate funding for social and cultural projects was to participate in any social and cultural project competitions that took place at the district or municipal level. As the project movement spread throughout the region, many lower-level administrative units ran their own competitions, and Lukoil-Perm would participate along the same lines that it had once participated in the region's competitions, as zakazchik (that is, by sponsoring a nomination category and helping to select the winners, with veto power over the results). The third funding track was by far the largest and most commonly used: the open corporate social and cultural projects competition, in which all sorts of organizations could apply for funding, so long as they were not engaged in overt political activity and complied with a set of auditing regulations

---

7. "Polozhenie o konkurse sotsial'nykh i kul'turnykh proektov OOO Lukoil-Permneft'," PermGANI f. 1206, op. 1, d. 203.

8. Although southern OOO Lukoil-Permneft had run its own internal corporate competitions for two years already, it was not nearly as closely connected with or invested in the broader social and cultural projects movement as were Buldashov and his team members from the northern ZAO Lukoil-Perm. Nearly all members of the unified Connections with Society office with whom I spoke recalled a period of significant adaptation as the two offices were combined. "We had to reeducate them," said one participant from the north, explaining that, in the south, these new grant competitions had served mostly as a thin veneer for the older form of patronage-based doling out of corporate funds to well-connected organizations and heads of districts, whereas in the north, an entirely new model of governance by expert commission was being contemplated and designed.

for the dispersal of funds. In all these categories, Lukoil-Perm expected substantial cofinancing to be obtained as a condition of the grant award, and often helped some of its favored applicants find cosponsors (usually included some mix of regional or district state budgetary funds, along with contributions from other local businesses).

Lukoil-Perm's competitions regularly included about two dozen of the Perm region's forty-six separate districts and municipalities, with slight additions and modifications that tracked with its own operations. When the company expanded exploration into a new district, for instance, organizations in that district became eligible to participate in the social and cultural projects competitions. The structure of Lukoil-Perm's judging commissions also changed slightly over the course of the decade. In the earliest years, the panel making final decisions often included region-level state administrators who were involved in the larger social and cultural projects movement. The first corporate competition run out of OOO Lukoil-Permneft, for instance, included commission members from the Perm Regional Legislative Assembly who represented southern districts: Sergei Chikulaev, Igor Pastukhov, and Robert Gabdullin (Chikulaev and Gabdullin had also been Lukoil employees at various times).[9] As the decade wore on, however, they were gradually replaced by a greater proportion of Lukoil-Perm's own employees, including representatives from local company workforces as well as district-level (rather than regional-level) state officials charged with coordinating social and cultural development initiatives in their territories.[10] In their official composition, if not in their language and overall effects, that is, the trend was in the direction of increasing corporate control.

Despite the organizational parting of the ways between state and oil company, however, the overall state-corporate field continued to coalesce around social and cultural projects in multiple ways, not least of which was the fact that both state and corporate competitions began asking the organizations applying for grants to seek cofinancing. In many cases, this meant making applications to both state and Lukoil-Perm competitions. By the middle of the 2000s, lists of cosponsors accompanied each project's featured page in the company's annual reports on the winners of the competitions—only serving to solidify Lukoil's joint role with the state as metacoordinator of diverse spheres of activity. No other company came close. In 2003, for instance, of the 132 projects sponsored in part by corporate contributions, at a total cost of 6.3 million rubles, OOO Lukoil-Permneft

9. Ol′ga Petrova, "Biznes i Sotsium: Glavnyi prioritet—chelovek," *Zvezda*, January 28, 2003.

10. See, for instance, the corporation- and district-focused list of commission members in "Polozhenie o VII-m konkurse sotsial′nykh i kul′turnykh proektov OOO Lukoil-Perm (2007)," http://old.lukoil-perm.ru/static.asp?id=116 (accessed March 26, 2012).

and ZAO Lukoil Perm accounted for 84.5 projects (or 64 percent) and 5.1 million rubles (or 80 percent of total corporate project expenses). Their closest competitor was PermRegionGaz, with a mere five projects and 83,000 rubles.[11]

In the decade between Lukoil-Perm's first social and cultural projects competition in 2002 and the tenth anniversary competition in 2012, the annual funding that the company allocated rose from around 2 million rubles to nearly 30 million rubles. In the 2004 competition, 84 projects were funded out of a total of 294 applications; by the time of the 2008 competition, the commission was receiving over 1,200 grant applications a year, and funding just over 10 percent of them.[12] Even in the context of the downturn in global oil prices and the resulting challenges for the Russian and regional economy in 2008–10, Lukoil-Perm maintained the same level of funding for social and cultural competitions (Lukoil-Perm 2008). The categories in which grants were awarded remained basically the same: Sports, Education, Culture, Health, Ecology and Religion, often with slight modifications that responded to larger Russian or regional projects, such as the Year of the Teacher in 2010 and the sixty-fifth anniversary of the end of the Great Patriotic War in 2010. In all years, the largest proportion of grants was awarded in the Culture and Spirituality category—in the neighborhood of 40 percent of the total number of grants since the very first competitions in the early 2000s.[13] I explore why and how this emphasis on culture arose—and its effects—in subsequent chapters.

The reunification of the two Lukoil production subsidiaries, combined with the turn to social and cultural projects as a primary zone of interface with regional state agencies, had immediate and significant consequences for the ongoing remaking of the spaces of the Perm region. If the key spatial differentiation of the late 1990s and early 2000s mapped financial and industrial capital onto the north and south of the Perm region, respectively, then after 2004 there appeared an increasing divide between oil-producing and non-oil-producing districts, inasmuch as the former had many more opportunities for development grants, and many more funded projects, than the latter. Already by 2003, the districts running their own competitions—separate again from regional, corporate, and federal district competitions—were entirely in oil-producing administrative units: the Barda, Uinsk, Cherdyn, and Krasnovishersk districts and the city of Kungur.[14] In 2004, setting aside Perm itself and the northern city of Berezniki (home to the

11. "Informatsiia po itogam provedeniia munitsipal'nykh konkursov sotsial'nykh i kul'turnykh proektov v 2003 g.," 7–8, www.perm.ru/files/mynic.doc (accessed May 2, 2013).

12. See Lukoil-Perm 2005b and "Tysiachi proektov," *Permskaia Neft'* 8 (232), April 2008.

13. "Sotsial'naia Politika OAO Lukoil v Prikam'e: Praktika primenenia sotsial'nogo kodeks v OOO Lukoil-Perm, 2003–4," 23; Lukoil-Perm 2010, 4.

14. "Informatsiia po itogam provedeniia munitsipal'nykh konkursov sotsial'nykh i kul'turnykh proektov v 2003 g.," 1.

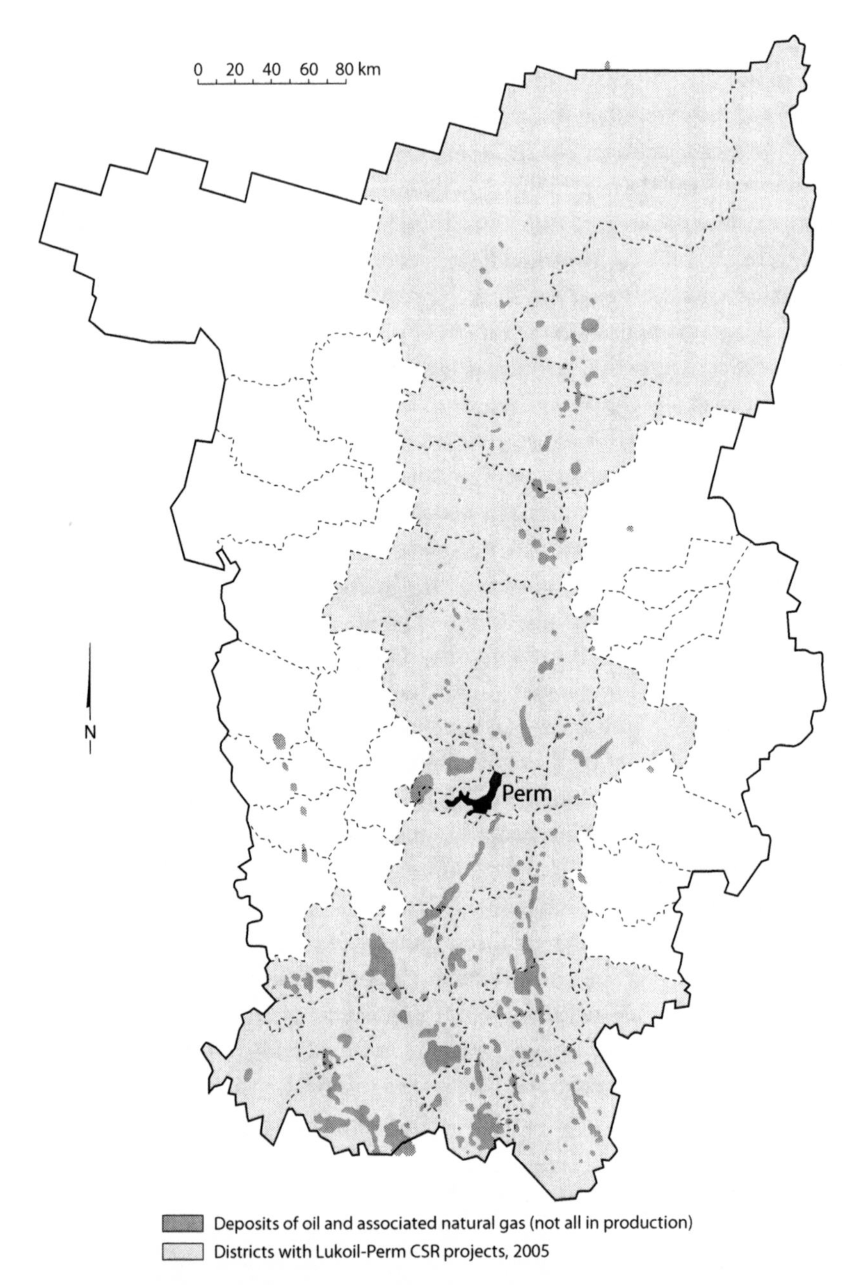

MAP 2 Spaces of oil and culture in the Perm region. The overlapping geography of oil deposits and districts that were home to social and cultural projects sponsored by Lukoil-Perm in 2005.

potash mining giants Uralkalii and Silvinit), both the largest number of grant applications and the highest sums of grant money across all categories were generally associated with oil-producing districts: Krasnokamsk, Dobrianka, Barda, Osa, Orda, and Usol'e.[15] Districts associated largely with agriculture such as the Vereshchagino district, and with coal such as the nearly abandoned mining town Kizel, barely register as participants, having submitted two and zero applications, respectively, in 2003. None were funded.[16]

As Trutnev and his team had hoped, the project movement was spreading from Perm out into the districts and small cities of the region. But it was spreading along corporate vectors as much as—and along with—state vectors, even after the initial unified corporate social and cultural projects commission was disbanded. Map 2 shows the spread of corporate social and cultural projects in the Perm region in 2004, a distribution that overlaps significantly with the highest amount of social and cultural project activity sponsored by the regional state administration. (The shaded area covers most of the significant cities of the Perm region, as well as most of the more highly populated districts.) Although the spread of these projects is tied to the location of oil deposits and oil production activities—something we find elsewhere in the world—the Perm region was home to a far more elaborate and intertwined corporate-state field than is often described in other studies of the ways in which extractive industry corporations take on parastatal functions. This is not just the revival of Soviet company towns or a postsocialist version of oil-sector overseas labor camps (e.g., Vitalis 2006). Nor is it another case of a highly defended, fairly small corporate enclave of the sort discussed by Ferguson (2005), Reed (2009), or Appel (2012b) in African cases and, in Russia, more commonly found in sparsely populated areas of Siberia or the Far North (Mitrofankin 2006; Stammler and Peskov 2008). For reasons stemming from the distribution of oil deposits across the length and breadth of the Perm region, the 1990s heritage of direct oil company participation in the means of regional exchange, the subsequent coincidence of basic state building and the oil boom in the 2000s, the absence of international companies, and the relatively small oil sector workforces as compared to relatively populated surrounding areas (at least in comparison to, say, Siberian oil regions), we find company and state intertwined to an especially high degree in the Perm region, at an especially

15. "Informatsiia po itogam provedeniia raznourovnevykh konkursov 2004 g.," www.perm.ru/files/itogikon.doc (accessed May 2, 2013). In 2004, there were also substantial funds awarded to organizations in the Ocher and Bol'shesosnovsk districts; on these grants, and their association with the regional gas distribution company PermRegionGaz, see chapter 7.

16. "Informatsiia po itogam provedeniia munitsipal'nykh konkursov sotsial'nykh i kul'turnykh proektov v 2003 g.," 4.

high administrative level, and on nearly the entire terrain of social and cultural development.[17]

CSR Effectiveness and Corporate Effects

The CSR world is steeped in the evaluation of effects and effectiveness. Corporate divisions, outside auditors, and other interest groups perpetually ask: Is one or another CSR project achieving its goals (i.e., actually responsive to local communities or actually sustainable)? Does it meet the industry or international standards to which the company has committed?[18] The world of CSR is very much a world of audit culture. I have in mind something different in considering the effectiveness and effects of Lukoil-Perm's CSR projects. To begin with, I do not assume unified or unitary corporate goals: the effects of CSR projects were various, overlapping, and often unplanned; they should be recognized and traced as such. Moreover, I look for effectiveness not just in terms of meeting plans or actually executing programs, but in the sense of "corporate effects" that I have been tracing. How, in other words, were Lukoil-Perm CSR projects involved in projecting the corporation's appearance as a unified and powerful entity and how were these corporate effects entangled with the region-level state effects discussed in the previous chapter?[19]

Electoral Politics

The Yukos Affair sent a clear signal to Russia's oligarchs to stay out of politics at the national level, and one element of this was the investigation and ultimate closure of the Open Russia Foundation, a Yukos-funded international organization dedicated to philanthropy, corporate sponsorship, and civil society building that was founded in 2001 and was widely seen as a vehicle for Mikhail Khodorkovsky's national political ambitions. In 2004, the replacement of direct elections for regional governorships with presidential appointments reduced the field of political possibilities for big business still further. But through all this, elections

17. In this respect, the configuration in the Perm region is close to what Rajak (2011) describes for Anglo-American Mining in South Africa, where the company's CSR projects—especially with respect to HIV/AIDS—have become an important template for South African governance as a whole. In the Perm region, however, this corporate influence extends not just into a single sector such as public health but into many domains—and their coordination—simultaneously.

18. See, for example, Frynas 2005 and, in the former Soviet world, Gulbrandsen and Moe 2005, 2007.

19. Jessica Smith Rolston puts this approach well in her study of Wyoming coal mining: "even if CSR initiatives are imperfectly realized, they nonetheless transform social relationships and imaginaries, providing a discourse through which miners and managers alike assess corporate practice and position themselves" (2014, 57).

remained open and highly contested at the district level in the Perm region; indeed, the foreclosing of possibilities for electoral politicking at higher levels only amplified its importance at lower levels, including along the vector of corporate sponsorship and philanthropy. In the Perm region, district-level political stakes were often high and not always in Lukoil-Perm's favor. Although the company was in most cases by far the biggest earner in any given district, it also had a small employment profile, especially because its production operations were increasingly reliant not on local workforces in the old company town model but on highly mobile brigades of specialists in the oil services companies it had spun off during labor-force restructuring in the late 1990s. District-level politicians could often assemble large and influential networks in support of their own candidates, who might or might not be friendly to the oil company (recall Osa's entrenched networks from the previous chapter), and the regional governor's office commanded a powerful political machine (see Hale 2003). Moreover, there were no opportunities to play districts off against each other as leverage: with a collection of old and depleted oilfields and ongoing targeted rather than wide-spectrum exploration, Lukoil-Perm needed very specific kinds of access to very specific sites on the ground. The subsoil deposits in which the company was interested were fixed by geology, and aboveground politicians had a very keen sense of where Lukoil production subsidiaries wanted to explore and drill.

One former Connections with Society member answered my question about why social and cultural projects were useful to the company as follows:

> What does Lukoil-Perm need to advertise? They aren't producing a product that requires competition. It's not cups or spoons or something. So they don't need that kind of an advertisement. Or not much. What they need is social calm (*sotsial'noe spokoistvo*) of the population, so that it's possible to work peacefully in the territories where oil production happens. That's the most important thing. . . . New deposits are needed [as old deposits run low]. New land is needed. And, if, earlier, it was possible to get that land with the help of central authorities, now it has to come from organs of local government. That's where the problem has to be solved, and you can't reach a separate agreement with everyone. So a system is necessary.

Social and cultural projects were, that is, an ideal route by which the company could seek influence in district-level politics. Although I have no direct and detailed evidence, my sense from talking to some of those involved in the decision-making process is that grants were awarded on their merits—as defined by the vision of social and cultural projects spreading throughout the Perm region under Trutnev—but that, at Lukoil-Perm, all projects were also quietly evaluated for their potential political implications in the district. Would a certain grant validate or empower a particular local coalition aligned with political factions friendly to Lukoil-Perm's operations? Would it help solve a social problem that might, unaddressed, draw votes away from a preferred candidate? Consider another interlocutor's reflection on these matters:

> Lukoil very correctly pursued the path of winning over to its side the intellectual elite in localities, the most authoritative people, who decide to a significant degree the politics of a district. Sure, the mass population votes, but two-thirds of them don't know the difference between the candidates. They go by personalities.

Lukoil-Perm, he said, can work to build up certain personalities, by providing resources through the social and cultural grants competition and through the media coverage and public relations that came with those competitions (especially awards ceremonies) and the resulting projects.

The company's district-level political strategy, that is, involved expanding its electoral coalitions outside of their natural (and in most cases decreasing) constituencies of oil workers to include a range of influential local elites: schoolteachers, culture workers, health-care specialists, social services workers, low-level state administrators, and others. The reemergence of a stratum of former Party officials, social sphere specialists, and culture works at the regional center—exemplified by figures such as Tatiana Margolina on Trutnev's team and Sergei Buldashov at Lukoil-Perm's Connections with Society Division—thus extended out into the districts as well. Said Buldashov in one of our interviews: "Precisely in those districts, those budget workers [i.e., state employees], those culture workers were the first to understand how to write grants. Absolutely."

This was not, however, an easy process. In the beginning, most members of district-level intelligentsias were skeptical that Lukoil-Perm would fund anything other than its own preferred interests. But when the money began to roll in, and projects began to succeed and make a difference, loyalty to the company began to grow. Part of the brilliance of social and cultural project competitions, another insider told me, was that the company did not even have to fund all the projects in order to have this influence. If members of the district-level elite attended only a single seminar on how to write grants, he said, they would still say to themselves, "even if I don't participate [by writing a grant], then at least I saw that I have that possibility. And they know that that possibility was provided to them by Lukoil-Perm."

Lukoil-Perm's preferred candidates had mixed results in local elections. By no means did the company's candidates always prevail. Other coalitions—supported by other networks, by the regional governor's office, or by other local businesses—

often came out on top.[20] When they did, however, they were still welcomed to the world of social and cultural projects by Lukoil-Perm: recall that all elected heads of oil-producing districts enjoyed their own subcategory within Lukoil-Perm's larger corporate social and cultural projects initiative. Incumbent district heads thus had access to one Lukoil-Perm grant each year with essentially no competition—not at all insignificant in the difficult budgetary environment in which most found themselves. After a few years of this, one former employee boasted to me, Lukoil-Perm's Connections with Society Division was as well or better plugged into the ins and outs of district-level politicking than was the regional state apparatus or any political party (including Putin's United Russia). "When all those district heads came to Perm for meetings," this acquaintance told me, "they stopped in our offices first, and then went over to do their work with the regional state administration."

Public Image

In a classic illustration of corporate response to critique by means of CSR, Lukoil-Perm also saw social and cultural projects competitions as an ideal context in which to change its public image. The company did not hesitate to deploy these projects and their results as a way to shift its image away from the associations of inequality and ecological harm that attached to it in both the Soviet and post-Soviet eras. "Social and cultural project competitions," Irina Marasanova told me, simply, "were a good model for the company's public relations."

The image-burnishing aspect of CSR was widely acknowledged and even welcomed as a motivating factor. Trutnev's deputy Tatiana Margolina, for instance, had no trouble seeing Lukoil subsidiaries' decisions to join the social and cultural projects competitions that her office was coordinating in this vein:

> It was very important for them to invest money in public opinion, to change it to the positive. If you take into account that in the early 1990s we had a very aggressively unpleasant Lukoil through PNOS, inasmuch as the company really, well, a whole micro-region had to be resettled because it was in an [ecologically] dangerous zone.... And so it needed to invest money in projects to establish a positive image ... and through those projects Lukoil-Perm basically solved its image problems.

Images and the production of images were, by all accounts, a very carefully managed part of the entire grant process, from applications through results and,

20. I am grateful to Petr Panov for his meticulous (but to my knowledge as yet unpublished) data on this issue, as well as for several conversations about the dynamics of district-level elections.

especially, publicity. During one of my trips to visit social and cultural project sites in an oil-producing district, the head of the local administration—a former Lukoil-Perm employee himself—took me to one of the company's social project sites that he was most pleased with—a school for handicapped children. He explained that it had taken an exceptional amount of effort to get this school included on the social and cultural projects list because Lukoil-Perm did not usually like to associate itself so directly with handicapped children or orphans. "It wants to sponsor sports, activities, festivals . . . things like that. What would it look like if publicity materials or media reports had the Lukoil-Perm logo and, next to it, a picture of a handicapped child?" It was, he said, a testament to the doggedness of the school's organizers—and, one presumes, his own contacts at Lukoil-Perm from the time he was employed there—that this school had earned a grant for itself.

Although Lukoil-Perm had an entirely separate and elaborate media relations operation, there was a natural spillover to the Connections with Society Division when it came to corporate image making. Indeed, this message was clear enough that some of the grant applications received were tailored to it quite specifically. A 2006 application from the Perm State Institute of Art and Culture, one of the premier institutes of higher education in the Perm region focused on training future culture workers (from librarians to staff for clubs and Houses of Culture), is illustrative if not typical. The Black Gold of the Kama River Region project that some of its staff proposed defined its problem as follows: "The activity of a major modern corporation is not possible without the propaganda promotion of that activity (both material and immaterial) in the consciousness of society." It went on to suggest that the Institute had just the resources that Lukoil-Perm needed, and proposed the formation of "mobile agitbrigades" dedicated to broadcasting the history, activity, and social commitment of Lukoil-Perm. Students and faculty from the institute, it went on, would work with Lukoil-Perm experts to compose educational materials and then to train culture workers throughout the Perm region in staging performances and other events that would improve Lukoil-Perm's image.

The proposal was quite self-conscious in its redeployment of early socialist-era propaganda techniques in the modern oil corporation.

> There exist many mechanisms for the promotion of a product on the market. But, amid the new collection of marketing technologies, the possibility for reviving traditional technologies, technologies that have lost their relevance, for a range of reasons, is often overlooked. . . . The genre of agitation-publicity (in the form of agitbrigades) was very popular in

> the twentieth century, and has its unique methods for affecting an audience.[21]

Moreover, the application went on, the formation of Lukoil-Perm agitbrigades would allow the performers and culture workers themselves to "realize their creative possibilities" and "open up the internal resources of their personalities," and "raise their status as people" by associating with a socially responsible business like Lukoil-Perm. I do not know whether this proposal was funded, and I am unaware of Lukoil-Perm agitbrigades, at least that understood or referred to themselves in that way. Indeed, I expect that most of my contacts in the Lukoil-Perm Connections with Society Division would have raised an eyebrow at such an unsubtle appropriation of Soviet propaganda techniques. In other words, the proposal is of interest not for its execution, but for the extent to which it renders so clearly that part of the Connections with Society Division's image-burnishing efforts relied on forging a new alliance with culture workers and other local elites across the Perm region's oil-producing districts in the service of shifting the company's image. I return to these issues in chapter 6.

Beginning in 2001, Lukoil-Perm was one of the first companies in the Perm region to systematically track its corporate image through professionally administered surveys. The results of these surveys were highly confidential and never circulated beyond corporate leadership, and I have not seen them. Some data from the mid-2000s were, however, described to me in some detail by a reliable source. They showed a steadily growing approval rating for Lukoil-Perm, approaching 50 percent by 2007, with only around 5 percent of the population holding a negative view of the corporation. Approval ratings for the company were highest in districts where Lukoil-Perm had production operations, and these districts also had the lowest percentages of the population complaining about unemployment and drunkenness. The details of the survey revealed that the corporation's approval rating varied slightly with several factors: the price of gasoline, the number and topics of media stories in the previous quarter, especially those having to do with ecology or oil spills, and with knowledge of the company's social and cultural projects. The company's favorability remained higher than all other big businesses in the region, including Sberbank and PermRegionGaz.[22] Overall, the survey's designers recommended the continuation of social and cultural projects as a key vector of corporate activity in the public sphere.

---

21. Proekt "Chernoe Zoloto Prikam'ia," Permskii Gosudarstvennyi Institut Iskusstva i Kul'tury, 2006. PermGANI f. 1206, op. 1, d. 912.

22. These surveys also provided a wealth of detail on local social and political circumstances, and fed the company's sense of what social and cultural projects were necessary and most politically advantageous.

Whereas my own queries were neither as detailed nor as systematic, the vast majority of intellectuals and political figures I spoke with in the Perm region in 2009 and 2010—corporate and state, regional- and district-level—judged Lukoil-Perm's social and cultural projects to have been, on the whole, successful in shifting the company's image. Here is my friend Oleg Leonidovich Kut′ev, a former Lukoil-Perm Connections with Society employee, reflecting on eight years of the company's social and cultural projects competitions:

> [Since the late 1990s,] Lukoil has done something enormous. It has, step by step, changed the way people relate to it. The most interesting thing is that, among much of the population, the evaluation of Lukoil is still negative . . . "if it's rich, that means it's stealing, it's taking our subsoil resources," and so on. . . . [But] all these [social and cultural] projects gave the possibility to the population to feel that they can, precisely with the help of Lukoil, do something themselves. And so the payoff for Lukoil was high . . . all this gives Lukoil the ability to work calmly in oil-producing areas.

Social and cultural projects, said one of his former colleagues, "became the mechanisms by which we brought people to a place at which they weren't offended that oil workers were living so well."[23]

General Partner: A Corporate-State Effect

Although they had post-Soviet specificities, the goals of winning over local elites and improving corporate image are part of the standard CSR toolkit worldwide (see, e.g., Coumans 2011; Welker 2014, 157–82). Lukoil-Perm's ambitions and practices, I have been arguing, went still further. Social and cultural projects, one Connections with Society employee told me, "killed two birds with one stone. Without much money, they improved the image of the company and also obtained a certain amount of influence over the distribution of state budget funds." Lukoil-Perm, that is, was engaged in a joint project of building a state-corporate field, one that extended far beyond securing its operations and managing its image.

One of the ways in which we see this effort is that Lukoil subsidiaries in the Perm region spoke much the same language as state officials when it came to preaching the importance of objective, expertly judged panels in reviewing

23. Survey data presented by Frye (2006) show that, in Russia more broadly, "good works" undertaken by a corporation have generally been helpful in reversing a negative public image, particularly an image associated with the "original sin" of questionable privatization practices in the early 1990s.

social requests for funding and contrasting this technique with patronage ties. In answer to a reporter's question about how the heads of district administrations had reacted to the introduction of project grant competitions in the south of the Perm region, Nikolai Kobakov, head of OOO Lukoil-Permneft at the time, stated,

> It's a competition. That means that there are winners and losers. I know that the heads of local administration have not immediately welcomed this idea. Of course, the old system of "asking and giving" was better understood and more familiar. . . . Many are now talking about building civil society in the country, about socially responsible business. In my view, the old times, when there was the "askers" side and the "givers" side, are now gone. . . . Our projects are one step on the path to civil partnership, to civil society.[24]

Under the old system, an OOO Lukoil-Permneft employee told a regional newspaper reporter, "it was not always clear how effectively the money was spent, and there wasn't a system of reporting."[25] Or, as another Connections with Society Division employee put it to me, with social and cultural grant competitions,

> there are no complaints about giving something to one and not to the other. It's a magnificent technology . . . it never gets dirtied by decisions about giving to one person and not the other. There is an answer [to that kind of accusation]: "folks, we brought you this competition, fight for the grants, go to war over them, defend your proposals effectively . . . [the best will get funded]."

These comments use some of the same framing that regional state agencies employed in their attempts to spread the project movement, especially the desires to extract themselves from district-level patronage networks and to serve as an objective, disinterested arbiter.

In the first of its annual reports on the results of its corporate social and cultural projects competition, issued in 2004, Lukoil-Perm deputy director Vsevolod Bel'tiukov—a Kuziaev deputy since the days of the Perm Commodity Exchange—contributed a lengthy article describing the ways in which the company had "started to use a new social technology for work with the local population" (Lukoil-Perm 2004, 2). Both Bel'tiukov and Tatiana Margolina (in an accompanying article), drew a contrast between this kind of social and cultural project

24. Ol'ga Nikolaeva, "Nikolai Kobakov: Pomogaem tomu, kto dostoin." *Zvezda*, September 17, 2002.

25. Petrova, "Biznes i Sotsium."

funding and the allegedly older practice of patronage. Margolina endorsed Lukoil-Perm's competitions on behalf of Governor Trutnev's administration:

> What was it like before? Every leader of a region met once in a while with the director of a large enterprise, talked about his problems and hopes that the corporation would join in solving them. But it was impossible to support each in equal measure, and to pick one or the other was completely subjective. In the [new] regime of project[s], everyone is put on the same level. . . . What the oil company is proposing is a modern, international approach and a worthy investment of resources in the development of the territories of the Perm region. There are new relationships here—not on the level of "asker and giver" [*prostitel'—daiushchii*] but on the level of equal partnership. (Lukoil-Perm 2004, 4)

The title of a commentary in that 2004 Lukoil-Perm report on its sponsored social projects phrased the company's aspiration with respect to the regional state aptly: "Involvement in a Common Affair" (Lukoil-Perm 2004, 4). Indeed, perhaps the best label for this element of what Lukoil-Perm sought to accomplish in its relationship with the regional state apparatus and with districts was plastered on banners at many public events: "General Partner—Lukoil-Perm." In practice, the general partner designation meant that Lukoil-Perm had given the largest amount of monetary support to making one or another program or event happen (other organizations might be listed for the same event as an "official partner" or "sponsor"). In 2009, for instance, at an international Forum of Regions hosted in the Perm region and intended to be a major meeting and planning event for those working in subnational government across Europe—an event that had little to do with oil—Lukoil-Perm was billed as the event's general partner and its executives were accorded corresponding prominence. Lukoil-Perm was the general partner for the First Diaghilev Ballet Festival in 2013, the general partner for the yearly Oil–Gas–Chemistry trade exhibitions of that took place at the Perm Exhibition Hall, and the general partner for the annual Folk Crafts Show, also held at the Perm Exhibition Hall. It was also the general partner for an international forum on "the Muslim world," held in Perm in March 2013 with a strong presence from Lukoil-Perm's predominantly Muslim production districts in the south, the general partner of the 2009 Perm Alive festival of contemporary art, and of the fifth exhibition of business-angels and innovators held in Perm in 2007, and so on. General partner was more than a level of funding: it indicated a level of generality and metacoordination on par with, and in collaboration with, the regional state apparatus. This style of partnership was very much in tune with increasingly common international business practices in which corporations enter into partnerships with state agencies and nongovernmental organizations with the aims of furthering their brands and countering critique (see, e.g., Foster 2014 on Coca-Cola). In comparative terms, if anything was exceptional about the case of Lukoil-Perm in the Perm region, it was the depth and breadth of this partnership, rather than its existence in the first place.

The Durability of Corporate Social and Cultural Projects

In Lukoil-Perm's social and cultural projects, as in their analogues in the regional state apparatus, we find both pervasive politicking and networking *and* the denial of that politicking and networking through aspiration to the status of an external, objectively judged, assertively apolitical arbiter of grant competitions. State building, corporation building, and their intersection in a state-corporate field in the Perm region of the 2000s relied simultaneously on the mobilization of very particular networks and the effort to cast those networks as their opposite: natural, external, forward-looking, and in the very nature of things as they should be.

I noted at the conclusion of chapter 4 that there was some irony in this because Perm's Capital of Civil Society campaign wound down with the departure of Governor Iurii Trutnev for the Ministry of Natural Resources in Moscow, and was therefore frequently recalled not so much as a successful and completed state-building project but as the political/electoral strategy of Trutnev's team, to be replaced in short order by Oleg Chirkunov's signature projects. This was not the fate of Lukoil-Perm's social and cultural projects. In fact, they continued in basically the same form through three separate governorships of the Perm region—Trutnev, Chirkunov, and Basargin—and through several leadership changes at Lukoil-Perm, both in the general director position and in the Connections with Society office. Lukoil-Perm was thus, over a period of years, on firmer ground than was the regional state administration in insisting that its social and cultural projects competitions stood outside of network and patronage politics. The durability and dependability of its social and cultural projects was, in fact, something the company frequently noted, drawing significant attention to the tenth anniversary of corporate social and cultural project competitions in 2011–12.

Lukoil-Perm's CSR efforts increasingly began to serve as models for other companies, with variously situated parties identifying it as a longtime leader and innovator in this field. When Buldashov, Marasanova, and other members of the original Connections with Society team departed Lukoil-Perm after a leadership change in 2006, for instance, many of them took up positions that continued these projects in other forms. Buldashov moved from Lukoil-Perm to become head of the regional Council of Unions, where he coordinated relationships between all the region's employee unions, their employers, and the regional administration. Marasanova moved from Lukoil-Perm to a position in the regional state administration's office of municipal development where, among other things, she helped municipalities write applications for new kinds of state and funding competitions. (She worked with many of the same district and small-scale municipal specialists she had trained in her time at Lukoil-Perm, and it should come as little surprise that organizations from oil-producing districts were, given their experience and existing connections, most successful in these other grant competitions as well.) In 2010, Buldashov and Marasanova collaborated on a region-wide conference directed at further developing the ways in which factories throughout the Perm region could help develop their workforces and surrounding communities. Although none of the companies included in this initiative operated on the region-wide scale that Lukoil-Perm did, the point was very much to adapt Lukoil-Perm's CSR strategies and practices for other, smaller companies (see Marasanova 2010).

Lukoil-Perm's social and cultural projects also became a model for big Lukoil. Although big Lukoil had been one of the first companies in Russia to issue its own Social Codex, in 2002, as part of its effort to raise its international profile, that document focused in on labor relations, ecological responsibility, and charitable contributions. In 2006, the company signed on to the United Nations Global Compact and pledged to adhere to international standards of transparency, reporting, and social responsibility. But the ways in which big Lukoil's various subsidiaries fulfilled the obligations outlined in the unified codex and in international agreements varied widely in both scale and method. Among its own subsidiaries, the company frequently held up Lukoil-Perm as an example in its national and international reporting on CSR. Over the course of the 2000s, an ever-increasing number of Lukoil subsidiaries began to model their CSR efforts on the social and cultural project competitions run by Lukoil-Perm.[26] According to a 2008 presentation by A. A. Moskalenko, head of big Lukoil's Personnel Department, at the European Organization for Economic Cooperation and Development, corporate social and cultural project competitions, held first in 2002 in the Perm region, began to spread slowly throughout other Lukoil subsidiaries, reaching the Volgograd region in 2004; the Astrakhan region, Komi republic, and West Siberia in 2006; the Nizhegorod region in 2007; and the Kaliningrad region in 2008.[27] The same general categories of grant giving were used in each region, but the proportions of funding and grants allocated varied. A steady stream of visitors

26. Lukoil, "Otchet o deiatel'nosti v oblasti ustoichivovo razvitiia na territorii Rossiiskoi Federatsii," 2003–4, Moscow.

27. A. A. Moskalenko, "Sotsial'no-ekonomicheskoe partnerstvo kak vazhneishee uslovie ustoichivogo razvitiia regionov rossii (opyt' OAO Lukoil)," www.oecd.org/cfe/leed/43389224.pdf (accessed March 19, 2013).

from other Lukoil subsidiaries came to Perm to better understand the workings of Lukoil-Perm's social and cultural projects competitions. In 2012, big Lukoil reported that its subsidiaries had received over 9,000 applications across 10 Russian regions, and funded just over 2,100 of them, for a total commitment of 360 million rubles (in the neighborhood of $12 million).[28] Big Lukoil even began to run its own grant competitions, open to organizations in all its operating regions.[29] Social and cultural grant competitions, which had begun as Western aid agencies' efforts to seed civil society in Russia had, transformed by the Perm region's oil complex and dense state-corporate field, returned to the federal level a decade later, as a main strategy of one of Russia's biggest corporations.

State and Corporation Beyond Social and Cultural Projects

Social and cultural project competitions were the most public and high-profile arena of state-corporate interaction in the early 2000s. There were, however, also a number of other important ways in which Lukoil subsidiaries and regional state agencies were entangled in social and cultural matters. These other interactions shifted in their relative significance over time, and became ever more significant following Trutnev's departure from the governor's office.

Some of these other dimensions of state-corporate interaction can be glimpsed by tracking the categories in which Lukoil-Perm itself counted its contributions to the Perm region. An "Informational Report" about Lukoil-Perm's social policy in 2004, for instance, listed a number of streams of support for the Perm region over and above the social and cultural projects competitions. "Charitable support" included various onetime donations to regional causes, including major contributions to Perm State University and Perm State Technical University, as well as contributions to celebrations of the seventy-fifth anniversary of the discovery of oil in the Perm region and to Lukoil-Perm's own sports club. Still another category in the 2004 report listed a series of festivals and seminars that made up a major initiative called Historical Cities of the Kama River Region and an array of projects supporting the revival of traditional folk handicrafts.[30] Five years later, in 2009, Lukoil-Perm reported that its contributions under the general rubric of

28. Konkurs sotsial'nykh i kul'turnykh proektov Kompanii Lukoil: 5 let v Nizhegorodskoi oblasti. Lukoil-Volganefteprodukt, 2008, 1.

29. Konkurs sotsial'nykh proektov, http:///www.lukoil.ru/static_6_5id_2259_.html (accessed March 26, 2013).

30. Informatsionnaia Spravka o Sotsial'noi Politike OOO Lukoil-Perm v 2004 godu. Confidential source, 2009.

corporate social responsibility broke down as follows: 28.8 million rubles in direct aid, 28.5 million rubles distributed through social and cultural project competitions, and 8.5 million rubles from its charitable fund for special causes.[31]

The year 2009 was also the eightieth anniversary of the discovery of oil in the Perm region, an occasion that Lukoil-Perm believed should be marked with a special gift. The company proposed reconstruction of a part of the central esplanade area of the city of Perm. Much of the esplanade was, indeed, in poor repair, including the section that lay between Lukoil-Perm's central offices and those of the regional state administration. It was a small-scale development project, to be sure, but the space in question was highly symbolic: the contrast between the carefully arranged and scrupulously maintained masonry, grass, and flowerbeds outside Lukoil-Perm's offices and the scrubby grass and battered concrete across the street in front of the offices of the regional state administration had often been pointed out to me by friends when we walked through the city center. If sleek and well-maintained offices were an indication of who had the upper hand in this state-corporate field—and to at least some minds they were—this central area of the city offered up a ready answer: Lukoil-Perm.

The offer was rejected by Governor Chirkunov's administration. Several sources familiar with the negotiations told me that it had been one symbolic step too far to allow Lukoil-Perm to make a contribution to the reconstruction and beautification of the Perm region right under the state administration's own windows. The Chirkunov administration made a counterproposal, suggesting that Lukoil-Perm might look a couple of blocks down the street to Cathedral Square, a smaller and less central area of the city, but one that was heavily trafficked as a primary route to the walkway along the Kama River. The Russian Orthodox Church, which had only recently reacquired the cathedral from its Soviet-era occupant—the Perm Regional Studies Museum—readily agreed, and the project fit neatly into Lukoil-Perm's many other efforts to sponsor religious revival in the Perm region. With funding from Lukoil-Perm, Cathedral Square was paved in handsome red brick, with neat fences and nests of benches and a new monument to Saint Nicholas the Wonderworker. An engraved plaque, roughly two feet by three feet and standing at eye level at one end of the square reads,

> The Cathedral Square on Sludke Hill was reconstructed in 2009 as a gift to the city of Perm and the residents of the Perm region from the oil company Lukoil in honor of the 80th anniversary of the discovery of Permian oil and the Volga-Urals oil and gas province (1929–2009).

---

31. "Blagotvoritel'naia i Bezvozmezdnaia Pomoshch'," Lukoil-Perm, Press Release, October 11, 2009.

The section of the esplanade that Lukoil-Perm had offered to repair and rebuild was left untouched.

The Cathedral Square renovation was a small and peripheral—yet highly symbolic—example of the ways in which state agencies and Lukoil-Perm bargained and struggled with each other even as they collaborated in a broader field of governance. These struggles leaked out only occasionally; indeed, it is not a coincidence that the details of this relatively small disagreement were the ones that were freely shared with me by a number of acquaintances. Higher-stakes negotiations on the state-corporate field were murkier and more inscrutable, especially when they were not accompanied by the commitment to transparency that was part of the Capital of Civil Society campaign.

One such higher-stakes negotiation came in the wake of major changes in Russian corporate income and natural resources tax laws in the 2000s, and the resulting deal came to stand alongside corporate social and cultural project competitions as a main feature of Lukoil-Perm's CSR efforts in the region.[32] A 1999 Perm region law, passed as the era of fuel veksels wound down, as Lukoil's regional subsidiaries were heavily focused on consolidating their operations, and as ZAO Lukoil-Perm sought to open new production fields in the north of the Perm region, stipulated that region-level taxes on oil companies would be paid into the districts where the production was occurring. District-level budgets in oil-producing districts ballooned in those years—particularly in places like the Usol'e district that had seen new, high-quality deposits open up. The rise of social and cultural projects over 2000–2004, then, proceeded in tandem with quickly rising district-level budgets, and the pervasive seminars and grant-writing classes were cast by both company and regional state apparatus as training opportunities for newly flush district-level administrators.[33]

In tandem with the Putin administration's efforts to recapture oil rents for the federal state, however, this arrangement changed dramatically over the course of 2000–2004. In 2004, a new law governing Lukoil-Perm's corporate income taxes stipulated an 80–20 split between federal and regional tax collection, with no corporate tax receipts accruing directly to production districts at all. Under this new tax code, districts still received property taxes from Lukoil-Perm, but the highly compressed geography of oil extraction meant that the oil companies paid taxes only on the vanishingly small amounts of land on which their production

32. On the convoluted issue of oil sector taxes in 2000s-era Russia, see Alexeev and Conrad 2009; Appel 2008; Jones Luong and Weinthal 2010, 138–44; and Gustafson 2012. These studies, however, focus almost exclusively on the issue of federal taxes. The regional and district level taxes I discuss below were much smaller in real terms but played a crucial role in the constitution of the Perm region and of federal-regional relationships (see esp. Votinova and Panov 2006).

33. On district-level budgets in the 2000s, see Ross 2009, 82–131.

infrastructure sat—an amount far less than the barely solvent agricultural enterprises that covered much of the rest of the districts. In the Il'inskii district, an old oil-producing district with little else in the way of business, taxes paid into the district budget by Lukoil-Perm subsidiaries declined from 18,432,000 rubles, or 24 percent of the district budget in 2004, to 1,787,000 rubles, or 4.7 percent of the total budget in 2005. Producing oil, that is, was no longer a direct boon to oil-producing districts, a change that placed enormous pressure on the leadership of these districts and forced them to compete with all other districts in the Perm region for regional funding—and to demand more of Lukoil-Perm in other ways.[34]

It was clear to all parties that a new kind of arrangement among districts, the Perm region, and the corporation was necessary, and it was one of the first major agenda items taken up by Governor Chirkunov in 2004. Instead of using this additional region-level revenue to fund further social and cultural projects competitions—as some of members of Trutnev's team (at that point out of office) advocated—Chirkunov and his associates took a different path, proposing to develop the Perm region not through projects but through tax cuts for businesses, which they argued would spur companies themselves to reinvest. In the mid-2000s, the Russian tax code allowed regions to reduce the amount of tax they collected on business income, and Chirkunov's proposal was to reduce the region's corporate tax rate from the standard 24 percent to 20 percent for all businesses, effective January 1, 2006. Although the proposed tax cut was to apply across the board, it made the most difference for the region's two major taxpayers: Uralkalii, the potash mining giant in the Perm region's north, and, of course, Lukoil-Perm. In the face of some early criticism, Chirkunov insisted that this was not a concession to the region's biggest businesses. It was, rather, a supply-side tax-cutting measure designed to spur investment and attract the main offices of other large corporations to the Perm region. Chirkunov's office successfully pressured regional lawmakers to pass this tax cut quickly, claiming that they had to beat other Russian regions in order to attract businesses.

There is reason, however, to doubt the public rhetoric and the breadth of the benefits of cutting taxes in this way. Indeed, an alternate reason for the introduction of the 24–20 tax program circulated in the Perm region, both among contacts I knew and in the financial press. On this reading, the entire 24–20 arrangement was designed at the behest of—or at least to accommodate—Lukoil-Perm. Reforms to the federal tax code in the early 2000s eliminated many of the strategies that Russian oil companies had used to avoid paying their taxes, including transfer pricing, in which substantial profits could be logged in so-called domestic offshore tax havens with lower federal tax rates. But these reforms did

34. Passport Il'inskogo Raiona, Il'inskii, 2006.

not fully eliminate the possibility of interregional transfer pricing as it affected taxes paid to regions. If, for instance, Lukoil-Perm sold the oil it pumped in the Perm region at a low cost to another division of big Lukoil located outside the Perm region, and that second division marked up the price to Russian or world prices, then the Perm region would not be home to the profits—and therefore not the taxes either—although the Russian federal government would still get its share.[35] Or, to put this in the terms of the spatialities of the regional oil complex: although Lukoil-Perm was very beholden to district interests in order to access land for drilling into very specific geological formations (and could therefore not easily play districts off of one another), it had a much greater degree of flexibility with respect to where and how it accounted for profits on sales and transfers of oil once that oil was out of the ground. On this score, the company could and did press its advantage with the Perm regional government even after federal tax loopholes had been closed.[36] According to at least one observer I spoke to, the 24–20 program emerged precisely as an effort to stop Lukoil-Perm from using a trading company based in Moscow for its oil exporting and accounting purposes, a switch that could have decimated the Perm regional budget right along with district budgets. (In the early 2000s, Lukoil-Perm's tax payments made up around 20–25 percent of the Perm regional budget.)

Under Chirkunov's plan, however, the Perm region did not simply lose that 4 percent of Lukoil-Perm's tax money, for the company still felt the need to attend very closely to its production districts and remained sensitive to the strands of critique that had helped launch its CSR projects years earlier. In an arrangement unique to Lukoil-Perm in the Perm region, a separate and additional understanding about the 24–20 program was built into the five-year and annual negotiated agreements between big Lukoil and the Perm region. The two sides agreed that the company's annual regional tax savings (that is, the calculated difference between 24 percent and 20 percent) would not be pocketed by the company but, rather, would be promptly reinvested in the region. In any given year, half of those tax savings would go to Lukoil-Perm's own reinvestments in the Perm region, including roads, pipelines, office buildings, and other necessary infrastructure (and including modernizing in ways that would improve the company's ecological footprint). The other half would go to social and cultural development projects—over and above the amount that was already being spent through the well-established

35. See "Ne ustupit' pribyl'," *Izvestiia*, August 20, 2007. On transfer pricing in the Russian oil and gas sector at around this time, see, for instance, World Bank 2004 and Jones Luong and Weinthal 2010, 152–56.

36. On transfer pricing and hidden subsidies in the Perm region, see Zakhar Zlobin, "Printsip Litsemeriia," *Permskii Obozrevatel'*, September 27, 2004. The possibility of Lukoil moving certain operations out of the Perm region was frankly acknowledged by Governor Chirkunov; see, for instance, "'24–20': Sbudetsia li prognoz?" *Permskii Dialog*, February 2008, 5.

corporate social and cultural project competitions. All these projects, it was further agreed, would be situated in Lukoil-Perm's production districts, where they would help make up for the zeroing out of direct tax receipts. Finally, these new projects would not be planned or allocated through open competitions, but by direct, annual negotiations between Lukoil-Perm and Governor Chirkunov's office, with appropriate consultation with district heads.

In 2008, the third year of the 24–20 agreement, Lukoil-Perm's 4 percent savings on its regional taxes constituted 1.5 billion rubles, or roughly $60 million, leaving half for reinvestment in the region's oil industry and half for social and cultural development projects. "That is *big* money," said one person I spoke with who knew some of the details of these negotiations but did not participate in them. "You can build small hospitals, new schools, all kinds of things."[37] One of the governor's representatives at the meetings at which this money was allocated told me they were largely amicable (although he certainly had an incentive to say so): "They [Lukoil-Perm] might come to us and say we want to build two new clubs, one in such and such a district and the other over here. We might reply, sure, build two clubs, but we'd like one of them to go in this [other] district." A partial list of the projects that the $30 million in 2008 went to includes: the construction of three fitness and health centers; the construction or reconstruction of schools in seven districts; reconstruction of libraries, clinics, cultural monuments, churches, and mosques up and down the region; and repair and reconstruction of the water and gas infrastructure for another set of cultural sites.[38] As should be clear from this list, funds from this pot of money were dedicated largely to physical reconstruction, repair, and restoration, rather than to the everyday problem solving, subject formation, and community building that featured in the annual corporate social and cultural project competitions. These capital projects, I was told, were particularly significant and prized by the company because they left a permanent mark on the central spaces of towns and cities, and because they associated Lukoil-Perm directly with a durable infrastructure of schools, hospitals, sports centers, cultural centers, and so on. Whereas state funds generally went first to the maintenance of the status quo or to expenses like salaries, Lukoil-Perm was able to claim association with new and highly visible projects. These negotiations represented an entirely new stream and vector of funding, a new way in which Lukoil-Perm became central to regional planning and metacoordination after the end of Iurii Trutnev's Capital of Civil Society initiative.

Once these agreements were reached in the region-level negotiations, the details were incorporated into the annual signing of agreements between Oleg

37. I do not have enough data to calculate whether the 24–20 tax break bargain that Lukoil-Perm reached with the Perm region meant more or less net money for the districts.

38. Anatolii Petrov, "Krizis delu ne pomekha," *Zvezda*, April 17, 2009.

Chirkunov and Vagit Alekperov of big Lukoil. Separate meetings between Lukoil-Perm leadership and all the heads of local administrations in the region's oil-producing districts then followed, where specific plans for using this money were signed. These agreements were tailored to each production district, and specified the obligations of both sides. The agreement between Lukoil-Perm and the Il'inskii district for 2007, for instance, committed the company to continuing its oil production operations, to on-time and complete tax payments, and to continuing the corporate social and cultural projects competition. The company also promised to collaborate with the district on a year's worth of events celebrating the 430th anniversary of the founding of Il'inksii. The Il'inskii district, for its part, committed to fostering an optimal tax regime for Lukoil-Perm in the district—something the district had very little flexibility on in any case—and to three additional points, all of which had to do with the one thing that district administration could exercise some leverage over the company about: access to land and land rights in and around its operations.[39]

The 24–20 program was not without its critics in the Perm Legislative Assembly, who pointed out, for instance, that the program's first four years did not bring any major new businesses to the Perm region. At the same time, over the course of those four years, the program subtracted an estimated 3.5 billion rubles from the regional budget. Although 1.8 billion was returned through the negotiated development projects, deputies who opposed the tax break were fond of pointing to all the other projects—especially bridges and roads in Perm and in non-oil-producing districts—that the regional budget could have supported in those years. Despite these objections, proposals to end the 24–20 program never passed the regional assembly, and the governor's office continued to strenuously oppose efforts to change the law.[40] The 24–20 program endured as a new phase and dimension of the state-corporate field in the Perm region, one that gave both sides significant extrabudgetary power to shape districts and localities and, once again, placed Lukoil-Perm in the role of joint metacoordinator of domains of life up and down the region.

It is certainly true that reforms of the Russian tax code in the Putin years diverted significant flows of oil rents into federal coffers, and from there (partially) into an array of state-led welfare programs (e.g., Cook 2007). But, once again, this is only part of the story, and corporations have played a larger and more direct role than is often noted. Indeed, because federal taxes were increasingly non-negotiable, correspondingly increased pressure and value were placed on budgetary, financing, and development initiatives that *could* still be negotiated. There

39. Confidential source, personal communication, 2010.
40. "Poka khomiak ne sdokhnet," *Permskii Obozrevatel'*, August 21, 2010.

was ample opportunity to carve out new modes of region-level collaboration among state agencies and corporations, including on the terrain of development and welfare provision.[41] The 24–20 program thus deepened the state-corporate collaboration and provided one more way in which Lukoil-Perm was configuring and reconfiguring public space and infrastructure in the Perm region. This region-level, extrabudgetary state-corporate bargain, it is worth underscoring, was sealed not in the context of federal state weakness in the 1990s or in the very early stages of state building in 1998–2001, but right in the midst of the massive oil boom, federal recentralization, and pervasive state building in the mid- and late 2000s. When it came to metacoordination of spaces and domains of life within the Perm region, corporate-state relationships in the oil sector remained just as mutually constitutive, if differently so, as they were in the eras of petrobarter and fuel veksels.

• • •

The Soviet model of industrial development in company towns, especially well represented in the Urals, created what were widely known as city-forming (*grad-oobrazuiushchie*) enterprises. The 2000s cast Lukoil-Perm as a different, but related, creature: a budget-forming (*biudzhetoobrazuiushchyi*) enterprise, whose taxes made up a significant part of the regional budget. The chapters of part II have argued, however, that merely counting these taxes or seeing them as essentially or even primarily constitutive of the relationship between regional state apparatus and region-based corporate subsidiaries elides the many other ways in which state and corporation were knit together into a state-corporate field. Even with respect to taxes themselves, Lukoil-Perm sought a much more direct role in governance and metacoordination among spheres of human life than is assumed in the literature on petrostates. Lukoil-Perm's development grants and taxes, Andrei Kuziaev argued on one occasion, were best when that money still "smell[ed] of oil"—that is, when it had not been fully abstracted into the monetary sphere but remained materially associated with oil and the regional corporation that specialized in it.[42] His reference to the smell of oil points to another range of material qualities of oil that rose to salience in the 2000s, above and beyond oil's geographical spread in the region and the ways in which the oil industry's production, refining, and transport infrastructure lent some shape to political and economic transformations.

---

41. Elena Chebankova (2010) persuasively argues that nearly all dimensions of the first Putin administration's attempts to recentralize federal state power, contrary to its original intentions, ended up creating new and diverse opportunities for state agencies and corporations to collaborate on informal grounds.

42. Tat'iana Vlasenko, "Bol'she nefti—bol'she deneg," *Zvezda*, July 13, 1999.

Part III
THE CULTURAL FRONT

6

OIL AND CULTURE

The Depths of Postsocialism

The Perm Regional Department of Culture's 1996 annual report did not open on an optimistic note. "The current stage of the development of society," it began,

> is characterized by deepening financial-economic crisis and instability in all spheres of activity, including the cultural field. Nevertheless, the organs of culture in the Perm region are doing everything possible not only to preserve what has been accumulated but to develop further.[1]

The full report showed that the museums and libraries for which the department was responsible were squeaking by on shoestring budgets and Houses of Culture stood empty—or were being converted into nightclubs, discos, or bars. In a line of work that had long featured ideological campaigns, it was not especially clear to anyone what the content or goals of state cultural production should be even if funding could to be found. Indeed, during my visits to the Perm region in 2000–2001, I found that culture work as a profession had a largely discredited past and, by common consensus, not much of a future. Employees in the Perm Regional Department of Culture and its district-level subsidiaries were quite demoralized. One, who had worked in cultural affairs in a rural district for most of her career, explained grimly that her department had just been merged with others to create a single department of Culture, Sport, and Youth. A member of the regional Department of Culture's central office confirmed these worries, telling me that, "Soon there won't be any place at all for culture at the district level." He

1. Perm Regional Department of Culture 1997, 4. See also Gasratian 2004.

himself was racing around trying to make ends meet by teaching courses and tutoring at a local institute in order to make up for his meager earnings from the state budget allocation for culture. Even so, he only picked up his supplementary salary once every few months, he said, when enough had accumulated from these odd jobs to make the trip the accounting office worth the time.

Culture was at the bottom of regional officials' list of priorities as well. In a wide-ranging interview about the regional economy published in 1998, economist Evgenii Sapiro—"godfather" of the Perm Commodity Exchange and, at the time of the interview, chair of the Perm Legislative Assembly—described at length several routes by which he thought the Perm region could extract itself from economic crisis and compete successfully with other Russian regions: a revival of the defense sector; the successful conversion of defense sector factories to other purposes; the export of primary commodities like oil, timber, and potash; and transportation—especially river transportation along the Kama. He concluded with a brief mention of culture, citing Perm's long history of distinguished universities, its large collection of theaters, and the lingering influence of artists and other cultural figures who had been evacuated to Perm during the Great Patriotic War years. Perhaps, he said, culture "is our 'ace in the hole,' our principle reserve for the future" (Liubanovskaia 1998, 81).

Within the span of a decade, Sapiro's list of priorities had been nearly inverted. Beginning with the rise of Lukoil's corporate social responsibility (CSR) projects in the early years of the twenty-first century, culture workers from the Perm region's oil-producing districts were in high demand. Many found themselves both busy and, to their own considerable surprise, influential. Given the tight relationships between Lukoil-Perm and the emerging regional state apparatus, an interest in reviving culture of all sorts soon spread to non-oil-producing districts as well, and the Perm region declared itself a Territory of Culture in 2007.[2] Independent cultural organizations—managed out of neither state nor corporate offices—sprang up as well, organizing their own festivals and other initiatives. By the time of my primary fieldwork for this project in 2009–12, culture could not have been more squarely in the center of public attention: In 2009, Governor Oleg Chirkunov announced that Perm would seek to dethrone Saint Petersburg as the cultural capital of Russia. Within a year, he had modified those ambitions into a formal plan for Perm to be named a European Capital of Culture in 2016. The Regional Ministry of Culture—no longer simply a department—grew exponentially, receiving massive infusions of cash from the regional budget to support a range of projects and programs, from a new museum of contemporary art to

2. The Territory of Culture program, which allocated state funds for cultural development in a small group of cities and towns each year, was still going strong in 2014.

an experimental theater to a constant stream of cultural festivals. It became the centerpiece of Governor Chirkunov's efforts to put Perm on the map of Russia and the world.

The reasoning behind Chirkunov's Perm Cultural Project, elaborated and executed by a carefully chosen inner circle and modeled on the global movements associated with creative cities and city branding, was that culture was the Perm region's best chance to extract itself from postindustrial torpor. The city's factories were not going to come back, at least not without unimaginable investment. As the oil boom leveled off, it was becoming clear that, for all of the wealth it created, the oil sector employed too few people and could not be counted on forever. Culture, Chirkunov proclaimed to audiences far and wide, would light the way by providing new routes to creativity and entrepreneurship that, coupled with international, national, and corporate sponsorship, would bring tourist money rolling in. It would not be much of an exaggeration to say that, by 2010–12, every taxi driver in Perm was talking about culture—if only to blast Chirkunov for building a contemporary art museum instead of better roads.

This cultural boom is the subject of part III. Although it would be productive to focus on any one of the domains of life among which the corporate-state alliances described in part II aspired to coordinate—health care, education, ecology and environment, and others—I take up culture for a number of reasons. First, in addition to its sheer omnipresence by the end of the decade (see fig. 5), culture was the starting point. Lukoil-Perm's earliest CSR initiatives were focused on cultural revival and folklore, and culture retained primacy of place in the company's CSR work for much of the decade. In 2004, 42 percent of Lukoil-Perm's social and cultural project funding was earmarked for culture, and in 2010 grants in the category of Culture and Spirituality were nearly a third of all total grants (49 out of 149) and nearly double the next highest (Health and Sport, with 26 projects) of the 8 categories.[3] Even as state agencies joined through their civil society initiatives, culture was a favored project: of the 120 social and cultural projects that the Perm region brought to the 2003 federal district civil society fair in Nizhnii Novgorod, 31 were in the category of A Unified Cultural Space: Strategies for Development—the largest of any category.[4] Second, whether these cultural projects were company-sponsored, state-sponsored, or independent—and they were usually some intricate combination thereof—they were tightly linked to the remaking and reconceptualization of the Perm region that I have been tracing across the decades. Culture became, in significant ways, the 2000s-era successor to the petrobarter and fuel veksel circuits of the 1990s as the domain in which the

3. Lukoil-Perm 2005, 23; Lukoil-Perm 2010, 4.
4. "Informatsiia dlia zhurnalistov," PermGANI f. 1206 op. 1, d. 197.

FIGURE 5 The Culture Taxi Company opened in Perm in 2012, hoping to capitalize on the city's high-profile campaign to become a European Capital of Culture.

Photo by author, 2012.

corporations and materialities of the regional oil complex became entangled with emergent senses of regional distinctiveness.

Finally, a focus on culture enables me to continue to situate the Russian case with respect to other parts of the world. We find, again, both significant similarities and differences. The transmutation of oil into cultural spectacles of various sorts has, for example, long been characteristic of oil-exporting states, from 1970s Nigeria (Apter 2005) to post-Soviet Central Asia (Adams 2010). Yet corporations have much less frequently placed cultural production at the very top of their agenda for relating to surrounding populations (although, to be sure, it often occupies lower ranks). As Stuart Kirsch (2014) demonstrates, science and ecology have been far more common domains in which corporations seek to shift public opinion and pacify critics. A combination of factors produced the configuration I describe for the postsocialist Perm region, including the socialist heritage in which firms were heavily involved in cultural work in company towns; the exceptionally close relationship between state and corporation in the post-Soviet Perm region; and the fact that Lukoil-Perm was engaged in CSR activities "at home," where treading on the terrain of culture did not have the imperialist overtones that it might for corporations with operations in European postcolonial contexts. (Especially after the nationalizations of the oil industries around the postcolonial world in the mid-twentieth century, showcasing an expertise in local culture was not usually an attractive or viable strategy for transnational corporations.)

It is commonly asserted that Russia's oil boom brought with it a range of cultural transformations—a resurgent, swaggering, state-sponsored Russian nationalism; a strong, centralized federal state heavily invested in the performance of its power; and a legal and censorship regime dedicated to suppressing alternatives and challenges in all realms, from the political to the artistic. These chapters considerably complicate this common portrait by charting the contested and morphing ways in which oil and culture were entangled in the Perm region during the Putin and Medvedev presidencies.

Abstracting Culture: From Crisis to Spectacle

Lukoil-Perm's CSR projects responded to a number of vectors of popular and political discontent. In the years when the projects were first being designed, the most pressing concern was a sharp and sudden rise in inequality in the Perm region's oil-producing districts. Indeed, for oil sector employees in the rural Perm region, the story of Lukoilization in the late 1990s is interesting in good part for what it does *not* feature: the tales of deprivation, shortage, and crisis that continued to be rampant elsewhere in the region. One former Lukoil-Perm employee I knew summed up his evaluation of the period from the late 1990s to the early 2000s—the time of Lukoil's final consolidation in the Perm region—as follows:

> [In the Soviet period] it was all simpler and easier to understand. There was no Lukoil, there was a state oil system and it didn't bother anybody. Oil was pumped for the people and the profits went to the people. Everything was fine. . . . [But, in the post-Soviet period,] a sharp divide began to arise between oil workers and other people living in those areas, in income and in the ability to obtain things—televisions, VCRs, refrigerators. Oil workers aren't the majority of the population in any district [i.e., this new wealth was spread unevenly]. And then these oil magnates appeared at the top. . . . [Ordinary] people's evaluation [of the oil industry] was about as negative as it could be.

Some of the largest and most evident inequalities in wealth in the postsocialist Perm region, that is, began to emerge most visibly in precisely those districts where Lukoil-Perm operated production facilities, as workers in those areas began receiving modest cash bonuses at the same time that workers in nearly all other sectors of the economy were experiencing salary delays, receiving in-kind compensation, or not working at all. Scholars such as Caroline Humphrey (2002), Jennifer Patico (2008), and Olga Shevchenko (2008) have shown that, at the level of everyday experience and conversations, postsocialist inequalities were often measured, as they were by my interlocutor above, precisely at the point of consumption. In the case of oil sector employees in largely rural districts, that is, we have a comparatively rare case in which a minority of workers suddenly *could* obtain refrigerators and VCRs, those iconic objects of early postsocialist desire, while their much more numerous neighbors could not. These inequalities grew sharpest in the northern oil-producing districts that were in the domain of the successful, dynamic, and wealthy ZAO Lukoil-Perm, for it was here that the bonuses were the largest, the links to national and international finance capital most evident, and the vast wealth being accumulated by the new oil elite the most clearly on display. Although not without their own versions of these dynamics, OOO Lukoil-Permneft's still-lumbering and debt-ridden operations in the south did not create quite as obvious inequalities and did not do so quite as quickly.

The kind of inequality emerging in these contexts was not only measured by wealth, income, or the access to the more prestigious set of consumption items that came with oil industry salaries. Recall from chapter 5 that Sergei Buldashov described the problem faced by Lukoil-Perm when he began work in the Connections with Society Division by imagining a resident of an oil town speaking about his or her neighbor: "there's Vasia, he got himself into Lukoil and now he's [got the money] to build a cottage." This phrasing bespeaks a concern about income inequality, in part. But it also points to a concern about exclusion from a powerful collective, about Vasia getting himself into Lukoil and leaving others behind. As Caroline Humphrey (2001) has shown with reference to a range of historical and contemporary examples, inequality in Russia has long been understood in significant part as being about exclusion from a hierarchical set of insiders' collectives: it is an issue of unequal access along with unequal wealth or income. We have already encountered worries about this kind of exclusion in the debates about whether or not the Lukoilization of the Perm region in the 1990s would leave the residents of the region outside of newly important circuits of oil and oil money. Worries of this sort only increased in volume and prevalence as Lukoil-Perm became more and more firmly ensconced at the center of regional political power and economic exchange. The company's response to these critiques aimed at both dimensions of inequality—income-based and exclusion-based—by proposing programs that would simultaneously raise incomes in oil-producing regions *and* emphasize that Lukoil's regional subsidiaries were, far from being exclusionary, deeply embedded in all dimensions of local life. Culture was the terrain on which the responses to these overlapping vectors of inequality met and mingled.

The task of corporate response to critique, as I have described, fell to ZAO Lukoil-Perm's newly minted Connections with Society Division and the former Party operatives tasked with running it. One former member of the division framed the question with which she and her colleagues wrestled in their first months as, "How are you going to relate to [the non-Lukoil employees] in oil-producing districts?" The very first answer they arrived at was in the domain of culture: the revival of folk handicrafts. In 2002, after a judging process that included ethnographers and museum experts, the company gave grants to a number of carefully selected artisans with the goal of enabling them to begin their own small businesses producing and selling folk handicrafts. I asked another former employee why ZAO Lukoil-Perm's initial focus was specifically on artisanal handicrafts.

> That [decision] was linked to the high level of unemployment in those districts. In the north of the Perm region, in the Cherdyn, Usol′e, and other districts, there was never any agriculture that could feed people. What was there to do? And so the idea of self-employment [*samozaniatost′*] was born [in the Connections with Society Division.] . . . Sit home . . . . sew, make pottery, do something else, and maybe you can get some sort of income. Miserly income, but at least you have something to do. In the beginning, there was, perhaps, an illusion [at Lukoil-Perm] that you could actually develop folk crafts, and develop whole towns on the basis of folk crafts, but then it became clear that [for this] you needed a strong local tradition, more than one generation. And there was the problem of materials—where to get them, where to sell them. . . . There were no great successes, but there was a lot of noise. The press was interested.

Some of the classic dilemmas of the first postsocialist decade are here: how to create employment and income in largely rural districts that were struggling to survive without socialist-era agricultural subsidies; how to acquire materials and to sell them on the market; and how, in general, to manage the turn to household production and petty trade across rural areas of the former Soviet world. Through grants to folk artisans, Lukoil-Perm understood itself, and certainly advertised itself, to be working hard to close the gap between oil workers and nearly everyone else through the revival of folk handicrafts. Folk culture would be, in short, an entrepreneurial lifeline for the rural Perm region (see fig. 6).

Whatever its success on the score of reducing income inequalities—modest, at best, by most accounts—Lukoil-Perm's grants for folk handicrafts production ac-

FIGURE 6 A billboard outside Lukoil-Perm's main offices in the center of Perm. As with many of the company's public relations and advertising materials in the 2000s, the billboard juxtaposes images of oil industry workers and infrastructure with images of widely recognizably traditional items—in this case, a church and a girl dressed in folk attire. The billboard reads, "Protecting Traditions—Lukoil-Perm."

Photo by author, 2008.

complished other things, in good part by mitigating critiques of and worries about that other dimension of inequality: exclusion from powerful collectives. The company's projects began to recast dilemmas of everyday getting by that were so prevalent in the 1990s as the concerns of a leading regional corporation, as matters of culture, and, on both counts, as something more than just family, village, or otherwise small-scale struggles. With these initial seed grants, corporation-sponsored cultural production quickly become a domain that pointed to larger-scale identities, networks, and projects than the situational and local struggles of the 1990s. The artisans, culture workers, museum specialists, and others who received these grants soon found themselves part, however small, of a regional corporation's field of vision. They began interacting in new ways with other culture producers from other oil-producing districts, all of them unexpectedly embarked on a collective and well-funded set of projects. Out of the 1990s travails of the rural turn to household production and the ruins of Soviet cultural construction, culture was, courtesy of Lukoil's production subsidiaries, gradually becoming a more abstracted domain, a company-sponsored way in which to include oil-producing towns in larger collectives and projects.[5]

The Vertical Integration of Culture

In the first weeks and months of their new folk handicrafts program, company representatives made frequent trips from Perm to the northern districts—three to four hours by car each way—to bring their new craft specialists raw materials with which to work. They would return to Perm with finished products for sale. Very quickly, though, these Connections with Society employees realized what so many residents of the rural districts of the Perm region had themselves learned a decade earlier: an entrepreneurial spirit was, by itself, hardly an antidote to involution, demodernization, and raging unemployment. It would not be possible to erase the deeply felt differences between oil workers and the remainder of the population simply by giving seed grants to artisans and watching revenues pour in. (If this seems unsurprising, it is only so in retrospect. The vast majority of Western advisers and consultants blanketing Russia at the time shared Lukoil-Perm's faith in the magic of markets and entrepreneurship. Indeed, they were one important source of it.)

Rather than abandon their emphasis on folk culture at that point, however, the architects of this program decided that seed grants to support entrepreneurship were simply not enough in the prevailing conditions. The corporation would need to become still more involved. The path they chose for this involvement stayed close to Lukoil's own expertise and experience at the time: vertical integration. If, that is, the 1990s were an extended effort to achieve the vertical integration of the regional oil sector, and if the 1998–2004 period saw a state-corporate alliance take the place of Western aid agencies and advisers in shaping social and cultural development in the region, then it should not be surprising that the strategies that Lukoil-Perm brought to the cultural sphere borrowed heavily from the vertical integration textbook of piecing together supply chains and combining and recombining previously distinct entities. Although I never heard the term "vertical integration" specifically applied to culture in my interviews and other conversations, a broader vocabulary about the benefits of unifying production, circulation, and consumption was pervasive in talk about Lukoil-Perm's intervention in the cultural sphere through CSR (cf. Guss 2000, 90–128; Stammler 2005).

5. Rogers 2014c frames this transformation as between two senses of the "everyday" in Russian historiography: *povsednevnost'* and *byt*. On the post-Soviet fate of culture workers, see also Luehrmann 2011a, 2011b.

In order to make their new grants-for-handicrafts program successful, for instance, Lukoil-Perm specialists decided that the consumption side of cultural production—a tourism industry—would need to be developed, so that residents of Perm and other cities would make their own ways to remote northern districts and boost local economies by buying products from the company's artisans. A new Lukoil-Perm program, launched not long after the first grants were distributed to folk artisans, sponsored small festivals and other events in areas that the company identified as "Historical Cities of the Kama River Basin." These festivals—all of them in oil-producing districts—were intended to offer contexts for Lukoil-Perm's chosen craftspeople to sell their wares and for specialists to gather and plan further festivals, exhibits, and ways to market crafts across the region.[6]

In addition to the mobile Historical Cities festival, Lukoil-Perm adopted other strategies to draw tourists to its production districts to spend money in ways that would help local economies. For instance, at the company's initiative, the annual Garden, Field, Farm trade show held at the Perm Exhibition Hall (a combination exposition center and trade show facility) each year included a new element beginning in 2002: a show and sale of local crafts featuring everyday gardening items. Under a banner reading, "Yes to Social Partnership!" and featuring Lukoil's red logo, artisans sponsored by ZAO Lukoil-Perm from eight northern districts sold handmade baskets, belts, tools, and other items. Oil company specialists helped them adjust their prices to the urban market and, on the sidelines of the exhibition, ran seminars designed to offer instruction in how to further develop a business.

The early Historical Cities festivals and Garden, Field, Farm trade shows sought to commercialize involution, to make some profit out of the turn to household production across those rural areas that were home to oil operations yet, due to the oil industry's high capital-low labor profile, provided salaries to very few local residents. Especially in the early part of the 2000s, these major events were managed almost entirely out of the Lukoil-Perm Connections with Society Division. Although projects often received cofinancing from regional or district-level state agencies through the kinds of collaborations described in the previous chapters, these cultural projects themselves were, from inception to conclusion, firmly managed by Lukoil-Perm. A former state culture worker I knew who was peripherally involved in the first Historical Cities festival in Chermoz in 2002 recalled that, although the planning for the festivals and workshops in each location

---

6. This kind of relationship among oil, tourism, and cultural production is not at all common. See Davidov 2013, 61–77 on oil and museums in neoliberal Ecuador and Büscher and Davidov 2013 on the "ecotourism-extraction" nexus.

included culture workers who were, like her, in state jobs, they were closely monitored at every stage by Lukoil-Perm employees. The script for the public parts of the festival in Chermoz, for instance, was written entirely at Lukoil-Perm, right down to the order in which representatives from each district paraded through the center of the town.

Despite the fact that they did little to reduce income inequality, these tightly supervised roving festivals and trade shows were nevertheless central to processes of abstracting culture, and they increased in frequency and variety over the 2000s. Some of the distance that had been traveled from those first ZAO Lukoil-Perm seed grants for folk artisans in 2002 to what was often called a regional festival movement by the end of the decade was nicely encapsulated for me as I stood in the audience at the conclusion of the 2009 Obva: Soul of the Riverlands festival. The festival, celebrated by four rural districts clustered around the Obva River in the west-central Perm region and supported in part by local Lukoil affiliates, had been a success by all accounts. The weather was glorious. Folklore ensembles had proudly presented the uniqueness of each district in costume and song. There was heavy traffic at a bazaar where local artisans, from blacksmiths to puppeteers, displayed and sold their wares. A day of friendly competition among districts in everything from volleyball to skits had gone off without a hitch. An elected official, representing the Il′inskii district, stepped to the microphone to deliver one of a series of ceremonial closing speeches. After a few words of thanks and some playful sparring with the heads of other districts, he concluded by inviting the gathered masses to yet another festival, scheduled to take place a few weeks hence, that would mark the 430th anniversary of the town of Il′inskii.

He stepped off the stage and, shaking his head and smiling, commented to a cluster of organizers and local politicians: "We just go from one feast day (*prazdnik*) to the next these days, don't we?" Had they heard, he continued, the story about a visitor to one of Russia's remote monastic communities? The visitor had tracked down one of the hermits in the forest and asked him, "How do you live like this, out here by yourself day after day?" The answer: "As soon as one feast day ends, I start preparing for the next one." That's what life was starting to look like in the Perm region, he concluded, and went off to shake more hands. Participating in the 2009 Obva: Soul of the Riverlands festival and speaking with its organizers and attendees, I was struck by how much had changed since the 1990s, when discussions on the topic of "how do you live like this" were much more likely to concern issues like how to feed one's family than friendly disputes about which district's folklore ensemble had performed better. It was, as this local official said, hard to miss the omnipresence of festivals and cultural spectacles in the Perm region by the late 2000s: culture was becoming an increasingly abstracted domain, one in which it was easy for Lukoil-Perm to claim that everyone was being included.[7]

Oil and Cultural Abstraction

Vertical integration is but one of a number of overlapping ways in which Lukoil-Perm's CSR projects assembled a more abstracted, generalized cultural field out of the concrete, situational, disconnected, and highly localized dilemmas that characterized the 1990s. Cultural festivals and museum displays, each a domain of abstraction in its own right, were other important ways. So, too, was an encompassing context of increasing monetization and commodification, in which rubles as a state-sponsored general equivalent finally took hold after years of barter and surrogate currencies. The following sections and chapters return to all these processes in greater detail. It is worth pausing, however, to situate these abstractions in the still broader context of theories of oil and culture.

Recall that in 1990s contexts of demonetization, involution, and parcelized sovereignty, sensibilities about oil in the Perm region gathered around its concreteness, its circulation through local networks and personalized relationships of mutual indebtedness. Petrobarter chains and fuel veksel circuits were, I argued, integral to links between Permian oil and senses of Permianness in that decade, and they gained their significance in part through contrasts with highly abstracted, unfamiliar domains of exchange associated with monetary circulation. For the 1990s, the centrality of oil to regional political economy notwithstanding, the theories of oil, money, cultural production, and imaginations of the state developed by Fernando Coronil (1997) and Andrew Apter (2005) in the cases of postcolonial Venezuela and Nigeria, respectively, did not shed much light on the early postsocialist Perm region. The increasing monetization and abstraction of cultural production in the era of Lukoil-Perm's CSR projects, however, invites us to consider the utility of these approaches anew for the 2000s. Do we find, that is, that the emergence of a generalized, monetized sphere of circulating oil wealth in the Perm region is centrally linked, as in many postcolonial contexts, to the emergence of cultural identities and imaginations of "the state" as an autonomous, powerful actor?

My answer will be both yes and no. To be sure, as is already evident from my discussion of vertical integration in the cultural sphere, cultural production

7. Hemment's (2015) discussion of state-sponsored youth movements in the 2000s also follows, in part, a similar trajectory of festivalization, with youth group activity increasingly gathered around elaborate summer camps and festivals.

during the 2000s did draw attention away from the concrete material exchanges and social relations of petrobarter and fuel veksels to more abstracted domains of both culture and monetized exchange. These cultural processes were some of the central ways in which the state-corporate alliance described in chapters 4 and 5 strove to legitimate itself, respond to critique, and consolidate its power. On this score, we can certainly see the dynamics of the 2000s Perm region as cousins of those described by Apter and Coronil in other oil-producing contexts.

But they are more distant cousins than might be presumed, for two linked reasons. In the first place, the kinds of cultural and political formations that were at stake in these projects owed a great deal to long-running Russian and Soviet models of nation building and state formation. These differed substantially from those common in Euro-American postcolonial contexts—enough so that it is crucial to treat them on their own terms. For instance, although I focus on museums and festivals, two modes of abstracting culture that have often featured in theories of Euro-American cultural modes of domination, I am mindful that Russian imperial and Soviet museums and festivals were significantly different in the kinds of expertise they deployed, the modes of abstraction and classification they preferred, and the ways in which they mobilized people (on museums, see Knight 1998; Hirsch 2005; on festivals, see Petrone 2000).

In the second place, the general analytical approach that I have adopted in this book inclines me to look for the ways in which material qualities continue to inflect even highly abstracted domains. Lukoil-Perm's role in fostering local cultural revivals was more than simply providing the funds that underwrote festivals or encouraging monetized markets in folk handicrafts. It mattered in a quite substantive and, indeed, quite material way that an *oil* company was carefully integrating and scripting these activities in response to critiques of its operations. The cultural dimensions of Lukoil-Perm's CSR, I will show, highlighted one particular material quality of oil—its geological depth beneath the Perm region—and associated this depth with the legitimacy and authority of something else that could be considered deep and valuable: culture. Neither the qualities of oil nor the particularities of local culture were washed away by abstractions, commodification, or translations into exchange value; indeed, those qualities were absolutely central to the way these processes worked at the regional level.[8]

8. I refer readers interested in this argument about the materiality of corporate social responsibility projects to Rogers 2012, which provides a somewhat more extensive theoretical discussion. For additional discussions of the ways in which attention to materiality can challenge common assumptions about abstraction in ethnographies of capitalism, see Yanagisako 2002; Keane 2008.

The Depths of Postsocialism

In July 2004, the town of Ashap, in the oil-producing Orda district, hosted the Perm region's first Festival of Children's Folk Handicrafts with support from Lukoil-Perm and the regional state administration. Nikolai Kobakov, the general director of Lukoil-Perm at the time, was in attendance, and he reacted as follows to a reporter's comment that many people at the festival were thanking the company for its aid:

> We are not providing aid (*pomoshch'*). We are returning our debt to the Perm region. There is the opinion, circulating with help of the media, that we are acting in the role of colonizers, that our task is to dig down to the bottom of the subsoil [*vycherpat' do dnia nedra*] of the Perm region and then happily live in some other separate Palestines. We are not aiming for that! Our homeland [*rodina*] is the same as yours—the Kama River Basin.[9]

Kobakov's response clearly frames the children's festival as part of a corporate response to critique. Note, however, that it does so in language that ties the cultural festival that he was attending both to Lukoil-Perm's subsoil operations and to the company's preferred view of the changing space of the Perm region—a space in which the oil company was not simply enriching itself in separate enclaves, but was fully engaged in the project of building a common, inclusive, regional homeland.

The head of the Orda district, speaking on the same occasion, echoed this language of depth:

> We are delighted that the oil industry organized [a festival celebrating seventy-five years of Permian oil] on our territory. It's not a coincidence that the festival was held in our district. More than four hundred thousand tons of oil are pumped from our territory every year, and our traditional folk crafts are recognized not only in Russia, but in many foreign countries as well.[10]

And consider the comments of the head of the Usol'e district, at yet another summer 2004 festival:

> I am especially delighted that Usol'e's birthday has also become a feast day for the oil industry, which is celebrating the seventy-fifth anniversary of Permian oil. In our district in 2004 we estimate that 600 million

---

9. Val'demar Pyrsikov, "75 let Permksoi neft'," *Zvezda*, July 29, 2004.
10. Ibid.

> tons of oil will be extracted—more than 16 percent of the Perm region's production. . . . All earlier distortions and excesses notwithstanding, residents of Usol'e are preserving the historical and cultural traditions of the past.[11]

Such links among the oil-rich subsoil, cultural revival projects, and the spaces of the Perm region became frequent in 2004 and only expanded in subsequent years. A newspaper overview of Lukoil-Perm's new and expansive commitment to social and cultural projects, for instance, concluded by riffing on an old Russian proverb: "Among the ancients there was a saying: 'Don't bury your talent in the ground.' The oil industry, as it extracts natural resources from the ground, is allowing the richly talented residents of our region to not 'bury their talents in the ground.' And that is good."[12] By 2009, it made eminent sense for the small city of Lys'va, southeast of Perm, to stage a yearlong celebration of local history and culture under the slogan, "Lys'va—A Deposit of Culture." *Mestorozhdenie*, deposit, literally "birthplace," is a term used for underground oil and gas reservoirs; it sounds odd, yet certainly intelligible, when applied to culture.

In all these examples, depth as a material quality of oil, along with the infrastructures built to extract and transport it, features prominently in oil's reincarnation as a sponsored cultural project in an oil-producing district. These were not simply the rhetorical flourishes of the Connections with Society Division's public festival scriptwriters or company-friendly reporters at the regional newspapers. They were closely entangled with the processes of cultural production and consumption sponsored and funded by Lukoil-Perm across its rural oil-producing districts, and they ran the whole gamut of culture projects: from folk handicrafts and festivals to museum and library exhibits.

Producing Oil, Producing Culture and History

Oil had been discovered in the Krasnovishersk district, in the far north of the Perm region, in the 1950s—recall the underground nuclear detonations that, it was thought, might alleviate socialist shortage by reshaping oil reservoirs—and the district was one of the most aggressive in pursuing control over its own oil deposits amid the parcelization of sovereignty in the 1990s. By the early 2000s, however, Lukoil-Perm had gained control of the district's oil operations, and the Krasnovishersk library submitted an application for a project titled Permian Oil: Past, Present, and Future. The application, ultimately successful, planned a series of visits to the district's oil production facilities to take pictures and talk with oil

11. Val'demar Pyrsikov, "Veselis', Usol'e," *Zvezda*, August 19, 2004.
12. Dmitrii Krasik, "Zemlia talantami bogata," *Zvezda*, August 3, 2004.

workers. These materials would then be featured in publications in the district newspaper and a display in the library about labor dynasties of oil workers. (Labor dynasties were families in which successive generations worked in the same industry, and they were a staple of Soviet rural library and museum displays celebrating socialist successes. In the Soviet period, they were typically selected by local representatives of the Communist Party, and featured in various efforts to stimulate labor productivity [see Rogers 2009, 136–46].)

In returning in new conditions to the once familiar project of displaying labor dynasties in the district library, the culture workers of Krasnovishersk not only worked through new procedures of corporate grant writing rather than the Communist Party structures that had previously identified model workers. They also had to confront the fact that the oil industry had never been particularly highly esteemed in the Soviet period, and it was not immediately obvious which local family might be suitable for designation as a labor dynasty. A sign hung in the library in the early stages of their project:

> Respected Residents of Krasnovishersk!
>
> Among those with the title "Honored Citizen of Krasnovishersk" there are people of various professions: paper factory workers, diamond-drillers, road-builders, teachers, doctors, and so on, but there is not a single representative of the oil industry. We therefore ask you to help us select the most worthy oil worker to be named an "Honored Citizen of Krasnovishersk."[13]

The library eventually selected the Antipin family as an oil sector labor dynasty, including a grandfather who was born in 1951 and "stood at the source of the discovery of oil in our district" and later worked for a Lukoil drilling subsidiary; a son, born in 1974 who was at that point an oil engineer at Lukoil's West Siberian Langepas field; and a granddaughter, who had finished grade school in 2004 and "dreams of enrolling at the Perm State Technical University in the Geology Department."[14]

In the case of the Krasnovishersk library's project, the mode of exploring the depths of oil and culture was genealogical: it grafted family history onto corporate history onto district history, aiming to displace both critiques of the corporation and other industries or professions that might have vied for Lukoil-Perm's central position in district affairs. Indeed, an application from the newly founded charitable organization Visher Heritage, submitted to Lukoil-Perm's Ecology grant competition that same year—2004—aimed to put together a

13. "Permskaia Neft′: Vchera, Segodnia, Zavtra," PermGANI f. 1206, op. 1, d. 529.
14. Ibid.

mineralogical display with funding from both Lukoil-Perm and Ural-Almaz, the local diamond-extraction company. The project was partially funded by Lukoil-Perm, but only, it seems, on the condition that the project's name be changed from that on the original application—from The Museum of Diamonds (which Ural-Almaz had agreed to support) to the more general Museum of the Krasnovishersk Subsoil (*nedr*).[15]

If a genealogical logic informed the display in Krasnovishersk's library, then other temporal languages of depth were possible, too. An official history of Lukoil-Perm, published in 2009, took a geological route. It was titled *The Permian Period: Vagit Alekperov and His Team* (Neroslov 2009). The title plays on the Permian geological period (the last period of the Paleozoic) that, as every local schoolchild knows, shares its name with the Perm region. By linking the Paleozoic to the postsocialist, the book's title nicely encapsulates its overall goal: inserting Lukoil-Perm as inextricably into the historical, economic, political, and cultural fabric of the Perm region as possible. Indeed, it projected the company's Permian characteristics back to a geological time when the earth's oil deposits were still forming: oil and culture, depth and depth.[16]

Versions of both genealogical and geological depths were on display in the regional studies museum in Kungur, a city south of Perm. The museum had been the recipient of generous grants from Lukoil-Perm for several years, extending beyond the renovation of the existing regional studies museum to include the opening of a branch museum dedicated to the history of the provincial merchantry. When I visited, the regional studies museum included the standard progression of exhibit rooms dealing with nature and archaeology, local peasant cultural traditions displayed through mannequins in traditional dress and traditional tools and implements, and an exhibition about Soviet-era industry in Kungur and the Kungur district (which made scant mention of oil). The final and most recently constructed exhibit room—bright and modern to the point of appearing sterile—was dedicated entirely to the district oil industry (see fig. 7).[17] It prominently displayed a miniature oil pumpjack right in the middle of the

15. "Ekologicheskii Turizm i Prirodnoe Nasledie," PermGANI f. 1206, op. 1, d. 532.

16. A similar strategy was adopted in *The Story of Permian Oil: Pages of History* (Gasheva and Mikhailiuk 1999), published with the support of OOO Lukoil-Permneft, which divided its chapters by historical decade, terming each a stratum (*plast*). In contrast to the oil temporalities reported by Limbert (2010) for Oman, in which a projected fifteen-year window of reserves placed an end to oil-funded cultural production in the not-too-distant future, Lukoil-Perm projected a bright future for many decades to come—even if that rosy view was often doubted by residents of the Perm region. Compare Rajak 2014 on the invention of corporate histories.

17. This museum design is strikingly similar to the representational strategy that Dolly Jørgensen discerns in aquariums on the United States Gulf Coast, where oil rigs and other industry infrastructure are frequently incorporated directly into exhibits of the oceanic environment, thus naturalizing the oil industry's presence in these spaces and telling a story of the Gulf of Mexico as a "harmonious meeting place of oil and water" (2012, 480).

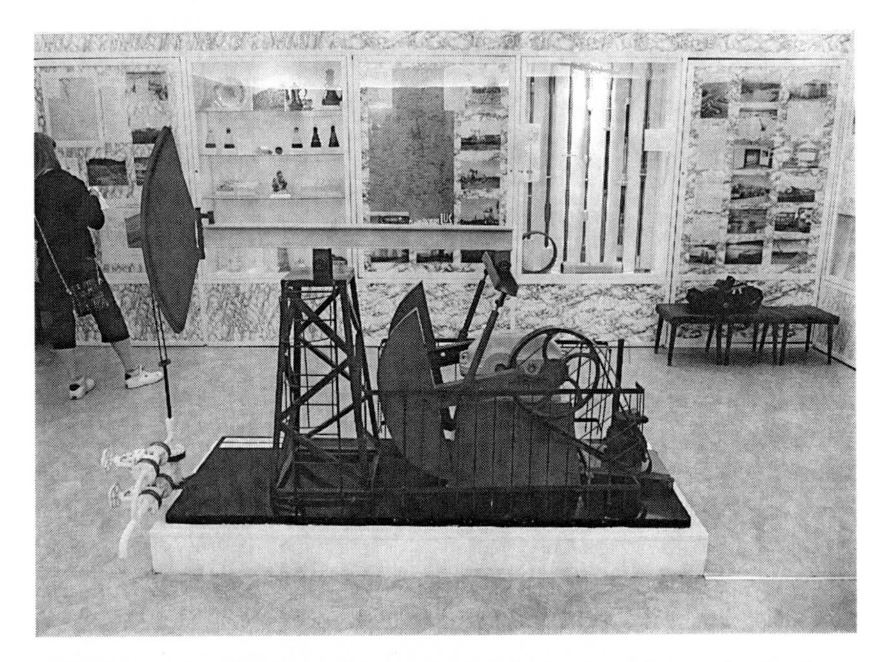

FIGURE 7 An exhibit about the district oil industry at the Kungur Regional Studies Museum. The exhibits on the back wall include samples of oil, lengths of drilling pipe, and maps and photographs of oil deposits and oil industry activities around Kungur.

Photo by author, 2010.

room, and the surrounding exhibition cases all displayed items from the depths: samples of oil from the region, lengths of drilling pipe, and cross-sectional maps of the geostratigraphy of local oil deposits. Although the exhibit avoided the language of labor dynasties—a term more commonly used in rural, less professional exhibits and displays—the history of Kungur's participation in the regional oil industry was crisply narrated in photographs, maps, and text alongside the objects on display.

To visit this museum was to experience Lukoil-Perm as dominating the most recent stage of history, the last in a succession from ancient artifacts to the present day, represented in successive rooms and display cases. As was the case with the festivals and others programs that brought visitors to the Perm region's oil-producing districts, it was to encounter Lukoil-Perm as chief regional specialist in geological, historical, and cultural depths—and in everything from their production to their consumption. This association of deep oil and deep culture was the product of a particular historical circumstance, one outcome of a broad range of corporate initiatives that worked to link oil deposits pumped by

Lukoil-Perm—a corporation registered in 1996—to all the historical depth and authenticity that "culture" in the context of festivals and museum displays could impart. Similar initiatives and their associated publications led to many of the regional oil industry histories that informed my reconstruction of the socialist oil complex in chapter 1.

Fixing Signs: The Corporatization of Postsocialist Signification

At the turn of the twentieth century, Standard Oil commissioned a memorial to Edwin Drake, whose 1859 oil strike in northwest Pennsylvania had opened the first U.S. oil boom. The heart of the memorial was a bronze sculpture, *The Digger*—a crouching, muscular man cast in an overtly classical style with sledgehammer poised high overhead, ready to pound a metal spike into a rock ledge. Art historian Ross Barrett (2012) identifies *The Digger* as the inauguration of a representational strategy that he calls "petro-primitivism," in which the U.S. oil industry sought to link itself to a deep historical past and to larger cultural discourses about timeless struggles between humans and nature. The Drake Memorial, Barrett goes on to show, was part of Standard Oil's response to late nineteenth-century critiques: ecological destruction in the oil-production counties in Pennsylvania; boom-and-bust cycles; and the company's emergent monopoly. The memorial's "archaizing vision"—and that of many other projects like it, both at the time and in later years—sought to "recode the cultural image of the oil industry" by drawing viewers' attention to much longer geological and historical time spans and casting the extraction of oil in a heroic, community-supporting mode (Barrett 2012, 397, 412). (An exhibit sponsored by Sinclair Oil at the Chicago World's Fair in 1933–34, Barrett notes, featured animatronic dinosaurs—a representational strategy similar to Lukoil-Perm's corporate history *The Permian Period*.)

If the version of petro-primitivism on display in Lukoil-Perm's sponsored cultural projects had some analogues in late nineteenth-century northwest Pennsylvania, its more proximate building blocks are to be found in widely circulating talk about depth in 1990s and early 2000s Russia. Indeed, things from the depths and the quality of depth more generally were central to aspects of the postsocialist experience before Lukoil-Perm arrived on the scene, in times when festivals and new museum displays were far from most people's minds. Former state farm employees I knew often joked about digging up and putting to use the items their great-grandparents had buried rather than allowing them to be collectivized in the late 1920s and early 1930s. Not long after the end of socialist regimes, phalanxes of dead bodies emerged from the ground—both literally and figuratively—to feature in struggles over history, land, and political power at all levels (Verdery 1999). So, too, did buried sacred objects, among them the texts and treasures that allowed what Anya Bernstein calls the "deepening of history" (2013, 96–98) through Buryat Buddhist religious revival. In those same years, as they were confronted with new uses for money and new kinds of commodities, many Russians agonized over the surfaces and deeper essences of things, finding in this distinction a rich language with which to apprehend and navigate a variety of social and cultural transformations (Lemon 1998). Depth also featured prominently in the proliferating talk about the Russian soul so evocatively traced by Dale Pesmen (2000). In talk of having a soul, invocations of depth were closely linked to widely circulating stereotypes of Russian national character and distinctiveness: the deeper the soul, the more authentic, the more authoritative, the more Russian.[18]

That "culture" can be treated as an object with qualities like depth is a common enough phenomenon in conditions of modernity (e.g., Handler 1988), and this process took on some particular shapes in the socialist project and its aftermath. As Bruce Grant put it in his study of the Nivkhi of Sakhalin, culture in the Soviet Union was "something to be produced, invented, constructed, and reconstructed" (1995, xi). After of a century of such projects that whipsawed between framing Nivkhi culture as socialism's other and as socialism's exemplar, between "their" culture as deep past and bright future, the Nivkhi Grant knew saw themselves as living among ruins—crumbled, discredited, meaningless bits of objectified and discarded culture. Much the same could be said of the 1990s state of cultural construction in the Perm region. The association of deep oil and deep culture in the Perm region was made in the 2000s out of building blocks—or ruins—present in the first postsocialist decade. But it took a capitalist corporation working hard to respond to critiques to create the conditions under which an acquaintance of mine could respond to my news that I had just come from one of Lukoil-Perm's folk handicrafts exhibitions with the comment, "Of course—where there are folk handicrafts, there is Lukoil." Her smile, like the minor awkwardness of the slogan "Lys′va—Deposit of Culture," points to the recently cobbled-together nature of this relationship.[19]

---

18. Surfaces and depths are, of course, also central to an important strand of interpretive social science (e.g., Geertz 1973; Alexander 2008). I leave a metatheoretical discussion of how this approach fits with the postsocialist context and the analytic approach I take in this book for another day, except to reiterate that my analysis does not rest on an account of how viewers or participants interpreted the cultural events they were attending. This is an extremely important issue, but it is not my focus here. For an incisive treatment of the semiotics and politics of depth in a quite different cultural context, see Collins 2011, and for a fascinating characterization of the relationship between oil and American culture as "ultradeep," see LeManager 2014.

19. There is also good precedent in Russian literary and cultural history for working with representations of *nedra* to imagine human-nature relationships (see, e.g., Costlow 2004 on Turgenev's *Poezdka*

We can trace this cobbling fairly precisely to the Lukoil-Perm Connections with Society Division. Recall that, in 2001–2, Andrei Kuziaev had hired well-connected players from the former party-state apparatus to facilitate the company's networking and relationships within the Perm region. As the idea of fostering folk handicrafts became an important aspect of those efforts, these members of the old factory-based Party elite looked, in turn, to experts in cultural construction, among them my friend Oleg Leonidovich Kut′ev. One of our first conversations took place in the fall of 2000, as we walked around Khokhlovka, the Perm region's open-air museum complex. We spoke about the state of museums in Russia, and I recall his telling me that Russia's new rich had not yet decided that it was time to leave a legacy behind in the way that the Fords and Rockefellers and Carnegies had. Until they did, he said, museum funding would suffer. In 2003, Oleg Leonidovich left the Department of Culture to take up a new position in the Connections with Society Division at Lukoil-Perm, where he helped to oversee the corporation's rapidly expanding grant competitions for social and cultural projects for several years. He moved to Lukoil-Perm in part because of his dissatisfaction with the sinking status of cultural affairs in the regional state apparatus. The future did not look bright. Lukoil-Perm promised almost the opposite—a new and interesting context in which to carry out his life's dedication to regional culture, and one that offered real, tangible, and, by comparison with state work at the time, massive resources with which to do it. The new rich of the Perm region were, at last, turning their eyes to culture.

Indeed, the opportunity to revive culture—rather than any corporate strategy or personal endorsement of Russia's nouveaux riches—drew Oleg Leonidovich to his new job. He struck many compromises along the way and was acutely aware that to work for a corporation meant toeing lines to which he was neither accustomed nor particularly sympathetic. But it was a job he relished. Oleg Leonidovich once told me that his favorite part of working for Lukoil-Perm was spending his days on the road, seeing and managing the cultural diversity of the Perm region. One day he would be in the frozen far north consulting about a planned summer festival that seemed far off; a day (and ten or twelve hours in the car) later, he might be in the south, evaluating a new museum exhibit and watching

v Poles′e) and for building links between extracted energy sources and cultural construction (e.g., Bird 2011, on early Soviet peat). Cultural relationships between geological formations and imaginations of community have long been noted in the anthropological literature on mining (e.g., Ferry 2005), where they often serve as ways to counter or divert corporate strategies. In the cases I explore, corporations themselves are forging these links at substantial remove from local experience at the point of extraction, and then placing them at the center of CSR programs. See Cepek 2014 for an intriguing case in which the idea that oil is the blood of subterranean gods seems to have originated in the context of environmental mobilization and to be perpetuated primarily by faraway environmental activists.

residents plant the first seeds in their gardens. He had never, he said, experienced anything like this in his many years of association with the cultural affairs departments of the regional state administration. In the end, the downsides of working for Lukoil-Perm and close association with the oil industry—including not a few lost and strained friendships ("he's an oilman now," one of his former colleagues said to me, rather disdainfully)—were well worth it for the opportunity to be at the center of the cultural revival he had sought for many years.

Scholars of socialist and postsocialist societies have paid ample attention to fixers and operators like Oleg Leonidovich and his superiors in Lukoil-Perm's Connections with Society Division (e.g., Ledeneva 2006). They have not yet, however, appreciated the extent to which, when transplanted into the world of corporate capitalism, these figures have devoted their skills to resignifying the meanings of material objects as much as facilitating business deals. For this is what Oleg Leonidovich and his colleagues did, now in oil-boom conditions of almost unimaginable plenty rather than the socialist and postsocialist shortage to which they were accustomed. For the four years Oleg Leonidovich worked at Lukoil-Perm, the new association between the depth of oil and the depth of culture was, in significant part, forged by him, his colleagues, and the networks they facilitated: in the grants for local culture they helped to write and award, in the museums they helped to build, in the festivals they helped to organize, and in the ways all this reshaped the politics of the Perm region's oil-producing districts.

This was, in fact, a three-dimensional move. The first dimension, as I have described, involved emphasizing the significance of oil against other salient objects, especially those hard-times agricultural staples that had become the hallmarks of postsocialist involution and demodernization. Nancy Ries shows beautifully, for instance, that the potato was a quintessential "ritualized mode of activity, exchange and negotiation within families, networks, and communities" (2009, 183) in and after socialism. Lukoil-Perm's cultural sponsorship sought to build new rituals, exchanges, families, networks, and communities—all of them gathered around oil. (Recall the labor dynasties of Krasnovishersk.) But oil already had other qualities and associations, not all of them positive. So the second dimension highlighted the geological depth of oil beneath the region as against toxicity, convertibility into massive wealth, and inequalities of exclusion, qualities of oil that had become central to critiques of the industry. And the third dimension, accomplished through omnipresent sponsorship, entailed borrowing the authority and legitimacy that might be associated with the depth of culture. The local intelligentsia in the Perm region's districts—those writing and receiving grants and executing new cultural programs—was the Connections with Society Division's crucial partner in this effort.

Oleg Leonidovich's first job as a young man was at the small regional studies museum in the town of Il'inskii. Decades later, under his watchful eye at Lukoil-Perm, Il'inskii became quite successful in grant competitions, receiving funds to remodel the library, start a folk crafts center, and support annual festivals celebrating local culture. Perhaps the most significant grant-funded addition to Il'inskii has been a newly renovated and expanded 1,600 square meter regional studies museum dedicated to the famous Stroganovs, whose large family estates in tsarist-era Perm province were administered from offices in Il'inskii. As we walked through Il'inskii with a mutual friend in 2008, Oleg Leonidovich was greeted solicitously at nearly every turn. He answered questions about when the library's most recent batch of construction materials would be appearing and inquired how sales were going at the folk arts and crafts center (founded with support from Lukoil-Perm). A town priest eagerly showed us the most recent renovations to his church, also supported in part by a grant from Lukoil-Perm. Our tour of the new regional studies museum revealed expensive glass cases, spotlights, and video screens the likes of which I had never seen in rural Russia and, I was told by our tour guide, had begun to win federal awards.

As in the examples discussed above, depth as a shared feature of oil and culture featured prominently in these projects. As a complement to the new regional studies museum, Oleg Leonidovich published a book to commemorate the town's 425th anniversary in 2004. *Il'inskii: Pages of History* charts the town's beginning in the prerevolutionary period and concludes with a chapter on "Il'inskii's Big Oil," an extensive account of the geology and labor history of oil production in the district. This book sought to accomplish much the same work as the official Lukoil-Perm history *The Permian Period*, mentioned above, but at a district rather than regional level. If, that is, the Permian qualities of oil featured as an element of critique of Lukoil-Perm as a corporation in the 1990s, a decade later they were redeployed in a much different way through CSR projects: Permian oil and Il'inskii's oil became chapters in the recounting of the best and deepest of regional and local cultural traditions.

My research has not turned up a deliberate semiotic branding strategy built around depth and culture in the Connections with Society Division's efforts, at least in the early days that I focus on here. Indeed, the story as recounted to me by many of its key players was one of improvisation, of experienced Soviet-era networkers seizing opportunities in a new corporate context, with massive funds and a mandate to remake the corporation's relationships with the region. Although a language of corporate social responsibility imported from Western corporations played a role, as did the technology of grant competitions borrowed from Western aid agencies in the Perm region's Capital of Civil Society era, the central focus on shifting the relevant qualities of oil that I have described here emerged largely from an improvised repurposing of what was at hand: the ruins of Soviet cultural construction and a pervasive concern with the surfaces and deeper essences of things—from souls to dollar bills—in the 1990s. Culture, that is, also had to be revived from its post-Soviet ruins and placed in a new position of authority; in the Perm region in the early 2000s, Lukoil-Perm's CSR projects were the primary vehicle for this revival, at that point far outstripping the efforts of the regional state apparatus even in those cases where state agencies also provided funding and staffing.

Studies of signification in the early post-Soviet period theorized the symbolic vacuums and indeterminacies of the 1990s, even going so far as to diagnose a condition of "aphasia"—a loss of metalanguages of identification coupled with a general absence of workable post-Soviet signifiers (Oushakine 2000). A subsequent set of studies charted ways in which states sought to reestablish themselves through the control and channeling of meaning, including through state-sponsored cultural production (Adams and Rustemova 2009; Denison 2009; Adams 2010). Lukoil-Perm's cultural projects demonstrate that corporations also powerfully shaped the field of signification in this era, seeking to bend it to their own interests and to those they shared with regional state agencies.

Deep Oil and Deep Culture on the Map of the Perm Region

Lukoil-Perm's culturally focused CSR projects increased in scope and scale, gradually coming to occupy the very center of the company's relationships with oil-producing districts of the Perm region (see map 2 in chapter 5). Especially because oil companies' taxes did not accrue directly to oil-producing districts after 2004, many district-level politicians came to see answers to their problems precisely in Lukoil-Perm's funding for social and cultural projects. With greater and greater frequency, complaints about joblessness were answered—however ineffectively—with Lukoil-Perm's attempts to create and then make profitable the tourist industry through museums and folk craft centers. Complaints that the Soviet period had left culture in ruins were answered with corporation-funded festivals, celebrations of local culture, and the reconstruction of churches and mosques. District head after district head relied on Lukoil-Perm's social and cultural programs to blunt critiques of their own past work, speaking ever more clearly in a language of cultural distinctiveness and revival.

Speaking with the elected head of one of the Perm region's non-oil-producing districts one day in 2009, I asked why his district did not follow others in applying to become a Cultural Capital of the Kama River Basin in order to bring in tourist money. "We can't be a cultural capital," he replied immediately, "We don't have any oil!" This comment is all the more striking because, by the time we had this conversation in 2009, Lukoil-Perm was no longer the only potential sponsor when it came to facilitating cultural production. Indeed, the Cultural Capital of the Kama River Basin program that I was asking about was run *not* out of Lukoil-Perm, but out of the Municipal Development Office of the regional state apparatus, and was therefore open to all districts in the region. The program was, however, administered by a former Lukoil-Perm Connections with Society employee who had moved from the oil company to the state administration. It may be that the district head with whom I was speaking presumed that this official would continue to make decisions that supported cultural producers in districts familiar from his time at Lukoil-Perm. Or he may have been all too aware that the local intelligentsia in oil-producing districts simply had a decade's worth of experience in writing grants that experts in his own district would find hard to outcompete. Whatever the case, the expectation that oil and culture were to be found together had become so strong that it outlasted Lukoil-Perm's initial near monopoly on cultural sponsorship.

This spatialization of culture, such that it rested largely and in a quite materially specific way on the region's oil deposits, was quite different from both the Soviet period, when unified party-state organs sought to apply a somewhat more evenly distributed set of social and cultural technologies to the population—and certainly one that did not closely track the spaces of oil production—and the early post-Soviet period, when neither major corporations nor state agencies were in much of position to embark on such explicit projects. These two earlier periods, however, provided some of the crucial ingredients—from objectified senses of culture to political networks to widely circulating talk about depth—out of which experienced operators like Oleg Leonidovich and his colleagues in the Connections with Society Division could, in new corporate contexts, usher in a new, more abstracted, and vertically integrated cultural field. Bruce Grant has shown how long-running myths and narratives about the Caucasus that circulate through genres of Russian popular and official culture work as an "art of emplacement." That is, they "generate a powerful symbolic economy of belonging . . . that naturalize[s] violence and enables diverse Russian publics to frame their government's military action [in the Caucasus] as persuasive" (2009, 16). Lukoil-Perm's cultural CSR projects, inasmuch as they sought to naturalize Lukoil-Perm's activities by inserting them deeply into the spatial and temporal coordinates of the Perm region, are usefully understood as younger, corporate cousins to the more diffuse and older arts of emplacement Grant describes.

"The Riches Are for the Fatherland, The Name Is for Ourselves"

We can discern still more about the workings of this state-corporate field by exploring more precisely which layers of Russian history and culture Lukoil-Perm and its grant applicants were most interested in extracting, processing, and making available for consumption. What, in other words, were the kinds of relationships and imagined collectivities into which Lukoil-Perm's culturally focused CSR invited residents of the Perm?

In the 1990s, with a weak federal center, local currencies proliferating, enterprise bosses more powerful than in anyone's memory, and millions of people feeling tied to the land, talk of a return to feudalism was common across the postsocialist world (Humphrey 2002, 5–20; Verdery 1996, 205–7). The Perm region was no exception (see, e.g., Rogers 2009, 223–45). We might expect that these reappropriations of the feudal past would begin to dissipate with the consolidation of a more unified state-corporate field beginning in around 1998–2000. Consider, however, a 2001 interview with Andrei Kuziaev. In response to a question about whether he was offended that many were beginning to call him an oligarch, Kuziaev replied:

> No. What haven't I been called in the last decade? An upstart, a Komsomol businessman, a bourgeois stockbroker, a New Russian . . . and now they're calling me an oligarch. Tomorrow, maybe, something else, maybe an oil baron. I'm not offended. I think a businessman doesn't have a right to take offense at anyone or anything, since that would make it impossible to work. . . . What is important is the final evaluation of someone, not the current one. So we should do everything we can now to ensure that, in the future, people recall us with the same respect that they now recall the Stroganovs, the Demidovs, the Liubimovs, and the Gribushins![20]

For much of the twentieth century, the Stroganovs and Demidovs—the land- and factory-owning noble families most closely associated with old Perm province—were not recalled with much respect in official Soviet discourse and were otherwise often forgotten. But when Kuziaev needed a name for the invitation-only club for the wealthiest and most influential residents of the Perm region that he founded in 1998, he chose "The Stroganov Club." The club's periodic meetings and other entertainment events at the boutique hotel Kama—paid for by a Lukoil subsidiary and built by a Swedish construction company—were thought by many

20. Grigorii Volchek, "Gol kak sokol," *Zvezda*, January 16, 2001.

observers to be the most important site for backroom decision making and deal making among the Perm region's industrial, financial, and political elite.[21] Nor was Kuziaev the only member of the region's new oil elite to latch onto the Stroganov name and reputation. Nikolai Kobakov, while serving as head of OOO Lukoil-Permneft and explaining the company's new interest in the sponsorship of social and cultural projects in 2003, explicitly cast the company's CSR projects as reviving a regional tradition of private sponsorship of arts, culture, and religion that had been interrupted by the Soviet period. "The Stroganovs," he said,

> had a marvelous model: "The riches are for the fatherland, the name is for ourselves: (*bogatstva—otechestvu; imia—sebe.*) And if, in recent times, we are opening up more and more [through sponsorship of social and cultural projects], then I hope it is all done for the benefit of thousands of people—our oil workers, whose labor helps the region and the whole of Russia bring market reforms into being.[22]

At least some recipients of Lukoil-Perm funding for social and cultural projects also drew on this comparison to pre-Soviet models of patronage. The final report to Lukoil-Perm from the Russian Orthodox Church's Perm Spiritual Academy, following the completion of a project called The Beauty of Orthodoxy: For the People of the Kama River Region that had aimed to revive the sixteenth- and seventeenth-century Stroganov School of icon painting, read in part as follows:

> Today, when secular art is spiritually and morally empty, it is especially important to revive and strengthen sacred art. . . . We would like to give our deep thanks to the leadership of Lukoil-Perm for the enormous help that was provided in realizing this project. The cultural heritage famous in the lands of the Urals was accumulated in large part thanks to the trusteeship (*popechitel'stvo*) of the Stroganovs. Traditions of trusteeship . . . also need to be revived in our days. The Russian Church and Russian culture need enlightened patrons. Today, with the means provided by charitable giving, churches are being restored, icons are being painted, bell towers are being raised. It is encouraging that the revitalization and development of the Stroganov School of icon painting came about with the help of charitable giving.[23]

---

21. See, for instance, "Personalii: Kuziaev, Andrei Ravelevich," *Neftegaz.ru*, October 18, 2009. See also OFPI Bastion 2000.

22. Sergei Zhuravlev, "Bogatstva-otechestvu, sebe—imia," *Zvezda*, February 2, 2003. This slogan was also the motto of the fifth in the Historical Cities of the Kama River Region festivals, held in Usol'e in 2006. See Maria Molchanova, "V Usol'e, u Stroganovykh," *Zvezda*, July 20, 2006.

23. Proekt "Krasota Pravoslavnaia—Liudiam Prikam'ia. Permskaia Dukhovnaia Uchilishche. 2006," PermGANI f. 1206, op. 4, d. 289. Although my project did not explore these issues in detail, I note that casting powerful corporations in the role of patrons—a register of moral claims-making that

Press releases and thank-you notes were only a small corner of the ways in which Lukoil-Perm's response to critique reclaimed the legacy of the Stroganovs. The Stroganovs' many factories, residences, and headquarters, spread throughout what are now the central and northern districts of the Perm region, overlapped significantly with the territory of Andrei Kuziaev's ZAO Lukoil-Perm, providing ample opportunities to link the depths of oil being exploited by the company to this particular stratum of regional cultural history. As exhibits and displays of Lukoil's own materials from the depths—oil rigs, beakers with petroleum samples, drilling pipes—began to appear in museums, so, too, did exhibits, and indeed entire exhibition complexes, dedicated to Stroganov-era objects, including both items of material culture and entire buildings that had been neglected in the Soviet period.

I have already noted my friend Oleg Leonidovich's efforts to foster cultural revival in the old Stroganov stronghold (and more recently oil-producing town) of Il′inskii, where a number of sponsored cultural projects were coordinated by The Stroganov Capital, a local noncommercial organization headed by an employee of the regional studies museum. Il′inskii, however, was far outpaced in this regard by Usol′e, which was both a historical center of Stroganov activities in the Perm region and the district seat for one of ZAO Lukoil-Perm's most productive oilfields—the Siberian deposit. The enormous profits being generated in the Usol′e district meant that, even though the district was the least populous of the entire Perm region—with scarcely fifteen thousand residents—it attracted some of the most lavish corporate sponsorship and highest number of social and cultural project grants.[24]

Beginning in the seventeenth century, Usol′e was a hub in the salt extraction and processing industry and, due to its prime location on the banks of the Kama, in river transportation as well. The Stroganov-era offices and churches in old Usol′e were "in ruins—in the direct sense of the word"[25] in 2000, but by the time I visited in 2010 they had been transformed into a major museum and architectural complex. The architectural sites were rebuilt in large part by ZAO Lukoil-Perm's substantial district taxes and direct contributions, and individual exhibits and smaller projects were supported by a series of dedicated grants for social and cultural projects.[26] Usol′e's restored Stroganov ensemble, as it was often

contests the logics of market and capital in which those corporations specialize—is a common strategy pursued by local communities around the world (see, e.g., Sawyer 2006; Shever 2012; and Welker 2014).

24. Usol′e was also helped in this regard by its proximity to the wealthy mining cities of Berezniki and Solikamsk, which were home to many former residents of Usol′e. Given the competition between ZAO Lukoil-Perm and business interests in the north (see chapter 3), Usol′e was also an important point at which Lukoil could project its munificent image through sponsored projects.

25. Dmitrii Krasik, "Usol′e na Kame," *Zvezda*, June 9, 2005.

26. On the reconstruction of ruins in the post-Soviet period, see Schönle 2011, 219–30.

called, boasted 44 separate architectural sites, the largest and most significant of them being the Stroganov Palace—transformed into the Usol′e regional studies museum—and the Transfiguration Cathedral. Usol′e's Stroganov ensemble rapidly become one of the chief contexts in which residents of the northern Perm region encountered their past in a museum context. The number of visitors jumped from 2,400 in 2003 to nearly 60,000 in 2009, and the museum was a common stopping point for exhibits traveling throughout Russia. It was also a major center for scholarship on regional history in the northern Perm region and on the Stroganovs more generally (see, e.g., Shekhmetov et al. 2004–2006).

The reconstruction of the Stroganov complex at Usol′e began in precisely those years that the Perm region as a whole was consumed with Governor Trutnev's efforts to transform it into the Capital of Civil Society. It became, in fact, something of a proving ground for the project movement that emerged from state and corporate collaboration in those years. The district's high tax revenues coupled with its low population meant that new projects had a higher chance of succeeding in the Usol′e district than in much of the rest of the Perm region. If the project movement was going to work as a form of state-corporate governance, that is, it would work first in a place in Usol′e. The district thus became home to some of the Perm region's most high-profile experiments with local project organization, councils and conferences, grant writing workshops, and others experiments. The number of societal organizations in and around Usol′e rose from four to eighteen between 2000 and 2006. 2006, the four-hundredth anniversary of Usol′e, was also the peak year for grant awards to local organizations in the district, with over 2.2 million rubles received—a more than fivefold increase from the 400,000 rubles received in 2003. The Stroganov complex itself was the site of a new experiment in cultural preservation in the Perm region: the establishment of an independent nonprofit organization called The Board of Trustees of the Historical-Architectural Complex Usol′e on the Kama, composed of the directors of district companies and specifically dedicated to cultural sponsorship, revival, and the development of the district's tourism industry.

If one of the guiding premises of Trutnev's Capital of Civil Society project was the search for a model of statehood and social development not generated entirely by international foundations and Western transitologists, then the Stroganovs were good candidates for becoming that model. Indeed, the quest for historical analogues to the kinds of state forms that emerged in the 2000s was very widespread in Russia, including in the Kremlin itself (Humphrey 2009). Where better to explore those possibilities for the Perm region than in Usol′e, surrounded by the restored walls of the Stroganov corporate offices, in one of Lukoil-Perm's most lucrative districts, amid exhibits composed of objects drawn from the Stroganov past? One of the first events held in the Usol′e museum complex, in fact, was a roundtable discussion about corporate sponsorship and corporate social responsibility hosted by Lukoil-Perm. The attendees—representatives of businesses, state agencies, and a variety of local organizations—took detailed historical-cultural tours of the restored site so that the "local atmosphere would sink in" before embarking on the business of discussing how to continue developing social and cultural projects competitions. A newspaper report on the gathering described this atmosphere in familiar terms of cultural and historical depth, noting that Usol′e was a place "where rich cultural traditions of singing, iconography, and architecture developed. And in addition to all this [the district] is second in oil production only to the Kueda district."[27]

Two years later, in 2006, the fifth in the series of Lukoil-sponsored Historical Cities of the Kama River Region festivals was held in Usol′e, complete with its standard components of new museum exhibits, folk crafts show and sale, and specialist roundtables dedicated to the ongoing improvement of social and cultural projects as a technique of district and regional development. The theme of the entire festival, held on the four-hundredth anniversary of the founding of Usol′e, was "From Stroganov Traditions of Management [*Khoziastvovanie*] to the Socially Responsible Business of the Twenty-First Century."[28] If it seemed that all that was missing in Lukoil-Perm's efforts to wrap itself in the legacy of the Stroganovs by linking the material depth of oil and the material depth of culture was a real, live Stroganov, this was a comparatively minor obstacle. For the 2006 Historical Cities festival in Usol′e, Lukoil-Perm flew in a special guest from Paris: the last known living member of the Stroganov line, the Baroness Hélène de Ludinghausen, herself a major supporter of architectural restoration in Russia through the Stroganov Foundation (Vikkel′, Fedotova, and Iuzifovich 2009, 289–90).

The origins of the Stroganov slogan "The Riches Are for the Fatherland, the Name Is for Ourselves" are not especially clear, but it only began to appear on the family crest in the late nineteenth century, an era as remarkable as the late 1990s and early 2000s for its critiques of inequalities and of the massive accumulation of wealth in the hands of a small elite. There were, to be sure, many differences between the imperial period and the post-Soviet period (the Stroganovs were not known for their grant competitions), but Lukoil-Perm's embrace of the Stroganov legacy points us to at least one important similarity: out in this province of the Urals, at any rate, the power of the Russian state has long been inextricably tied to the power of regionally influential notables—to such an extent that it makes little sense to talk about one separately from the other. It was that legacy, as much

27. Krasik, "Zemlia talantami bogata."
28. Molchanova, "V Usol′e, u Stroganovykh."

as a legacy of philanthropy, that Lukoil-Perm's revival of the Stroganov name in the 2000s bespoke.

Lukoil-Perm and the "Friendship of the Peoples"

My conversations with Oleg Leonidovich about Lukoil-Perm's social and cultural projects usually began with examples from his native Il′inskii or from other northern oil-producing districts with which he was most familiar. But they often moved quickly south, either by way of comparison or in tracing his own increasing familiarity with cultural projects across the full spread of the Perm region's oil-producing districts. "In the south," he explained on one occasion, "folk handicrafts were important from the perspective of multinationality (*mnogonatsial′nosti*), that is, self-identification. There are Tatars, Bashkirs, Udmurts, Maris, and others there. And they all eagerly applied for those grants." Indeed, Lukoil-Perm's grant competitions always included nomination categories designed to encourage projects aimed at reviving and preserving non-Russian as well as Russian ethnic and national identities. In addition to the large Culture and Spirituality category—which funded both Christian and Muslim communities—each grant competition year featured an entire category dedicated to national cultural traditions. In 2005–9, for instance, The Preservation of the National-Cultural Uniqueness of the Peoples of the Kama River Basin category funded ethnographic expeditions, museum displays, festivals, folklore ensembles, and handicraft operations focused on studying "the peoples of the Perm region."[29]

If resonances between the depths of oil and one particular stratum of the deep Russian past—that occupied by the Stroganovs—were particularly dense and compelling in the north of the Perm region, they were far less so in the south, outside the Stroganovs' old domains and where Tatar and Bashkir communities' relationships with the imperial Russian state and its noble families often gathered around the politics of civilizing missions to the *inorodtsy* ("aliens," those of other origin [see Werth 2001]). Given this history, Lukoil-Perm's sponsorship of handicrafts and other forms of cultural production in the south frequently

29. Soviet nationality policy made elaborate distinctions among different classifications of "peoples," depending on a whole host of variables and shifting considerably over time (see, e.g., Shanin 1986; Slezkine 1996). Although these classifications and their post-Soviet trajectories continued to matter a great deal in some domains of state policy and cultural politics, I have not discerned any appreciable impact on the shape of Lukoil-Perm's funding priorities, and so I do not emphasize these distinctions (e.g., whether or not a group was a "numerically small" nationality) in my discussion here.

eschewed the imperial period altogether, recalling instead those epochs of Soviet nationality policy that sought to create a "friendship of the peoples" working in harmony on a common project of advancement (cf. Adams 2010, 141–44). To be sure, gone were an overarching Marxist ideology (critiques of bourgeois nationalism were certainly out of place at Lukoil-Perm) and an encompassing socialist political economy. Gone, too, was any suggestion that differences among nationalities would be eliminated in the course of future development. Although in more muted ways than the celebration of the Stroganov legacy, we find in this subset of CSR projects, as Francine Hirsch did for the pre–World War II Soviet state building, the importance of expert ethnographic knowledge, the showcasing of abstracted cultural "ways of life" (*byt*) of "the peoples" through museum display, and the use of mobilizing, nationality-themed festivals in the service of consolidating power and legitimating rule.

CSR projects also came to the south by a somewhat different route than they had to the north. The Perm region's southern districts, recall, were home to the region's heritage oil company, OOO Lukoil-Permneft—the poorer, less self-consciously dynamic, and less globally oriented sibling of the northern ZAO Lukoil-Perm. Whereas the state-corporate alliance that grew up around grant competitions in the early 2000s was hatched largely at the intersection of ZAO Lukoil-Perm's Connections with Society Division and the regional state administration under Governor Trutnev, these techniques were initially unfamiliar in the south. OOO Lukoil-Permneft had, of course, long provided resources to a number of cultural organizations on its territories, usually through informal contacts and local patronage ties. If money was needed to help rebuild a mosque in the 1990s, kinship, workplace, or other channels were the main routes along which company funds could be unlocked. Corporate funds were available, that is, but their dispersal along patronage networks kept them largely local affairs, not yet incorporated into larger, systematized, and region-wide techniques of funding, sponsorship, and corporate-state governance.

At the encouragement of its northern sibling and Governor Trutnev's office in the early 2000s, OOO Lukoil-Permneft began to participate in grant competitions, and some of their first grants were focused on the preservation and display of traditional culture. "Ancient Russian customs and rituals are being revived in Chernushka thanks to the help of the oil industry," declared an article in the Perm newspaper *Zvezda* in 2002. One of the employees at the Chernushka regional studies museum was getting married, the article went on to relate, and her coworkers decided that an appropriate gift would be a full traditional wedding held in the museum, using the displays of material culture to help recreate old wedding practices. The event proved so popular and attracted so much attention that the museum allowed other couples to celebrate their own traditional weddings in the museum, to the point where the events were beginning to strain the budget. The solution was a successful grant in the very first round of applications for social and cultural projects sponsored by OOO Lukoil-Permneft, in the nomination category of The Harmony [*sodruzhestvo*] of the Peoples in the Territory of the Perm region.[30] The grant bought new costumes and decorations and allowed the celebration of a number of new holidays in the museum, in addition to weddings. Instead of closing on holidays, the museum began the practice of staying open and serving as a gathering place for those interested in reclaiming old traditions and rituals.[31] Based on his long-term research in Bulgaria, Gerald Creed (2002) has drawn attention to the ways in which the increasing financial burdens of postsocialist transformations in rural areas were associated with a decline in ritual activity and an erosion of the kinds of sociability associated with it. Although I do not have the ground-level ethnography to engage specifically with this question in the Perm region, it is not too big a leap to surmise that OOO Lukoil-Permneft's indirect financial support for life-cycle and other sorts of rituals through grants to museums like the one in Chernushka would have made the company especially popular with local residents.

As cultural projects gradually spread to the south, so too did the experts who had developed them and who were, like my friend Oleg Leonidovich, largely unprepared for what they would find. One Lukoil Connections with Society employee with experience largely in the north recounted to me that he was initially skeptical when a cultural affairs worker from the Barda district—a majority Tatar-Bashkir district in the far south of the region—suggested that, with the help of a grant from Lukoil-Perm, there might be as many as thirty thousand Tatars at the district's 2003 summer Sabantuy festival. Somewhat over thirty thousand participants turned out in the end, astonishing the Connections with Society employees from the north and immediately reaffirming their decision to focus the company's sponsorship attention on local cultural identities.

The revival of these national traditions in the southern districts of the Perm region had, of course, been under way since the end of the Soviet period—particularly because of the strong links between Tatar and Bashkir populations in the Perm region and the Russian Republics of Tatarstan and Bashkortostan, two of the county's most vocal in asserting culture-based autonomy. In the first years of grant competitions, when Lukoil's operations in the Perm region were still divided into two subsidiaries, there were two complementary associations of

30. See also "Sodruzhestvo narodov Chernushinskogo Raiona," PermGANI f. 1206, op.1, d. 329.
31. Ol'ga Iakovleva, "Svad'ba v . . . muzee," *Zvezda*, November 30, 2002.

folk artisans sponsored by the company, one in the north and one in the south (Lukoil-Perm 2005, 19). But with the merging of ZAO Lukoil-Perm and OOO Lukoil-Permneft in 2004, these associations merged as well. Houses of Culture and museums in the south found themselves part of the same region-wide festivals, grant competitions, reports, displays, corporate advertising, and other events as those in northern oil-producing districts. *Permskaia Neft'*, the unified Lukoil-Perm's internal newspaper, began to dedicate a substantial number of column inches to the company's social and cultural projects. Reading though its pages of descriptions of festivals and museums up and down the Perm region is not, in fact, a bad guide to the diversity of cultural production in the Perm region in the second half of the 2000s. As they were integrated into the region-wide social and cultural projects competitions run out of a unified Lukoil-Perm after 2004, projects in the south came to share some of the same goals as those in the north, including folk handicrafts production aimed at alleviating unemployment in districts hit hard by economic crisis. Lukoil-Perm funded blacksmiths in the Orda district, Mari traditional dress in the Suksun district, honey production in the Uinsk district, and many others—as well as museums and festivals at which these items were displayed, bought, and sold. A 2004 report on the handicrafts industry from Barda reported:

> Modernity is when people, having studied old folk crafts, can earn some money for themselves. This is how the project "The Folk Artisan," which was chosen in Lukoil-Perm's Third Annual Social and Cultural Projects Competition, was born. It's well known that the oil industry, beginning with the very first competition, is reviving historical traditions in those territories where they work, making good on the slogan, "Lukoil: For the Good of the Perm Region." The goal of the project is to teach folk crafts to the poor and unemployed residents of Barda.[32]

In the southern districts, Lukoil-Perm devoted as much attention to the reconstruction of mosques as churches. A winning project in the Barda district in 2004, The Rebirth of a Mosque Is the Rebirth of Spirituality, included the building of a small new mosque. The applicants described their goal as "the revival of cultural and spiritual tradition of the Tatar and Bashkir peoples, the spreading of the idea of spiritual unity (*dukhovnoe edinstvo*) and international (*mezhnatsional'noe*) agreement." Religious communities in the south would also write their thanks to Lukoil-Perm for its support of revitalization efforts, but these took on a somewhat different valence than in the north (leaving out, for instance, references to the legacy of Stroganov patronage of the Russian Orthodox Church). The orga-

32. "Narodnyi umelets," *Rassvet*, July 28, 2004.

nizers of another mosque-building project in that same year, for instance, wrote in the grant report:

> November 3 is a national feast day for Muslims—Uraza-Bairam. But for residents of the town of Kar′evo that day in 2005 will be a double holiday—a mosque will be opened. Young and old alike await that day—how long they have waited for a mosque in Kar′evo. Hopes and dreams come true. Thanks to grant funding from Lukoil-Perm, believers in Kar′evo will now have the ability to visit a mosque in their own town. The problem of how to preserve national culture, spirituality, and customs is solved.[33]

Although many grant proposals coming from these southern districts were concerned with one or another national cultural tradition, a significant number, echoing their Soviet predecessors, concerned the relationships among them. One of the more intriguing of these came, in 2004, from the Chernushka district, an oil-producing district in which around three-quarters of the population were Russian. A group of cultural workers based in Chernushka, with support and promised cofunding from a range of region-wide organizations, applied for funding to support a Festival of Ethnocultures that would coincide with Chernushka's celebration of the seventy-fifth anniversary of Permian oil. The organizers envisioned an expansive, multiday festival that would "widen interactions among the various peoples of the Kama River region, improve interethnic interactions, and preserve the national-cultural uniqueness of the peoples of the Kama River region." The proposal identified interethnic relationships as one of the most important issues of contemporary life in the Perm region, "one of the most polyethnic regions in Russia," and suggested that the Chernushka district, long a site of interaction among Russians, Tatars, Bashkirs, Udmurts, Chuvash, and Maris, would unquestionably be the best place to hold such a festival.[34]

The festival organizers placed a particularly high value on ethnographic knowledge as central to cultural revival. Their plans built explicitly on their 2003 grant, the Harmony of Peoples in the Chernushka District—the project that had supported holding weddings in the Chernushka museum. In this new, much more expansive project, the organizers envisioned that a number of the Perm region's most established and influential ethnographers would be present to help achieve the festival's mission of being "not just a presentation of the cultural traditions of the peoples of the Kama River region and visiting guests, but a big dialogue-discussion about how we will live in the world in harmony and together preserve

33. "O dukhovnosti i verovanii, s. Kar′evo Ordinskogo raiona," PermGANI f. 1206, op. 1, d. 573.

34. "Chernushka festival′naia," PermGANI f. 1206, op. 1, d. 717.

and respect each others' traditions." A book called *Ethno-Folk of the Prikam'ia* was planned for publication in conjunction with the festival. It would serve as an informational compendium on creative collectives, national centers, museums, and other cultural resources throughout the Perm region, and would allow all these groups to make new contacts among the growing number of cultural producers throughout the region. Finally, the event was projected to help develop tourism in the Perm region and solidify the region's reputation as a multiethnic and harmonious territory.

In the language I have been using here, the Festival of Ethnocultures was a classic case of the vertical integration of culture, uniting production, circulation, and consumption under the banner of Lukoil-Perm. Indeed, in addition to interethnic dialogue and display, the applicants listed among their goals advancing a collaborative relationship among Lukoil-Perm, state agencies, social organizations, and creative and scientific communities in promoting the social and cultural development of the region. Although one of the major components of the festival was The Festival of the Land, holding it together with the major celebration of seventy-five years of Permian oil and with major sponsorship from Lukoil-Perm meant that the focus was just as much on the subsoil as on the land itself. In a familiar juxtaposition of the depth of oil and the depth of culture, Chernushka would simultaneously assert itself as a major oil producing district and a center of interethnic cultural celebration and dialogue.

In Chernushka and other southern districts, even support for Russian cultural traditions was usually phrased in multinational terms. A 2004 project in the Kueda district that supported the construction of a Russian cultural center dedicated to "the revival, preservation, and propaganda of Russian national tradition" noted that the Kueda district was one of the most multinational of the entire Perm region, with more than twenty nationalities represented and only about two-thirds of the region's thirty-three thousand residents identifying as Russians.[35] The need for a specifically Russian cultural center was established in part by noting that there were already functioning Tatar-Bashkir and Udmurt cultural centers. The project envisioned ethnographic expeditions, museum exhibits, scientific seminars, festivals, and other events. A later grant, in 2007, supported the Kueda regional studies museum in a project dedicated to exploring all the districts' nationalities, "from their origins to the twenty-first century" by teaching students about the foods and traditional attires of Russian, Udmurt, Tatar, Bashkir, and Chuvash communities in the district (Lukoil-Perm 2007, 59).

---

35. "Sokhranenie national'no-kul'turnoi samobytnosti narodov Prikam'ia. Kueda," PermGANI f. 1206, op. 1, d. 548.

Through these projects, and their direct and indirect funding and deployment of ethnographic knowledge, Lukoil-Perm's CSR helped to shape the perceptions, assumptions, and careers of an entire generation of intellectuals and managers in the cultural field—museum specialists, folklore ensembles, university ethnographers, tour guides, and others. As their work became increasingly important in the 2000s, these specialists performed, sold their wares, researched, and wrote in contexts that took the cultural diversity of the Perm region as a basic and largely uncontested good, one that had the clear support of the region's most powerful corporation and its allies in the regional state apparatus. Although many factors have contributed to the Perm region's generally calm interethnic relationships in recent years, among them is surely the subsumption of so many local cultural practices and expectations into Lukoil-Perm's version of a corporate harmony of the peoples. The revival of this stratum of Russian history was, perhaps, less explicit than the revival of the Stroganov name in the north, for it was a much trickier thing to associate oneself with the socialist-era past. Indeed, the very Soviet-sounding "friendship of the peoples" was usually eschewed in favor of other, similar yet not identical labels: tolerance, multinationality, preservation, harmony, or collaboration. But the resonance with certain dimensions of the Soviet party-state's efforts to foster harmony among its constituent peoples was clear. Yuri Slezkine famously described Soviet efforts to foster ethnic particularism within an encompassing socialist state through the metaphor of the "USSR as a communal apartment" (1994). Efforts to foster ethnic particularism endured in the Perm region into the 2000s, although we might say that the communal apartment was replaced with a tourist bazaar featuring folk arts and crafts and the USSR with a corporate-state alliance centered on the oil industry.

When scholars write about oil and the formation of Russian identities, the tendency is to see a generalized Russian nationalism fostered by a resurgent central Russian state flush with oil rents. This sense of national identity and purpose undergirds Russia's newfound assertiveness abroad, whether in armed conflict in the Caucasus or Ukraine or in hardball natural gas contract negotiations with European consumers. This is not inaccurate. It is certainly true, for instance, that big Lukoil worked hard to cast itself as working "for the good of Russia" and that Vladimir Putin's centralization of federal state power is associated in various degrees with resurgent civic pride and a sense that Russia has returned to superpower status in the world.

Links between oil and Russian national identity came up only very occasionally during my fieldwork, and, as a rule, only in those rare conversations that were specifically about this aspect of international relations. This tracks well with survey data and elite interviews suggesting that the Russian public is somewhat ambivalent about Russia's status as energy superpower on the global stage and worried about being a simple "raw materials appendage" of the West, while federal political elites generally reject the idea of building national identity around the energy sector (Rutland 2015). Rather, the links among oil, oil companies, and various categories of identity that I found were discussed, displayed, and debated in quite local terms—at the level of the Perm region and its constituent districts, subdistricts, and territories. My findings about the relationship among oil and ethnic, national, and territorial identities are thus in line with a range of scholarship that has pointed to the crucial significance of regional and local identities in the post-Soviet period (Liu 2005; Dickinson 2005). They also speak to a much broader literature that sees local contexts as crucial sites for the creation of national sensibilities in ways that cannot be reduced to, subsumed into, or fully explained by global or state-level processes (e.g., Confino and Skaria 2002). In their grant application to Lukoil-Perm for support for a project to revive traditions of blacksmithing, applicants from the Orda district expressed a very common opinion on this matter: "the collective of the School of Folk Handicrafts is convinced that the rebirth of Russia will begin not in the capital, but in the provinces."[36]

The Depths and Spaces of Oil Culture, 2010–2011

The points I have made in this chapter—about abstraction and vertical integration, about the entwined depths of oil and culture, and about the legacies of the Stroganovs and the Soviet-era friendship of the peoples—were all on display at the Perm region's annual folk handicrafts fair. The fair, held at the Perm Exhibition Center, had grown from its origins on the margins of the 2002 Garden, Field, Farm exposition into a separate large event beginning in 2007. The days when Lukoil-Perm's Connections with Society Division anticipated that folk handicrafts would provide a company-sponsored path out of agricultural involution, that is, were largely over. The display and sale of handicrafts had become its own event, more closely tied to abstracted displays and district identities than to the practical dilemmas of getting by in difficult times.

By the years I attended (in 2010 and 2011), the annual fair was drawing scores of artisans from all over the Perm region and over fifteen thousand visitors over the five days it was open. Signs of Lukoil-Perm's influence were everywhere, beginning with advertisements for the exhibition scattered throughout Perm.

36. "Kuznets—vsem remeslam otets, s. Orda, 2004," PermGANI f. 1206, op. 1, d. 572.

The company not only funded many of the exhibitors through its ongoing grant competitions dedicated to handicraft production but also was a general partner of the event, providing a significant portion of the funds necessary to stage the exhibition each year. In 2009, the exhibition even opened a day early to accommodate the travel schedule of a very important guest: Vagit Alekperov, president of Lukoil, who was visiting from Moscow to sign the company's yearly agreement on cooperation with the Perm regional government.

On the exhibition floor itself, the organizers arranged exhibitors by district of origin within the Perm region, often with more than one artisan from each district occupying a booth. Significantly, the display stands of the oil-producing districts of the Perm region were grouped together in the center of the exhibition space, near the main stage, under banners featuring Lukoil-Perm's logo (see figs. 8 and 9). Other exhibitors from the Perm region were arrayed around the periphery of the exhibition hall, farther from the main stage and the heaviest foot traffic. Their plain white-and-blue painted identifying labels contrasted sharply with the ornate, folklore-inspired lettering and background shared by the booths

FIGURE 8 A view of part of the main floor at the Perm Exhibition Center's 2011 Folk Arts Show and Sale, taken just as the exhibit opened. Booths with Lukoil-sponsored artisans are grouped together near the main stage, and independent artisans or delegations from non-oil-producing districts are set up around the periphery.

Photo by author, 2011.

FIGURE 9 The Uinsk district's booth at the 2011 Folk Arts Show and Sale, sponsored by Lukoil-Perm (note the company logo in the upper right). The sign at the back of the booth reads "Uinsk: Honey Capital of the Kama River Region, a project of the M. E. Igoshev Museum."

Photo by author, 2011.

of oil-producing districts. The centrality of Lukoil-Perm and Lukoil-Perm's production districts to the burgeoning trade in folk handicrafts could not have been asserted more forcefully, or mapped onto the space of the Perm region more precisely.[37]

A representative of the Perm regional government opened the fourth annual exhibition in 2010 by succinctly bringing together local cultural identities, the depth of traditions, the Perm region's multiethnic composition, and the abstracted kinds of cultural interaction enabled by the fair:

> We often ask ourselves how to preserve our identities, how to preserve our ethnic-national values. We ask questions. But there are artisans in the Kama River Basin who, with their own hands and souls, assert our

37. See B. Gustafson 2011 for a useful typology of spatializing practices in the context of hydrocarbon extraction; for another instructive reading of identities, territories, and remappings of post-socialist space, see Reeves 2014, 65–100.

> ethnic-national uniqueness, who show how we are very different from other regions in the richness of our diversity. I'd simply like to thank those in this auditorium who came from the depths (*iz glubinki*) of the Perm region and usually work one-on-one, in small groups, or simply by themselves. To present themselves and their craftsmanship in the regional center, to see their fellow artisans, to enter into what are, for many, new market relationships—these are important opportunities.

The chief facilitator of these relationships took the stage next, as the master of ceremonies turned the microphone over to a representative of Lukoil-Perm. He greeted the participants "on behalf of all the oil workers" in the region and continued,

> Probably more than half of the participants here are from . . . districts where our company extracts oil. This isn't a coincidence. In collaboration with the heads of districts, the company has actively supported the revival of traditions for many years, including folk crafts, in the districts where it works. The words, "Lukoil—Protecting Tradition" are not just a slogan. Brands such as Elokhov fish and Uinsk honey have become known far outside of our region. This is all the result of our work with you.

At least in the staging of the event for the public, that is, the 2010 exhibition was a collaborative company-state project dedicated to showcasing and selling the cultural distinctiveness of the different districts of the Perm region, and many of the displays were staffed by or had other support from local museums as well as artisans themselves. Socialist states commonly turned to the celebration of rural populations and folk traditions as a method of shoring up their faltering legitimacy and responding to critiques (Kligman 1988; Rogers 2009). In Russia's oil-boom years in the Perm region, this was a joint state and corporate project, with Lukoil-Perm very much in a leading role.

When I returned to the fair again in 2011, this company-state relationship had deepened still further, and the exhibition had begun to overlap with that other classic domain for the display of abstracted culture: festivals. The state office charged with administering a program of Fifty-Nine Festivals for the Perm region—discussed in more detail in chapter 8—had a prominent booth of its own, featuring a large map of the Perm region with dots showing the location of many festivals and a long list of the rest. A large-screen television played clips from a year's worth of video footage of festivals on a continuous loop. Moreover, an entire program of folk songs had been added to each day of the exhibition, arranged so that representatives of many of these festivals would appear briefly, showcase their traditional attire, sing a song or stage a skit, and then invite those present to visit an upcoming festival in their district.

The delegation from Gubakha, a former coal mining town that had seen its fortunes (and population) decline rapidly with the closing of the Perm region's coals mines in the post-Soviet period, was one of these performers. It took the stage to invite those present to Gubakha a month hence for a theater festival named Secrets of Krosovaia Hill after an unusual local rock formation popular with tourists. As part of their pitch, they had prepared a mini-quiz for the audience, with small prizes awarded for correct answers. The topic of the quiz was the history of the Stroganovs. As exhibition attendees milled among the booths and grouped at the front of the stage, the head of Gubakha's delegation asked questions over the amplified sound system, cupping her hand to her ear to catch the shouted responses. "What new status did Maksim Iakovlevich, Nikita Grigor′evich, and Andrei and Petr Semenovich Stroganov receive from the tsar in 1610?" "What honor did Aleksandr, Nikolai, and Sergei Stroganov receive from Peter the Great in 1722 in recognition of their ancestors' service to the fatherland?" The questions turned to the Perm region.

> Q: All generations of the Stroganov family were occupied with creative activities on their lands. What kind?
>
> A: Embroidery!
>
> Q: Yes, embroidery—pictorial needlework, with gold and silver threads. Correct. What else?
>
> A: Metallurgy!
>
> Q: Of course, metallurgy, building metallurgical and salt-processing factories. This is where that industry started in our lands. And what else? Yes. Theater, of course. The Stroganov serf theaters. Our theater [in Gubakha] can be considered a kind of descendant of the serf theater. And what else? Apart from factories what else did the Stroganovs always build, wherever they were?
>
> A: Libraries?
>
> Q: Libraries, of course! . . . And what else? What else are the Stroganov lands rich in? In Usol′e and others? Monasteries, churches, of course! The Stroganovs did a great deal for them. . . .

The quiz ended on that question and the head of the Gubakha delegation summed up her message for the audience: "Well, we've identified the Stroganovs' great philanthropy [*blagotvoritel′nost′*]. God grant that all those with great power in our days [*nashim velikim*] support our population in the ways that the Stroganovs did."

7

ALTERNATIVE ENERGIES

Lukoil-Perm in Corporate and Cultural Fields

By the middle and later 2000s, Lukoil-Perm was not the only company in the Perm region engaged in corporate social responsibility projects, nor was it the only sponsor of projects aimed at fostering cultural renewal. Attending to some of these other projects helps to illuminate both the influence and the limits of the conjuncture of corporation and culture that I have described thus far. In other words, one useful way to evaluate the significance of Lukoil-Perm's integration of corporation and culture is to situate it within a broader field of imitators and critics—to ask how and to what extent the company was able to set the terms by which other corporations and other cultural producers went about their own work. My discussion moves first along a corporate vector, exploring CSR projects in another energy sector company in the Perm region: PermRegionGaz, the regional subsidiary of the Russian state gas company Gazprom. Adding a second energy company to the analysis allows me to highlight the ways in which different energy sector materialities in the Perm region have become wrapped up in different corporate projects of social and cultural transformation, with quite different fates. My second example moves along a cultural vector, considering another organization—the Kamwa ethnofuturist festival—that was heavily engaged in the production and celebration of local and regional cultural identities in the Perm region in the 2000s. Although different from Lukoil-Perm in many ways, the independent Kamwa festival favored some quite familiar strategies of cultural production and consumption, notably the contention that things extracted from the depths of the Perm region—especially archaeological artifacts—could be a powerful source of energy with which to reimagine the region's past and future.

These larger corporate and cultural fields were well understood by those who moved within them. Nearly everyone I introduce here knew everyone else and was eager to talk about different trajectories, past and potential collaborations or disagreements, the ins and outs of politicking across domains of energy and culture, and the implications of all this for the Perm region as a whole. It was entirely common for my interlocutors to make unprompted comparisons, for instance, between Lukoil-Perm and PermRegionGaz, or between Lukoil-Perm's Stroganov-focused projects in the Il′inskii district and cultural projects embarked on by the Kamwa festival in and around Perm itself. I draw attention to these "indigenous" comparisons whenever possible in order to highlight the ways in which multiple intersections of energy and human social and cultural life constituted a field of shifting possibilities for the stratum of intellectuals and cultural managers that became so influential in the Perm region's second postsocialist decade. These intersections, in turn, powerfully shaped the ways in which everyone in the region encountered energy, corporation, culture, and the regional state.

Socializing with Pipelines: Gas Sector Corporate Social Responsibility

In one of our early conversations, Oleg Leonidovich told me that he suspected he would not be at Lukoil-Perm for much longer. Management changes were coming and the new regional leadership would, he said, probably prefer to put its own team in place. I asked where he would go next. "I don't know," he replied, "maybe to the gas industry." Although I was not aware of it at the time, PermRegionGaz had also begun to award grants for local projects, in part after observing the publicity successes of Lukoil-Perm's Connections with Society Division. Indeed, of the corporations of regional scope in the Perm region, PermRegionGaz was the only one that had adopted a program of corporate social responsibility on anything like the scale of Lukoil-Perm. Although Oleg Leonidovich did not ultimately end up at PermRegionGaz, the lateral move in the energy sector that he briefly contemplated was part of a family of comparisons and juxtapositions of oil and gas that I frequently encountered, one that had much to do with the material subsoil and surface of the Perm region.

Like crude oil, natural gas is pumped from the depths of the subsoil. There are, however, no significant deposits of natural gas beneath the Perm region.[1]

1. Several oil deposits managed by Lukoil subsidiaries included natural gas—so-called associated gas—that was simply flared into the atmosphere in the Soviet period. At the time of my primary fieldwork, Lukoil-Perm was gradually increasing the amount of this gas that it processed and sold.

Natural gas extracted in West Siberia flows through the region in pipelines, the largest of which run from the Siberian supergiant fields southwest through the Perm region and into the Druzhba network, which transports gas as far as Europe. Branches from these trunk lines feed the Perm region's municipal gas networks, factories, and heat and power cogenerating plants (see map 3). The material qualities of gas as it moved through this pipeline infrastructure—rather than the geology of subsoil deposits, as in the case of Lukoil-Perm—was central to corporate critiques and responses.

Many of the critiques to which the regional gas industry responded in its CSR initiatives were, for instance, pegged quite closely to the transient qualities of pipeline gas. The fact that the gas companies in the Perm region specialized in buying and selling something produced elsewhere made them vulnerable to a genre of criticism quite common in the postsocialist period: that the corporations were engaged in speculation, making money purely by means of moving a product around, marking up the price, and trying to eliminate any competition to drive that price still higher. One relatively high-ranking regional state employee drew the following distinction for me between Lukoil-Perm and PermRegionGaz: "Look at what hard work it takes to pump oil out of the subsoil, how much money and labor. But the gas industry (*gazoviki*) . . . they're just sitting on the pipes, turning the spigot on and off." A newspaper reporter opened an interview with a gas company executive rather bluntly: "Who loves *gazoviki*? Everyone loves gas. But gazoviki? No one. Gazoviki sit on the valve and threaten to shut off anyone who doesn't pay."[2] Although the consumption-driven boom of the mid-2000s certainly dampened the efficacy of speculation as epithet substantially as compared to the mid-1990s, the transient quality of gas remained available as a critique of the regional corporations that specialized in it.

Gas's transient qualities in the region also meant that it could not easily be said to have the Permian qualities that were commonly attributed to oil extracted from the depths. The potent set of place-based associations so central to early critiques of Lukoil-Perm and, later, to the company's CSR projects aimed at diluting those critiques, did not characterize the gas industry. Whereas Lukoil-Perm worked hard to associate itself with the production of *Permian* oil—in part through the production of Permian local cultural identities—the fact that PermRegionGaz largely managed transient gas meant that the company was far more often associated with laborless circulation and speculation. This association had real political consequences, at least at the level of rhetoric and deal making. In the late 1990s, for instance, the governor of the Perm region, Gennadii Igumenov, reported that he sat down with executives at the region's main gas-processing

2. "Pedal′ Gaza," *Svetlyi Put′*, October 2, 2001.

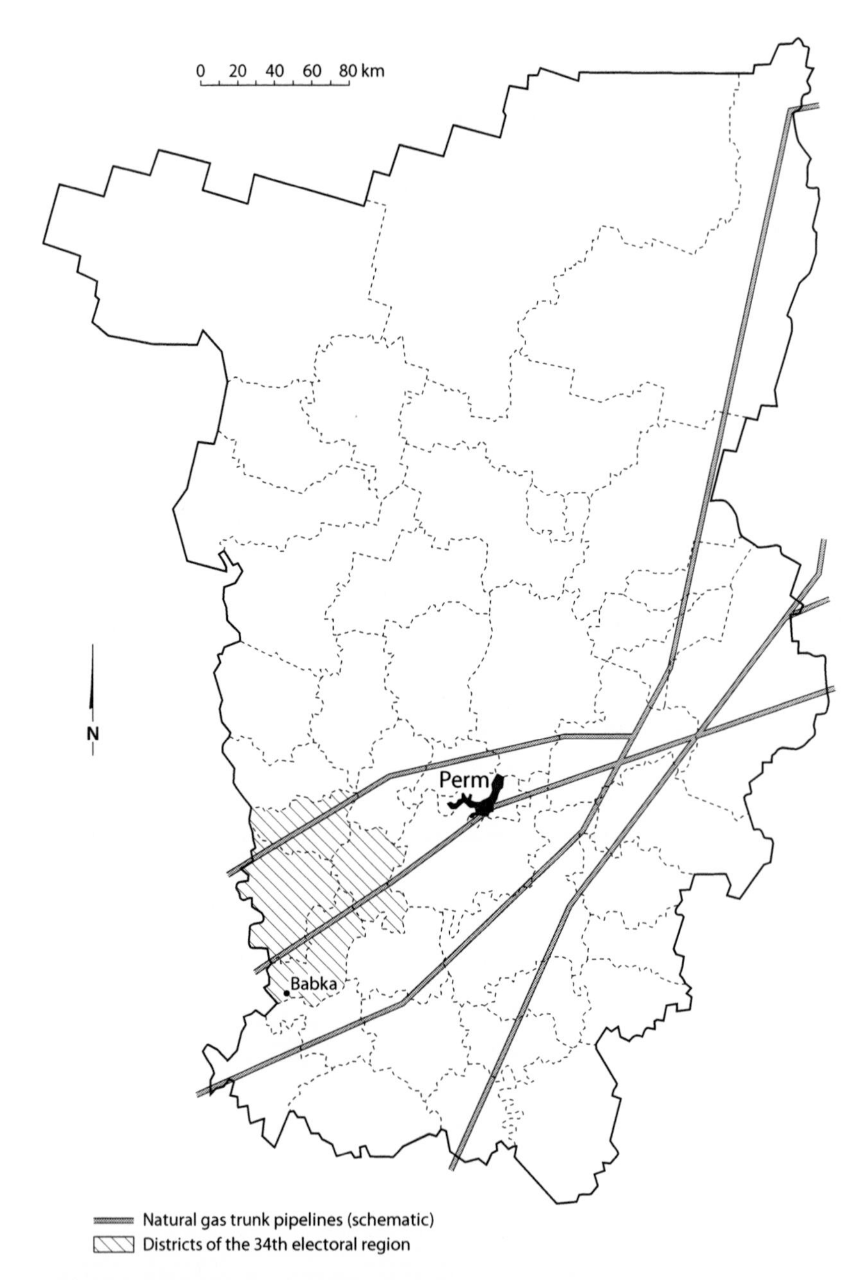

MAP 3 Natural gas trunk pipelines running from east to west through the Perm region. The pipelines intersected with the Perm Legislative Assembly's thirty-fourth electoral region, represented by Igor Shubin from 2001 to 2005.

plant to inform them that simply processing transit gas—and therefore paying only federal taxes—did not excuse them from contributing to the Perm region. If the company did not recognize that it "is located in our territory," he had the means to ensure that it would soon encounter "a certain discomfort in its activities" (Kurbatova 2006, 547).

Such was the set of qualities associated with natural gas—transience, non-Permianness, and, of course, convertibility into massive wealth—that Perm-RegionGaz sought to shift as it followed Lukoil-Perm into the practice of giving grants for social and cultural projects in the early 2000s.[3] Its sponsored projects, administered through Cooperation, a Perm-based noncommercial organization, picked up on another quality of pipeline gas in the region—connectivity—and linked it to widespread postsocialist longings for sociability. Like talk about depth, dollars, and souls, discussions of sociability (*obshchenie*) circulated everywhere in the first postsocialist decade, articulating worries about the amount and intensity of fulfilling personal interactions as compared to the Soviet period. Whatever goods, consumer and otherwise, that the postsocialist era had brought, the lack of sociability often hovered in the background as a casualty of the speeded-up paces, pervasive disorientations, and rapid social stratification of the new age. Cooperation's projects, with funding from PermRegionGaz, took aim at this lack.[4] What emerged was a corporation-sponsored variant of a common phenomenon tracked by anthropologists and others, in which material infrastructures—roads, bridges, sewers, or, in this case, gas pipelines—embody a collective "fantasy of society" (Larkin 2013, 329).[5]

Consider, for instance, the small town of Babka, which distinguished itself in grant competitions in the mid-2000s even though it did not sit atop oil reserves. Reconstruction of Babka's central square began in 2004 with financing from PermRegionGaz. The new square featured a garden, a fountain, new benches for sitting and talking, paved paths designed with stroller-pushing mothers in mind, and even a small monument to the town's namesake—a grandma (*babka*). "We call it 'Babka Arbat,'" one of the out-of-town specialists who helped secure grant funding for the square told me with a smile and some pride, referring to the old

3. On the possibilities for rent capture afforded by transit gas in post-Soviet space, see Balmaceda 2008.

4. On postsocialist longings for sociability, see Paxson 2005, Pesmen 2000, Ries 1997, and Yurchak 2005. For another postsocialist case in which the difference between local notions of culture and sociability became a political and ethical fault line, see Rogers 2009, 246–68.

5. The embedding of ideology into material infrastructures was a significant aspect of the socialist project as well (Humphrey 2005), although the meanings produced through interactions with this infrastructure were far more diverse than their architects intended. On urban heating infrastructure as a socialist "index of modernity," see Humphrey 2003. Bouzarovski and Bassin (2011) show how energy sector materialities figure in Russian national discourses and imaginations of the country's "hydrocarbon superpower" status in the Eurasian region.

Moscow district known for its strolling pedestrians, resident intellectuals, and air of sociability. Projects like those in Babka, it was thought by their gas company funders, might provide new ways to make connections, to get to know one's fellow townspeople. Indeed, the benches and paths lining Babka's new square were explicitly designed and promoted to provide a new place for the sociability that many Russians longed for after the tumultuous and alienating 1990s.

Well before I met her in 2010, the director of Cooperation, Nina Nikolaevna Samarina, had begun to hold up Babka's central square as a model of collaboration between a corporation—in this case PermRegionGaz—and a local community. She expanded on her choice of exemplar in several interviews with me. The new spaces of Babka, she said, meant that everyone had a place to go and socialize now, that far more people were "included in the life of the town" than had been previously. Indeed, one of the main goals of Babka Arbat was to address the alienation and declining sociability that so many experienced in the proliferation of markets and inequalities after socialism. This was no backward-looking socialist project, for Cooperation's efforts were wrapped tightly in the language of capitalist entrepreneurship, not least in competitive grant writing of the sort that had come to dominate the region in the early 2000s. But they held out the possibility of regaining and reassembling a sociable world through this self-actualization. Here, it was not the postsocialist ruins of cultural construction that the corporation sought to reassemble through its CSR work, as was the case for Lukoil-Perm, but the postsocialist ruins of sociability.

The fruits of engagement in one's community, of actively engaged citizenship (another word appearing throughout Cooperation's work), would be new kinds of valued connections. A photograph of inviting benches arrayed around Babka's modest fountain featured, for instance, in Cooperation's 2009 annual report and in any number of PowerPoint presentations showcasing the foundation's activities. The little desk calendars that Nina Nikolaevna handed me after each of our interviews tended to feature examples of society work (*obshchestvennaia rabota*): children repairing a fence, teenagers building a spare soccer field near a ramshackle school, or volunteers building Babka's new central square. (Lukoil-Perm's annual reports and desk calendars, by contrast, featured month after month of children and elderly women in assertively traditional outfits, or perhaps small handicrafts—all items that evoked the depths of culture.) One particularly striking image in a Cooperation calendar featured schoolchildren stacking firewood for elderly townspeople, a classic Soviet-era activity of the Komsomol in this cold region. Nina Nikolaevna's socialist-era networks were, in fact, largely in the Komsomol—a point to which I return shortly.

Like Oleg Leonidovich, Nina Nikolaevna was an accomplished networker and skilled operator newly recruited, in her case, into a contractual relationship with an energy company (rather than direct employment). She had seized on an opportunity to continue the "work with people" to which she had long dedicated herself, in the process rethinking it for a new era. Cooperation's sociability-focused CSR projects did not originate as part of a conscious gas industry branding strategy any more than did those run out of Lukoil-Perm. They were made in a corporate context that facilitated the improvised combination of widely circulating postsocialist discourses about sociability and its lack, resurgent social/political networks, and massive new wealth. The material qualities of pipeline gas and its connective infrastructure played no small role.

Gas Wars: From Social Sphere to Sociability

My discussion of the CSR projects sponsored by both oil and gas companies in the Perm region has focused on the ways in which shifts in the significant material qualities associated with certain objects can originate in corporate contexts and shape political, social, and cultural processes. Thus far, the shifts I have been tracing have been largely at the level of resemblances. Through the work of well-funded and skilled operators, that is, oil and culture in the Perm region came to resemble each other through a shared quality of depth, gas and society through a shared quality of connectivity. In each case, the shared material quality was the bridge by which corporations sought to respond to critiques and local needs. At the same time, these new resemblances diverted attention from other qualities—more physically harmful or morally suspect—that had been associated with the objects in which each corporation specialized. In the case of pipeline natural gas, we find, in addition to these resemblances, a set of physical connections in play: pipeline grids, and especially the extension of gas pipelines to communities who had long demanded them.[6] Indeed, these aspects of pipeline gas were crucial to the gas sector long before PermRegionGaz began funding sociability. A brief excursion into the 1990s history of the Perm region's "gas wars" demonstrates how pipeline grids became crucial to regional notions of society in the post-Soviet period and, at the same time, helped assemble the political coalitions and networks that later came to administer PermRegionGaz's CSR projects.[7]

6. In the terminology of Peircean semiotics, natural gas pipelines, like wires and other aspects of built infrastructure, are indexical icons, working through both resemblance and connection.

7. On the imbrications of political and technical/infrastructural networks on the terrain of development projects, see also Anand 2011 on water in Mumbai; Bouzarovski 2010 offers a useful framing of post-Soviet pipeline networks and other infrastructural issues at the scale of EU–former Soviet Union relationships.

The story revolves around a contentious split between federally and regionally managed pipeline networks. In the Soviet period, gas production and the management of region-crossing trunk pipelines fell under control of the Ministry of the Gas Industry (and then, from 1989 to 1991, the Ministry of the Oil and Gas Industries), while the Ministry of Housing and Utilities controlled the municipal pipelines feeding urban housing blocks. In the Yeltsin era, the federal monopoly Gazprom emerged out of the former structure. It initially owned and controlled only the trunk lines passing through the Perm region from its West Siberian fields. The Perm region's municipal gas distribution networks were privatized in 1991 into a corporation called UralGazServis, which was controlled by regional businessmen and politicians.

UralGazServis bought natural gas from Gazprom's east-west running trunk pipelines and sold it to industrial and residential customers in the Perm region. In conditions of 1990s demonetization and barter, however, most of those customers were paying in kind or simply not paying. UralGazServis thus accumulated astronomical debts to federal Gazprom: it was collecting almost no cash payments but was legally prohibited from turning off the gas supply to 90 percent of its customers, including electricity companies, heating facilities, factories of national significance, and, crucially, the social sphere—residential apartment buildings. As Gazprom repeatedly tried to collect on these debts, regional coalitions in Perm did everything they could to avoid paying. They had powerful leverage against the center. Federal Gazprom could not shut off the supply to UralGazServis either, especially during the winter months, without prompting a public outcry and, in all likelihood, a national crisis of political legitimacy. So the gas continued to flow and UralGazServis continued not to pay, instead circulating all manner of non-cash payments, favors, and influence that were an important element of the regional politics discussed in earlier chapters.

Recall that, in the 1990s, the assertion that regional oil had important Permian characteristics was central to critiques of Moscow-based Lukoil's attempt to take over the regional oil industry. In the gas industry at around the same time, the hard-piped material connectivity of pipelines supplying the post-Soviet social sphere played a not dissimilar role, and was an important basis for regional coalitions to accumulate power and influence at the expense of the federal center.[8] Several factors, however, kept gas from taking the central place in the regional economy that oil did at the Perm Commodity Exchange and later at the Perm Financial-Productive Group: the lack of a regional gas production or processing

8. As Collier (2011) shows in the service of a different argument, the routing and technical specifications of pipes and wires were crucial elements of the brand of social modernity that characterized the Soviet Union; the "intransigent materiality" of this infrastructure in the post-Soviet period powerfully shaped the ways in which post-Soviet reforms could and could not be realized.

industry analogous to Permneft and Permneftorgsintez in the oil industry; the centrality of refined oil products to the struggling agricultural sector, which was such a large part of the petrobarter and fuel veksel exchange circuits; and the region's inability to export gas to other regions or abroad for either cash or other goods.

Later episodes from the gas wars show how the material connectivity of the social sphere in the 1990s morphed into sociability-focused CSR projects funded by a gas company a decade later. In the first Putin presidency, beginning in around 2000, Gazprom again sought to collect on debts and consolidate control over regional pipeline networks, this time with more success. As part of this effort in the Perm region, a new company, PermRegionGaz, was created to replace UralGazServis. A power-sharing compromise between federal center and region allocated the majority of shares in PermRegionGaz to Gazprom, but allowed Governor Iurii Trutnev to appoint the director of the new company. Trutnev chose Igor Shubin, the Perm region's vice-governor for energy affairs from 1994 to 2001 and a veteran of the gas wars.

Shortly after his appointment as head of PermRegionGaz in 2001, and in a move common among influential businesspeople, Shubin ran for—and handily won—a seat in the Legislative Assembly (*Zakonodatel′noe Sobranie*), the Perm region's chief lawmaking body. No residency laws governed the eligibility of candidates, and Shubin selected an open seat in the thirty-fourth electoral region, encompassing the Bol′shesosnovsk, Okhansk, Ochersk, and Chastinsk districts in the southwest of the Perm region (see map 3). Although some of Gazprom's trunk pipelines passed through these districts, they had no other special connection to the regional gas industry; indeed, many of their constituent towns and villages were not on the gas grid at all. Not surprisingly, then, Shubin campaigned on a promise to bring municipal gas pipelines to the districts—to include them in the social sphere linked by pipeline grids. By 2004, he had fulfilled at least part of his promise, and a newly constructed set of pipelines brought piped gas to the Bol′shesosnovsk district, one of those Shubin represented, for the first time ever.

It is important to note that the materialities of the region's oil industry did not permit Lukoil subsidiaries to pursue this particular kind of response to the critiques and demands of a local community. Oil, oil pipelines, and refined oil products can, in principle, also work through connection; their signifying capacities are not limited to the domain of resemblance. However, several factors pushed against this possibility at the regional level of analysis on which I focus here. Most notably, gasoline refined from crude oil was highly mobile, easily trucked around the entire region by tanker and widely available at Lukoil and non-Lukoil service stations. By contrast, the hard-piped nature of municipal gas supply placed a premium on connectivity that gas companies could provide only through the much more expensive and time-consuming extension of pipelines; towns, villages, and districts were either linked into this network or, like many residents of Shubin's districts, desperately wanted to be.

Gas lines were not all that Shubin's constituents wanted, however. Requests and complaints about social and economic problems poured into his office from these largely agricultural and impoverished districts, and critiques revolving around gas's transience did not disappear. Indeed, they had started during Shubin's electoral campaign, and one of his responses was to invoke his connections as head of PermRegionGaz and promise to imitate Lukoil-Perm's already-running social and cultural project competitions to provide the most up-to-date kinds of support for the population. He told a local newspaper, "We have, for instance, Lukoil, an influential, popular, and respected organization. [PermRegionGaz's] goal is to become just as influential, popular, and useful to the people of the region."[9] In 2003 and 2004, then, PermRegionGaz entered this new field of social and cultural projects, not in all districts of the Perm region where gas flowed, but, rather, only in those represented by Shubin in the Legislative Assembly.

Initially, the effort was run internally at PermRegionGaz, with Shubin himself as chair of the commission selecting projects to fund. Nina Nikolaevna of Cooperation served in an advisory role at first, taking over the full administration of the grant program in 2005 and steering grants farther in the direction of fostering sociability. Her relationship with Shubin was not at all coincidental. In fact, it ran through their joint membership in the large and powerful regional coalition headed by Governor Iurii Trutnev, who had appointed Shubin to his position at PermRegionGaz and also installed Tatiana Margolina and a number of other former Komsomol members in the upper echelons of his government to administer services and projects for the population. Indeed, as I outlined in detail in chapter 4, these former Komsomol members were some of the earliest regional adopters of the strategy of running grant competitions to fund local initiatives. Many continued to think of their work as society work and work with people, now recast as helping the population of the Perm region adapt to new capitalist circumstances.[10]

I am arguing, then, that we can understand the focus on sociability epitomized by the CSR projects in Babka's central square as emergent from two linked networks: first, a material infrastructure of pipelines connecting the all-important post-Soviet social sphere, and second, a political and economic coalition that, under Governor Trutnev, included both the powerful regional manager of the pipeline infrastructure (Shubin) and former Komsomol affiliates doing society work in a state administration context. Here, as in the oil industry, the ongoing

9. "Pedal' Gaza," 4–5.
10. On society work and work with people, see Rogers 2009, 248.

significance of a material quality of pipeline gas infrastructure—connectedness—into PermRegionGaz's CSR projects was neither a mere coincidence nor a careful, prehatched plan. It emerged from the active recombination of a whole range of postsocialist processes in the crucible of a corporation, and it was a small corner of the state-corporate field that grew up initially around Lukoil subsidiaries and the regional state apparatus.

In sharp contrast to the increasingly entrenched district-level politics swirling around Lukoil-Perm's culture-focused CSR programming, the relationship between PermRegionGaz and Nina Nikolaevna's Cooperation turned out to be short-lived. Shubin was appointed acting head of the City of Perm in November 2005, and relinquished his dual positions at PermRegionGaz and in the Perm Legislative Assembly. The district's new representative to the Legislative Assembly did not have anywhere near Shubin's networks and the resources of PermRegionGaz at his disposal. Andrei Agishev, the new director of PermRegionGaz, won election to the Perm Legislative Assembly, but from a different electoral region than Shubin. His own interests in social, cultural, and charitable work also lay in other districts, and the resources that had flowed from PermRegionGaz, through Cooperation, and into Babka and other towns and villages in the districts of Shubin's electoral region promptly dried up. Cooperation, of course, pressed on. Nina Nikolaevna continued to skillfully wrangle other connections, especially in regional government and local businesses, but they were never as good (and the funding never as substantial) as they were during the alliance with PermRegionGaz under Shubin's leadership.

Hydrocarbons, Histories, and Materialities in Global Perspective

> The last millennium in the history of the [Perm region] has been a period of expansion and continuous territorial development—by the Novgorod Ushkuiniks and the missionary bishop Stephen of Perm; by the Stroganov and Demidov dynasties; by Timofei Ermak and Vasilii Tatishchev; by the colleagues of the OGPU and the victims of the GULAG; by Communist Party members and nonmembers; by Lukoil and Gazprom.
>
> "The New Perm Period,"
> *Novyi Kompan'on*, December 26, 1999

Casting its eyes back at the turn of the millennium, one of the Perm region's leading newspapers characterized the present age as defined by two massive corporations that, it suggested, had become as central to life as had the Communist

Party and the Stroganov family in earlier ages. Both corporations were in the energy sector and both would soon extend their influence in the region by taking up the new (for Russia) practice of corporate social responsibility. Although Lukoil-Perm's efforts on this front were earlier, more comprehensive, and ultimately much more durable than PermRegionGaz's, considering the field occupied by both corporations permits some broader observations about the intersections of energy, corporations, and states in and beyond the post-Soviet Perm region.

Wealth accumulation, marketing and image making, the commodification of culture, the very visible prominence of the oil and gas industries, and the perceived social and cultural lacunae that CSR projects in the Perm region were calibrated to address are all, as I have presented them, important and intersecting dimensions of specifically postsocialist transformations. But, I have also argued, attending to the history and materiality of energy-sector CSR projects in the Perm region demonstrates that Russian corporations have been very much in step with their counterparts around the world as they have responded to criticisms that their products and other associated material things and effects (from built infrastructures to environmental pollution) have often engendered. Through their CSR programs, both Lukoil-Perm and PermRegionGaz emphasized qualities of those same products and things—qualities, such as depth, that are not necessarily directly related to the use value of their products. Part of this process involved working to shift entire semiotic fields of resemblances and/or connections with the intended result that the newly emphasized material qualities of the corporation's favored objects appear to address precisely the problems that other material qualities were understood to create in the first place. These corporate representational strategies played out on some of the most basic fields of human social and cultural life in Russia, from the historical consciousness prompted by museums and festivals to the promise of sociability held out by a new town square, and they have been enormously influential shapers of the ways in which residents of the Perm region have navigated the second postsocialist decade. They have helped remake lives, expectations, and sensibilities as much as—and in ways very closely bound up with—the increasing income inequality and social stratification that is most often noted in Russia's boom years. They are, moreover, a far cry from the early years of the Second Baku, when the project of constructing a new socialist society was imagined in metallurgical images and in Stakhanovite factory labor. A full seven decades after the discovery of oil, that is, hydrocarbons were becoming firmly lodged in the symbolic and representational landscapes of the entire Perm region, and they were doing so as one member of a worldwide family of corporate social and cultural responses to anticorporate critique.

Among these transformations and convergences with global trends, we also find some striking continuities with and reproductions of Soviet-era practices.

Not only did former Communist Party members become some of the key players in developing and implementing CSR projects in both the oil and gas segments of the Perm regional energy sector, but aspects of the overall former party-state hierarchy were reproduced as well. In the 1990s, as I showed for the Perm region in chapters 2 and 3, shifting and often contentious alliances of red directors and new financiers were the main players, and the pacts into which they entered shaped the formation and early interaction of financial and industrial capital in the regional oil complex. As both wealth and anticorporate critique began to accumulate in the late 1990s and early 2000s, necessitating a new dimension of corporate work—legitimation and public relations rather than primitive accumulation, mergers and acquisitions, and labor-force restructuring—the corporate and financial management of these new companies turned to some of the former leadership of Party organizations once attached to Soviet-era Perm's flagship factories. They became some of the leaders of the new CSR movement. In new, public-facing divisions of energy companies, these former Party members reconstituted, through the medium of grant applications, an old alliance with the former low-level Party operatives who ran educational initiatives, clubs, museums, and libraries out in the districts of the Perm region, whether in oil-producing districts such as Usol′e and Il′inskii or in districts located in Shubin's electoral region such as Babka.[11] Out of these alliances, museums and town squares were rebuilt, each with its own ties to the materialities of an energy source. Moreover, as these experts in social and cultural construction reentered their old field through energy corporations, they reproduced the internal Party hierarchy of the Soviet era. Former members of the region's leading defense factories moved laterally, to the newly prestigious oil industry, while those hailing from what were once lower ranks of the regional Communist Party apparatus—the Komsomol—moved into somewhat lower rungs of the hierarchy of energy-sector CSR, working for the less highly respected gas company or the regional state administration itself.

The socialist past was also important in another, related sense: it generated an overall shared expectation that firms should be maximally embedded in local communities. Although there have long been corporate company towns in the West as well, of course, the Soviet model extended firms' influence into social and cultural life far more intensively and materially, and at higher levels of party-state administrative coordination. Recall from chapter 1 that, in the standard Soviet model of political and economic organization at the level of small cities, towns, and even villages, the leading edge of governance was located in farms or factories. State organizations served in an administrative and secondary role, with the most

11. To be sure, some members of this new CSR network were not Party members, like Oleg Leonidovich, and some had begun work after the Soviet period.

powerful Communist Party offices attached directly to firms, not to state agencies (see Humphrey 1999, 300–373; Rogers 2009, 107–46). The Soviet-era Perm region was thus governed by official committees and informal networks that knit together Party structures and factory leadership, while in oil towns like Chernushka, Permneft's oil and gas production administration (NGDU) presided over what Lukoil-Perm executive Nikolai Kobakov remembered as a "natural economy." Despite the large-scale reconfiguration of the state field in the 1990s and 2000s, the expectation that firms were heavily involved in designing and administering social and cultural life—well beyond their own workforces—was reproduced in new forms in twenty-first-century CSR, often enough by the very same former Party members moving into energy sector jobs or contracts. Indeed, one of my interlocutors who was well-informed about Lukoil-Perm's CSR programs argued to me at length that there was nothing distinctively post-Soviet about them at all; they were, he suggested, entirely an outgrowth of Soviet-era firms' mandate to attend to all aspects of life in their territories. His was certainly a minority view—far more common were emphases on flexible capitalist entrepreneurship and highly marketized, commodified culture in the age of CSR—but it points usefully to the inheritance of an expectation about the governing responsibilities of firms that was characteristic of socialist organization (see also Peregudov and Semenenko 2008, 248–338).

Why, though, do we find hydrocarbon materialities—such as the depth of oil and the connectivity of natural gas pipelines—so prominently reincarnated in energy-sector CSR in the Perm region? And why do we *not* find this phenomenon reported elsewhere in the world, at least on the level, scale, and scope that I have described? I would suggest that the twentieth-century capitalist and socialist oil complexes discussed in chapter 1 bequeathed twenty-first-century CSR programs differential inheritances of institutional expectations and representational practices. In the classic petrostates of the capitalist periphery, states collected royalties, taxes, and other rents from national or international oil corporations. After these petrodollars cycled through the international banking system and domestic federal budgets, they were put to work on development projects dedicated to building independent, modernizing, postcolonial states, with the kinds of cultural identities thought to go with them. It should not be surprising that, in this international monetary circuit and state-centered representational field, the qualities of oil itself had receded to a significant extent by the time oil was reincarnated as an on-the-ground development project. Abstractions of this sort enabled, for instance, the "tricks of prestidigitation" that conjured Venezuela's magical state (Coronil 1997, and see chapters 1 and 5).

By contrast, in the socialist world, especially at subnational levels, *firms* took the lead in projects to transform their local communities, and those communities were never linked to the international and highly abstracted circulation of oil and money in ways that were very noticeable to them. Socialism, that is, tended to generate relationships between local communities and firms that were closely mediated by the material qualities of the objects in which those firms specialized: metallurgical subjects of heroic socialist factory labor (see chapter 1), relationships between peasant and factory workers mediated by the very fruit they grew and processed into baby food (Dunn 2004, 94–129), and many other examples. In the Perm region's oil complex, this meant that Permneft—and its oil—retained a prominent presence in efforts to shape socialist subjects in the oil towns and cities of the Second Baku. When CSR became popular in the 2000s, both its corporate architects at Lukoil-Perm and PermRegionGaz and the districts that were on the receiving end of sponsored projects were very accustomed to corporate caretaking that was closely linked to the objects in which those corporations specialized. It made eminent sense to link the materialities of oil and gas to the cultural and social transformations envisioned by CSR projects. In sum, although post-Soviet CSR is certainly in the same extended family as CSR operations elsewhere in the global extractive industry, the different representational strategies that characterized mid-twentieth-century capitalist and socialist oil complexes—a subset of capitalist and socialist state–firm relationships more broadly—has meant that we find the material qualities of oil and gas themselves afforded a much more prominent place in a postsocialist context than we do elsewhere in the world of energy-sector CSR.

It is useful to recall, in this connection, that Lukoil-Perm's CSR projects took place *at home*, in Russia, at the initiative of a Russian corporation—rather than in an overseas extractive enclave (as is much more often the case for corporations based in the West).[12] This point became especially clear to me in a conversation with one of Lukoil-Perm's Connections with Society employees in 2010. We were talking about Lukoil-Perm's early innovations in CSR and grant-making, and the fact that representatives from other Lukoil subsidiaries were still, a decade later, coming to the region for advice on developing their own similar programs. It was, she specified, only for subsidiaries of Lukoil based in the former socialist world that these kinds of CSR projects made sense, both to the company and to local communities. They would never be useful for Lukoil's expanding operations in Iraq, Africa, or elsewhere. In those places, she said, relations between local populations and the company required fences rather than social and cultural grant competitions. The mutual intelligibility of speaking in terms of the depths of oil and the connectivity of gas, I understand her to have been saying, was a post-Soviet, and decidedly postsocialist, phenomenon, a way in which corporations

12. I am grateful to Pauline Jones Luong for pressing me to think through this issue.

that were built in the aftermath of socialism could respond to and divert critiques leveled by communities that shared the same socialist history. This strategy was not necessarily transportable as Lukoil expanded abroad. For similar reasons, it is not likely to be a representational practice pursued by oil multinationals based in the West, such as Chevron or Shell, in their overseas CSR efforts. In these contexts, we are more likely to find invocations of the material qualities of oil linked to cultural imaginaries in ways that challenge or offer alternatives to the claims of multinational companies (e.g., Adunbi 2013). We do find alternatives of this sort in the Perm region as well, but they have lain largely outside the oil complex and its representational possibilities, although not outside the energy sector—at least not if we grant, along with Stephanie Rupp (2013), that the sorts of energies that humans can harness extend considerably beyond hydrocarbons.

Kamwa: Ethnofuturist Energies

> The energy of the primitive life [of the Finno-Ugric tribes] . . . is a living source that fills the existence of a human being . . . and gives the feeling of a deep connection to the past and the future. . . . Where do we meet Permianness [*Permkost′*]? . . . During an ethnofuturist festival, when the shaman and the artist, in ritual practice, make the border between the everyday and the extraordinary imaginable.
>
> Ivan Riazanov, "The Cultural Archaeology of the Permian Myth"[13]

One of the booths on the periphery of the exhibition hall at Perm's annual Folk Arts Show and Sale was labeled, in simple blue letters, "Kamwa: An Ethnofuturist Festival." The booth's proprietors sold CDs of folk music from the Perm region as well as an array of well-made items designed largely for tourists—postcards and handcrafted jewelry, dresses, shirts, and blouses featuring traditional designs, and a smattering of books on the ethnic and folk traditions of the region's districts. The booth also featured large-format ads for Kamwa's signature annual event: a multiday summer festival that was attended by many thousands. Founded in 2006, Kamwa was the largest, longest running, and most successful of another new breed of sponsored cultural producer in the post-Soviet Perm region: an independent cultural organization whose vision and events originated neither with corporations nor with the local or federal state apparatus. Examining the shape and fate of the Kamwa festival in the mid- to late 2000s serves the same sort of analytic purpose in my overall argument as considering PermRegionGaz's CSR projects:

13. Ivan Riazanov, "Kul′turnaia Arkheologiia Permskogo Mifa," *Shpil′*, November 2010, 8.

by extending outward from the particular conjuncture of oil and culture that ran through Lukoil-Perm's Connections with Society Division—in this case along the vector of culture—we are able to better understand both that conjuncture's uniqueness and its place in a larger field of possibilities in the Perm region.

The moving force behind Kamwa was Natalia Shostina. Shostina, born in 1967, was a native of Perm but had grown up with few links to the official Soviet culture industry (such as Oleg Leonidovich's at Lukoil-Perm) or to Communist Party structures (such as Nina Nikolaevna's at Cooperation). Her education was in a technical field—the design of aircraft equipment—and she spent the 1990s involved in a variety of small-scale experimental theater productions in Perm. In interviews with news media and with me, she located her formative background in Perm's artistic, musical, and cultural underground and to a period of immersion in late Soviet and post-Soviet punk culture. After an extended trip to Moscow and Estonia in the mid-1990s, Shostina returned to Perm and, with several collaborators, began a series of small-scale projects that, a decade later, culminated in the founding of Kamwa. In other words, Kamwa's history is that of a formerly underground form of cultural production that gradually gained new legitimacy as an officially registered and highly visible organization over the course of the late 1990s and into the 2000s—a trajectory quite different from those animating the cultural production sponsored by Lukoil-Perm.

Kamwa was, of course, heavily reliant on sponsors of various sorts, and Shostina was an accomplished fundraiser. Kamwa drew financial support from a range of small and mid-range companies in Perm, as well as some funding for its festivals from the regional Ministry of Culture. Its major corporate sponsor was Ekoprombank, a Perm-based bank that was said to have close ties with the potash-mining operations in the north of the Perm region.[14] In some ways, Ekoprombank's sponsorship of the Kamwa festival resembled Lukoil-Perm's cultural sponsorship, particularly in its effort to use cultural programming to broadcast its solid ties to the Perm region. In its sponsor's introduction to one of the pamphlets distributed at a Kamwa festival, for instance, Ekoprombank explained that its reason for supporting the festival was that both the bank and Kamwa were Permian institutions, independent of and not beholden to Moscow-based financial structures.

But there were also differences, notably the fact that Ekoprombank shared its sponsorship of Kamwa with a wide range of independent and semi-independent organizations involved in the media, tourism, advertising, and cultural production sectors. Radio stations, newspapers, fashionable clothing stores, and many

14. These ties between Ekoprombank and Uralkalii were not generally well known and not an especially visible part of either the bank's operations or the Kamwa festivals.

other groups had some billing on Kamwa's materials. In sum, as an independent, nonprofit cultural organization specializing in cultural and youth projects and an active seeker of sponsorship from multiple sources, Kamwa was a quite different kind of operation than those corporate or state agencies that summoned cultural producers to work on CSR projects. In their instructive exploration of the dynamics of ethnicity in neoliberal times, Jean Comaroff and John Comaroff (2009) discern a dialectic between corporate and entrepreneurial paths to the commodification of ethnicity and territory. Kamwa was decidedly on the entrepreneurial side of this dialectic.

The name Kamwa is itself significant. In some written texts, the organization went by *Kamva*, but its official name for marketing purposes was KAMWA, with the "w" pronounced as a "v". Most observers read the name as a variation of Kama—the river snaking through the Perm region. They were correct, and the organization's marketing materials and house style encouraged this reading with a soft, undulating light-blue design and font. But the organizers also chose their name by inventing an origin myth that combined the Finno-Ugric words for shaman (*kam*), and water/river (*va*). Kam, this new myth went, was the name of an elderly and powerful shaman who, ages ago, solved his tribe's problem of repeated crop failures by striking his foot on the ground and calling forth a mighty, magical river—the Kama. Both the name Kamwa and the process of its creation—a myth and its making—are important clues to the kinds of transformation that Kamwa's cultural projects invited. These transformations took place through artistic engagement with shamanism, with the earth and sun, cosmos and rivers, and with the ancient, sometimes mythical denizens of the Perm region and the artifacts they had left behind for archaeologists to unearth centuries later. Festivals, concerts, folk arts, souvenirs, lectures, and other activities concretized these transformations by attempting to create new forms of relationships and local cultural identities. Often enough, they did this in a language of energy; that is, oil was not the only source of underground energy for cultural projects in the Perm region.

Ethnofuturism emerged among a group of poets, artists, and other creative intellectuals in early 1990s Tartu, Estonia, as an attempt to revive senses of ethnonational identity in the Finno-Ugric diaspora as the Soviet system disintegrated. "It carried," wrote one admirer, "the spirit of liberation and optimism of the 'Singing Revolution' of 1988 which signaled/heralded the collapse of the Union" (Treier 2003, 1–2). Through ethnofuturism, its earliest proponents believed, numerically small Finno-Ugric populations in and beyond the Baltics might find a way to reenvision themselves for a post-Soviet era. Ethnofuturism distinguished itself from other modes of fostering ethnic and national identities by combining a firm belief in an authentic ethnic past with a determination to experiment widely with the ways in which that past might be reimagined in the most contemporary—even futurist—of forms. Fashion, experimental theater, avant-garde art, and electronic music joined experimental poetry and fiction as favored ethnofuturist genres. As one account put it,

> Archaic mythopoetic tradition, folklore, and national romanticism plus modern forms of expression—that is the composition of ethnofuturism. . . . Ethnic culture serves as a base, a source for the creativity of the artist who, integrating tradition, introduces society to a new artistic product and renews its values. (Iuri 2008, 116)

Ethnofuturism in its original Estonian incarnation, in sum, sought to experiment creatively and artistically with some of the classic tensions of Soviet nationalities policy—form and content, deep past and bright future—in the service of charting new kinds of cultural identity in the early Soviet aftermath.

By most accounts, ethnofuturism had more or less run its course as a developed and active movement in Estonia by the time Natalia Shostina lived there in the 1990s, but it had also migrated to other corners of the Finno-Ugric diaspora in Russia and had attracted a following among Mordvin, Mari, and Komi intellectuals. The Udmurt republic, the Perm region's neighbor to the west, was home to a particularly strong ethnofuturist movement in the late 1990s. As Shostina considered the kinds of cultural projects she might embark on in the Perm region, the fact that the northern reaches of the region and the neighboring Komi-Permiak Autonomous District were also part of the Finno-Ugric diaspora, combined with ethnofuturism's willingness to experiment with genres that were quiet alien to official late Soviet cultural production—as she herself was—seemed promising. She began pitching an ethnofuturist organization to potential sponsors and to cultural officials in the Perm regional state apparatus in 2006, and received enough funding to officially register Kamwa and begin a series of cultural projects.[15] Although no longer as closely linked to immediate post-Soviet articulations of identity as was the original Estonian version of ethnofuturism, Kamwa retained the movement's initial focus on the zone between ethnic authenticity and modern genres of cultural production as a generator of artistic creativity, possibility, and new kinds of identities and identifications.

On the authentic side of ethnofuturism, Shostina and her collaborators at Kamwa, including a number of prominent scholars, were major contributors to the ethnography and ethnology of the Perm region. Dr. Georgii N. Chagin at Perm

15. Shostina and her collaborators had already been successful in applying for funding for other projects, during the earliest years of the project movement. See, for instance, "Izucheniia mifologii drevnei Permi s pozitsii etnofuturizma," PermGANI f. 1206, op. 1, d. 200.

State University, dean of regional ethnographers, contributed the introductory chapter to Kamwa's *Almanac*, in which he gave brief sketches of the major ethnonational groups in the Perm region, beginning with Russians and running through Komi-Permiaks, Komi-Iazvins, Maris, Udmurts, Tatars, and Bashkirs (Chagin 2008). Kamwa also ran its own internal ethnographic expedition, headed by Aleksandr Chernykh, a noted ethnographer based in the Perm region's branch of the Academy of Sciences. Kamwa's expeditions were, in many ways, classic salvage ethnography: they sought whatever remainders of authentic tradition and folklore they could find in the region. They collected, for instance, high-quality audio recordings subsequently released as a five-CD set, *The Golden Archive of the Perm Region*, featuring folksongs sung by representatives of the region's major national groups. Through these and other activities, Shostina and her group gained a great deal of respect among academic specialists in the Perm region and participated in the ethnography-based abstraction of national culture. Oleg Leonidovich, who never hesitated to point out frauds in the cultural field, told me on one occasion that Kamwa was "one of the most successful projects in the Perm region from the perspective of supporting interest in traditional culture." Shostina, he said, was "a very serious researcher, who understands everything."

In their efforts to locate and revive authentic cultural identities, in part through the deployment of ethnographic expertise, Kamwa's projects were not dissimilar to those undertaken by the Connections with Society Division at Lukoil-Perm. But Kamwa put new and old knowledge about the Perm region to a wide range of uses from which scholars—and the far more cautious Lukoil-Perm—usually distanced themselves. Far from resting at folk handicrafts or scholarly publications, Kamwa regularly plastered the city with advertisements for a range of activities that used these materials in new ways: Ethnomedia, Ethnofashion, Ethnomusic, Ethnoparty, Ethnotheater, Ethnoatmosphere, and others. "There Is an Ethno-Boom in Perm," read the posters advertising the Kamwa festival throughout the city in the summer of 2009. Two examples give some flavor of the ways in which Kamwa's cultural projects sought to foster new kinds of cultural identities and group them into an overall understanding of the Perm region as a whole.

Kamwa's signature event was an annual ethnofuturist festival, held each summer from 2006 to 2010, which brought together contemporary explorations of local cultural identities across a wide range of performative genres. Each festival brought together groups from around the Perm region and the Komi-Permiak Autonomous District with other performers from around Russia and the world. The 2008 Kamwa festival, for instance, included nearly two hundred guest performers from Ukraine, Petrozavodsk, Saint Petersburg, Estonia, and beyond—many of them specializing in electronic musical adaptations of traditional ethnic music from their own regions. The festivals included everything from all-night raves in disused factories to folklore ensembles from rural villages, and from fireshows on the banks of the Kama River to displays of traditional cultural attire drawn from the collections of Perm's museums. In fact, some of the same folk ensembles and artisans that were supported by Lukoil-Perm participated in the Kamwa festival; at the Kamwa, however, the next act was just as likely to be wielding didgeridoos or synthesizers as folk repertoires from another oil-producing district of the Perm region.

It is notable that the main event of the annual Kamwa festival took place at an outdoor museum complex in the town of Khokhlovka, some forty kilometers from Perm. Khokhlovka was a distinctly Soviet creation, a thirty-five-hectare open-air ethnographic museum that sought to collect and display the architectural and ethnographic styles of the peoples of the Perm region. Walking through the sprawling museum, one could visit representatively traditional houses, churches, barns, and other structures that had been dismantled and shipped, log by log, from villages that had been closed in the course of Soviet centralization and modernization drives up and down the Perm region. Khokhlovka was a classic case of Soviet displays of traditional culture in the service of Soviet nationalities policy. Opened in 1969, it languished for much of the 1990s, along with other museums and the cultural field as a whole. During the Kamwa festival, however, its staid Soviet-style presentation of the friendship of the peoples was transformed into an ethnofuturist spectacle, a multicolored, multiethnic reenvisioning of what cultural identities might look like in the 2000s.

In Kamwa's productions, rigid classifications were replaced with hybrids, unexpected juxtapositions, and all manner of experiments and improvisations on an ethnic theme. In 2009, for instance, the part of the Kamwa festival that took place at Khokhlovka began with the recreation of the ancient rituals associated with the Feast of New Wheat, in which a special section of the museum's territory, carefully selected and sown months earlier by students from the Perm Agricultural Academy, was harvested. The Feast of New Wheat did not draw on any one ethnic tradition, but mixed and matched to create what the organizers discerned as a common element in many cultural traditions: rituals surrounding the new harvest. The events at Khokhlovka that year concluded with an evening gala concert that featured singing and dancing groups from ten regions of Russia as well as guests specializing in African and Brazilian music and dance. At their peak, Kamwa festivals regularly attracted over ten thousand participants to their outside summer festivals, creating massive traffic jams in and around Khokhlovka. "The governor [Chirkunov] came to visit once," Shostina told me, "he was in shock that I had ten thousand people enjoying themselves at a festival—and no drunks."

Although the yearly festival at Khokhlovka was Kamwa's most well-attended event, part of the organization's success lay in its ability to maintain the momentum of its various cultural projects through the whole calendar year—something that no other independent cultural organization in the region could match, at least on Kamwa's scale. For a much smaller and more refined audience in Perm, for instance, Kamwa organized a series of fashion shows—billed as Etnomoda—in some of the city's central exhibition spaces. Etnomoda shows featured fashions that were inspired by the traditional attire of the ethnonational groups of the Perm region, the Finno-Ugric diaspora, and, to a more limited extent, beyond. This was fashion for northern climes: layers upon layers of thick woven fabrics, hats of all shapes and sizes, and elaborate colors and designs. Eschewing the standard runway format, Etnomoda shows featured slow-moving, often dark scenes and tableaus set to music and video selected from the folk repertoire of the ethnic group whose attire was featured, or, alternatively, to gentle electronica or world beat.

Strikingly, and entirely in keeping with the general stance of ethnofuturism, all Etnomoda shows featured not only collections by small-scale fashion designers but also collections of a different sort—those from the Perm Regional Studies Museum. The first Etnomoda show, in 2008, for instance, featured a number of segments in which elderly women modeled items of clothing that, decades earlier, had been collected from their districts by Soviet ethnographic expeditions. As was the case at Khokhlovka, items collected during the Soviet period in the service of the study and display of national traditions gained new life in commodified and spectacularized display. Kamwa's academic wing soon joined in this process as well: the 2010 Etnomoda show, titled "The Other Shore," included a segment on "Authentic Costumes of the Upper Kama Region" that featured items of clothing accumulated in the course of Kamwa's own recently launched ethnographic expeditions. The models on stage were accompanied by musical selections from the Kamwa's recorded folklore collections and projected black-and-white photos (likely taken during Soviet-era ethnographic expeditions) of men and women from around their district posing in traditional dress. This segment was followed by a collection of stylized cowboy outfits, presented to the rhythms of the British electronica band Faithless. In a perfect illustration of the apparently incongruous juxtapositions that were central to ethnofuturist experiments, a series of giant, close-up photographs of beetles filled the screen in the background.

More Energy from the Depths

Kamwa's many projects, then, formed an important way outside of Lukoil-Perm's and the state administration's cultural projects that residents of the Perm region—

especially younger, more urban residents—encountered the revival of ethnic or national identities. Although there was certainly a similar vision of tolerance, mutual engagement, and interethnic understanding to be found in Kamwa's events, these elements were routed neither through a lightly modified Soviet-style sensibility about the friendship of the peoples nor through the kinds of cultural politics that animated corporate social responsibility. But on other counts Kamwa's projects bore some significant resemblances to Lukoil-Perm's cultural sponsorship, most notably its focus on energy sources beneath the subsoil of the Perm region that might be tapped for their transformative powers.

Running through most of Kamwa's cultural projects—festivals, fashion shows, and otherwise—was an engagement with myths and artifacts of the Perm region. In some cases, as in the case of Kamwa's own name, these were new myths about ancient times: "We are all mythmakers," Natalia Shostina was fond of saying. Some myths had circulated for some time, such as the myth of the ancient Chud tribe, forerunners to the Komi-Permiaks. The Chud, it was said, were the inhabitants of what became the Perm region long before the arrival of the first Russians in the fourteenth century and well before the mighty Stroganov and Demidov dynasties. (The Chud are mentioned in the twelfth-century Primary Chronicle as inhabitants of the Urals.) Rather than be colonized by Russians, it was said, the Chud went into hiding in caves far below ground; in some places in the northern and western areas of the Perm region, mysterious groans and calls could still be heard echoing from below the earth. The Chud left behind, however, a great number of ritual objects and representations—bronze castings of animals, humans, and human-animal hybrids that were collectively known as "Perm Animal Style" (*Permskii Zverinii Stil'*).

Artifacts in Perm Animal Style (see figs. 10a and 10b) had long been collected and studied by archaeologists and other scholars, and they had featured frequently in regional museums since the nineteenth century. The possibility that more such artifacts would be discovered occupied a nontrivial place in the imagination of many in the Perm region during the 2000s. On one occasion, when I was standing at the Perm Animal Style display in the Perm Regional Studies Museum, a schoolteacher appeared out of nowhere and told me with great enthusiasm about the summer trips she had taken her students on in search of Perm Animal Style artifacts. They had not found anything, she said, but the students had learned a great deal about archaeological techniques and the history of the Perm region. Among culture producers and managers, rumors of valuable private collections and a vibrant black market for figurines circulated.

Recent scholarship dates most of these castings from the seventh through the twelfth centuries. They were likely made by a number of different Finno-Ugric

FIGURES 10A AND 10B Drawings of Perm Animal Style archaeological artifacts of the sort that became popular in the Perm region in the first decade of the twenty-first century. On the left, a human-elk (*chelovekolos'*). On the right, a three-headed goddess. Perm Animal Style artifacts were generally interpreted as conveying connections among three worlds: the underworld, represented by the burrowing pangolin upon which each figure stands; the earth, represented by humans, human-animal hybrids, or humanoid deities; and the sky, represented in the three-headed goddess figure by eagles.

Sketches by April Wen.

tribes and, indeed, professional archaeologists are more or less united in their opinion that "Perm" is a somewhat misleading adjective for this kind of Animal Style. Although there do seem to be certain distinctive qualities to some of the figurines found in the north of the Perm region, broadly similar figurines are part of the archaeological record across much of the Eurasian landmass. The scant record of these early tribes has made scholarship on Perm Animal Style a particularly challenging topic of archaeological investigation (see Ignat′eva 2009), and it is exceedingly difficult for scholars to confidently associate particular artifacts with particular tribes and their movements.

For Shostina and her ethnofuturist colleagues at Kamwa, however, scholarly precision about the artifacts classed as Perm Animal Style was not the only goal. They were particularly drawn to theories of Perm Animal Style positing that these bronze castings conveyed the worldviews of Finno-Ugric shamans, especially their relationships with animals and their ability to mediate among the lands of sky, earth, and underground.[16] Recounting her pre-Kamwa experiences in cultural production to me, Shostina traced some of her key inspirations to Perm Animal Style: "I began to see how much information these figures have for me." She offered a similar recollection to journalists who asked about the origins of the Kamwa festival:

> The Chud, for me, are a parallel, living world, which really exists, which allows us to know something about it, although sparingly . . . and we entered into that Finno-Ugric world as a theater project that was dedicated to the Chud and to Perm Animal Style. . . . I understood that it was my scientific and creative line, for which I am responsible, not because I like it, but because it can tell the world something that can only be picked up in Perm."[17]

As the Perm Parallel Project—Shostina's experimental theater group—grew into the much more ambitious Kamwa festival, the centrality of myths and Perm Animal Style remained. In the 2008 Etnomoda show, for instance, the mystical figures of Perm Animal Style were featured on everything from hats to dresses to shawls and more.

With a boost from Kamwa's cultural projects, Perm Animal Style began to rise in prominence throughout Perm. Replicas and various art forms based on Perm Animal Style, many of them sold at the annual folk handicrafts fair or Kamwa's year-round booth in Perm's central store, began to appear in several of my friends' apartments. One year I was surprised to see that the door numbers on the rooms of the hotel in which I stayed on my shorter trips to Perm had been redone in

---

16. In her treatment of the history and study of Perm Animal Style and its collections, O. V. Ignat′evna argues that, in the 1960s, the dominant mode of studying Perm Animal Style shifted to structural-semiotic, with scholars interested in the field of symbols and signs in which these representations were embedded; see also Dominiak 2010. In its authenticity mode, Kamwa fostered and popularized this approach to interacting with the objects of Perm Animal Style. See, for instance, Oksana Ignat′eva and Anna Panina, "Kuda Ushli Chelovekolosi?" Kamwa website, May 5, 2007, http://old.kamwa.ru:443/publications/1/ (accessed December 8, 2014). The Kamwa website was frequently updated with reprints and links to new scholarship about Perm Animal Style.

17. Aishat Temirova, "Vremia Ponimaniia," *Vestnik actual′nykh prognozov: Tret′e tysiacheletie* 2009 (11).

Perm Animal Style. The local chocolate factory began to produce boxes of chocolates in Perm Animal Style suitable for gifts and, especially, souvenirs for tourists and visiting businesspeople. Perm Animal Style was the starting point for the introductory video at the 2010 Forum of Regions, and the gathered state officials and businesspeople from around Russia and Europe received small figurine replicas as part of their registration packets. Perm Animal Style figurines also served as the cover and internal images for the Perm Regional Administration's annual report of the Department for the Development of Human Potential in 2010. These examples could be extended indefinitely (see fig. 11).

This was not all the work of the Kamwa festival, of course, but Kamwa's ongoing presence and its overall determination to resituate old artifacts in new, experimental, and commodified contexts were important vehicles by which Perm Animal Style rose to prominence in Perm. It helped, too, that the very name Perm Animal Style sounded quite modish and was not easily associated with any particular corporation or sector of economic activity. When city or local state officials, tourist boards, and others sought a brand for Perm or the Perm region, Perm Animal Style was an easy fit—vaguely exotic, obviously connected to age-old

FIGURE 11 Perm Animal Style images adorn the front of Perm's Central Store (TsUM).

Photo by author, 2011.

traditions, stylish by its very name, evocative of unspecified but meaningful powers of transformation, and Permian in the most general of senses.

In sum, Kamwa's ethnofuturist projects, as well as their broader reverberations, relied centrally on a myth about a tribe hidden in caves far beneath the Perm region and on millennium-old artifacts that were unearthed by archaeologists, still sought by schoolchildren in the summer, and increasingly reproduced and displayed throughout the city. It was not uncommon for Shostina and her collaborators to talk about the cultural transformations and revitalizations that these objects from the depths could bring about in the language of energy. Kamwa's outdoor festival at Khokhlovka, I was told, was designed to be an event that released "positive energy." So, too, were the handmade clothes that incorporated themes from Perm Animal Style on display at Etnomoda. To buy and wear such clothes, Shostina told me, was to receive "positive energy . . . quality, love, and ecology. . . . It's not a product that kills." Indeed, according to one participant, the migration of the entire ethnofuturism movement from Udmurtia to Perm, could be summed up in terms of energy: "the river of ethnofuturism brings its energy from Udmurtia to Perm" (Iuri 2008, 117).

The energy to accomplish personal and cultural transformation was, in these framings, precisely what could be accessed by engaging with the signs and symbols of Perm Animal Style and other ancient artifacts and myths within the context of ethnofuturist experiments. Ethnofuturism, one of its proponents wrote, has two wings, one in the past and one in the future; "life energy is formed at the junction of those wings."[18] The "energy of the primitive life" of the Finno-Ugric tribes, another Kamwa enthusiast wrote, is a "living source" that "fills out the existence of a person who so little and so rarely returns to himself and gives himself a deep connection with the past and the future." Where, he went on, do we meet Permianness (*Permkost′*)? The answer: "During an ethnofuturist festival, when the shaman and the artist, in ritual practice, make the border between the everyday and the extraordinary imaginary."[19]

As in Lukoil-Perm's handicraft projects and Historical Cities festivals, we find in this framing the transformation of the everyday in the context of cultural spectacle, and again it is energy drawn from underneath the Perm region that makes this transformation possible. The energy of the Kamwa festival was the energy of the shamans of the ancient Perm region, an energy that came not from reviving shamanism per se, but through artistic creativity that unleashed ancient powers in new ways. That this transformative, shamanic energy underground could be accessed without the intervention of massive corporations and their

18. Tina Karrel′, "My tol′ko tvorim, a nazvanie pridumaiut iskusstvovedy," January 20, 2009, http://properm.ru/news/afisha/8927/ (accessed December 8, 2014).

19. Riazanov, "Kul′turnaia Arkheologiia Permskogo Mifa," 8.

techniques for extracting, refining, and selling hydrocarbons was a point left largely unstated. But it did occasionally appear. A journalist from Moscow visiting the 2009 Kamwa festival at Khokhlovka reported that his guide, a Perm-based culture worker who was volunteering at the festival, told him, "when the oil runs out in Russia, we will be able to trade in historical experience . . . and the Permian lands will serve as a distinctive staging area for the movement of Russian civilization."[20]

Energy was not the only, or even the dominant, discourse of transformation in which Kamwa's organizers and enthusiasts spoke. The word "energy" made no appearance on Kamwa's major marketing materials or festival advertisements, despite its relatively common appearance in lower-circulating publications and interviews. It is hard to analyze a silence, of course, but one reason for this absence seems to be rooted in the fact that various mystical sources of energy had been a major and controversial part of Russian public discourse in the 1990s. Despite its ubiquity (Lindquist 2005), the talk of bioenergy in the 1990s was firmly associated with cults, sects, and new-age healing, and was roundly condemned by Church, state, and most intellectuals. Indeed, in the first years of Shostina's Perm Parallel Project, she and her collaborators were often accused of running a cult. Recalling her early theatric attempts to uncover the history and power of the Chud, she told one interviewer:

> In the first two years, I was accused of being a sectarian: white people [i.e., full face and body makeup on stage], white clothes, strange dances. Perm is, in general, one of the epicenters of extrasensory zones in Russia, as there are a lot of [scientific] experiments performed here. . . . In fact, what we were doing was not shamanism but art—the reconstruction of an ancient people through the tools of art. So we stood firmly on our feet and felt that the time when we would be understood would come.[21]

If, in other words, one wanted the intellectual and social legitimacy as well as a degree of state and corporate sponsorship that came with large-scale festivals and ethnic revival movements, then artistic creativity rather than mystical energies, bioenergetic fields, and actual shamanic revival was a far better choice, at least for the purposes of public presentation.

Nevertheless, the construal of energy from the depths as fuel for cultural transformation was common enough in the literature surrounding the Kamwa festival that it calls for some explanation. The most helpful pointer in this regard is Stephanie Platz's (1996) remarkable ethnography of Armenia at the time of the

---

20. Aishat Temirova, "Vremia Ponimaniia."
21. Ibid.

Soviet collapse. The Armenia in which Platz did her fieldwork was suffering through widespread energy crises: shortages of fuel and electricity were endemic in conditions of an Azerbaijani energy blockade and resulting economic demodernization, with major implications for the organization of everyday life. In these conditions, Platz unexpectedly found herself drawn to studying a social movement concerned with UFOs—unidentified flying objects—and especially with the scientific documentation of their visits to Armenia and the transformations of energy they heralded. UFOs, Platz found, were said to be recharging depleted energy reserves in Armenia and offering access to alternative points of energy, including the bioenergy of Mount Ararat. These visits, scientists demonstrated through soil samples and other methods of analysis, extended far back into the past and affirmed Armenia's special place in the world far into the future. UFOs' close association with Armenian pasts and futures, Platz was able to show, made them integral parts of the imagination of the Armenian nation in a time of crisis and energy blockade. They did this precisely by providing alternate sources of energy.

We might see the links between energy and culture fostered by Kamwa as part of the same family as those discussed by Platz for Armenia—a family in which fields of energy production and fields of culture interact to articulate a particular kind of past and project it into a future. Although Platz wrote of times of energy crisis and the Kamwa festival emerged at a time of energy boom, both were contexts in which the sources and transformations of hydrocarbon-based energy were controlled by politically powerful and increasingly unassailable others. By claiming access to alternate sources of energy and alternate ways to transform them and, moreover, by linking those sources to an alternate history and an alternate materiality of the nation/region, both Armenian ufologists and Kamwa festival organizers refused to allow the terms of power to be dictated exclusively by the powerful. There were, Kamwa posited, alternate sources of energy that were just as closely related to the identity of the Perm region as those of hydrocarbons. These sources were to be accessed through an alternate geology and stratigraphy of the region, and they were indexed to the artifacts, myths, and rivers of a much older civilization—a time between the formation of oil deposits in the Permian, Carboniferous, and Devonian geological periods and the coming of the Stroganovs and the Soviets. There was, in the phrasing of Shostina's original project, a parallel Perm, or, in the name of one of her later projects and title of a collection of articles she edited some fifteen years later, "Another Perm."[22] Festival, spectacle, and artistic creativity in all its forms, when trained on things from the depths, were the routes into this other Perm.

---

22. Natalia Shostina, "Drugaia Perm," special issue of *Shpil'* 7 (51), November 2010.

A Field of Energo-Cultural Production

Many of Lukoil-Perm's primary corporate slogans, adorning everything from the foyer at the company's headquarters in Perm to its gas stations, were versions of "The Energy of the Subsoil—for the Good of People" (*Energiia Nedr—Vo Blago Chelovka*) (see fig. 4 in chapter 5). The previous chapter described the ways in which Lukoil-Perm accomplished this transformation not only through extracting, refining, and selling oil products but also through increasing participation in, and indeed the vertical integration of, the cultural sphere. By expanding outward from these projects to another center of cultural production in the Perm region in the 2000s, this chapter has shown that Lukoil-Perm was not the only organization intent on cultural transformations that prominently featured energies and objects extracted from beneath the surface of the Perm region. The field of similarities and differences between Lukoil-Perm's and Kamwa's cultural projects facilitates some broader observations about the place of energy and culture in the 2000s in the Perm region.

While both Lukoil-Perm and Kamwa were interested in energy from the depths as a vehicle for cultural transformation, their preferred objects led to quite different profiles. Lukoil-Perm's invocations of the depths of oil and culture, for instance, associated the corporation with the Stroganovs, powerful patrons of the Russian past, and with the fostering of a friendship of the peoples reminiscent of certain periods of Soviet nationalities policy. The Kamwa festival concentrated on the much shallower depth of archaeological finds, but linked it to a deeper, hazier, pre-Russian history.[23] Although both were concerned with producing region-level cultural identities, there were notable spatial differences. Whereas Lukoil-Perm's cultural projects concentrated on oil-producing districts of the Perm region, most of them at some distance from Perm, the Kamwa festival's cultural projects unfolded mostly in and just outside Perm itself, and were staffed and attended largely by an urban population. Neither Lukoil-Perm nor the Kamwa festival shied away from the commodification of ethnicity and territory, and both were heavily invested in the use of cultural spectacles to transform the dilemmas and uncertainties of the 1990s into the more assertive displays of the following decade.

Exploring these intersections of depth, energy, and culture in the Perm region shows, once again, the great extent to which Russian national imaginaries are often produced and experienced first and foremost as regional and subregional

---

23. In an instructive account of the geographical metanarratives of Russian national imaginaries, Marlene Laruelle (2012) points to three dimensions in which Russia's unique place in the world is often framed and debated: larger, higher, and farther north. "Deeper" deserves a place on her list.

imaginaries. For both Lukoil-Perm and the Kamwa festival, a crucial shared expectation about cultural projects was that they should proceed through a semiotic recoding and/or decoding of material objects extracted from beneath the Perm region. Chapter 6 showed how Lukoil-Perm went about resignifying oil as a material substance. One of Kamwa's projects dedicated to Perm Animal Style proclaimed that it was about revealing the "secret origins and incompletely decoded symbolism" of these figurines.[24] To engage in these acts of decoding, of symbolic interpretation, was both to better understand the past and to reposition oneself in the present and future. Kamwa's cultural projects were, that is, largely an invitation to treat regional culture as a decodable text—a play of symbols and meanings that ordered past, and therefore, at least potentially, present regional lives.

As might be expected, the independent and noncorporate Kamwa organization had a much lighter hand in this regard than Lukoil-Perm, whose efforts went beyond invitation to careful scripting. A Kamwa publication on Perm Animal Style summed up this position succinctly: "Is the information encoded in these bronze cult objects, masterfully created by ancient casters, only intended for specialists? Now anyone can try to 'read' what they find in museum showcases or old books."[25] A good example of this stance is the New Faces of Perm Animal Style project, sponsored by Kamwa and a number of media and state cultural organizations. New Faces featured sketched outlines of Perm Animal Style faces on a large glass panel (see fig. 12). Members of the public were invited to pose behind the glass so that, viewed from the front of the display, their own faces were overlain with those of the figurines. Photographs of visitors posing in such a manner were hung nearby. It would be hard to imagine this kind of interpretive openness in projects sponsored and tightly managed by the Lukoil-Perm Connections with Society Division.

To be sure, Kamwa organizers and affiliates were not above pushing their own interpretations and policing those of others. The very same article that suggested that anyone could interpret artifacts went on to argue that a particularly popular Perm Animal Style casting of a hybrid of an elk and a person (*chelovekolos'*; see fig. 10a) had a very specific message: "Our deep ancestors are trying to say to us: 'Live in harmony with nature.' The human-elk is the embodiment of that harmony."[26] And certain interpretations were simply deemed wrong or worthy of contempt. In one story I heard, a government official interested in contributing to the branding of Perm obtained a large Perm Animal Style piece to hang in the

24. "Novye Liki Permskogo Zverinogo Stilia," Kamwa press release, February 26, 2009.
25. Stepanova, "Kuda Ushli Chelovekolosi?" 10. See also Erenburg 2011.
26. Stepanova, "Kuda Ushli Chelovekolosi?" 11.

FIGURE 12 The New Faces of Perm Animal Style station at the White Nights in Perm Festival in June 2012. Hanging nearby were large displays of printed photographs of the sort being taken here—photographs in which Perm Animal Style faces appeared superimposed on smiling guests at the festival.

Photo by author, 2012.

waiting room outside his office. He proudly called the Kamwa offices to consult about its meaning, only to find out that it was a death mask—not the ideal greeting for out-of-town dignitaries. According to another story, perhaps apocryphal, when Governor Chirkunov began to learn about Perm Animal Style and its potential as a regional brand, he summoned a number of archaeologists and, entirely unaware of the technologies used to make these figures nearly a millennium ago, asked them, "Can you find me a gold one?" The Kamwa volunteer telling me this story rolled her eyes hopelessly at this point. Apocryphal or not, the story speaks to the kinds of transformations that were appropriately thinkable from the Kamwa perspective. Unlike oil, Perm Animal Style castings extracted from the ground were not useful for their monetary value but for their interpretive value: looking for a gold one was just the kind of mistake a corporate or state agency representative would make. Even as they skewer state functionaries, however, both of these stories speak to the power of just those state representatives and their corporate allies in Lukoil-Perm, and to the dominance that their transformations of the cultural and geological depths of the Perm region had gained in recent years.

I noted above that PermRegionGaz's CSR projects did not last for long in the Perm region, dissipating after a few short years even as Lukoil-Perm's projects became ever more central to regional culture and politics. The same turned out to be the case for Kamwa, at least in the form I have discussed it here. Already by the time I interviewed her in 2010, Natalia Shostina had declared that, after five successful years, the time of major Kamwa festivals in the Perm region was over. When I asked why, she merely nodded out the window in the direction of the main offices of the regional state administration, which was, at the time, in the early phases of its new campaign to transform Perm into a European Capital of Culture. In the face of this massive new state-sponsored cultural production, most of which she considered substandard at best, it was time to move on. She and her crew were, she said, headed "back to where we came from—underground [*ander-graund*]."[27] My notes show that I felt disappointed for her at the time, but I now understand this comment somewhat differently: for Kamwa's organizers and close followers, there was no more powerful and generative place than the underground.

---

27. See also her similar phrasing in Iulia Batalina, "Chuzhie Zdes′ ne Khodiat," *Novyi Kompan′on*, January 3, 2011.

8

"BILBAO ON THE KAMA"?

The Perm Cultural Project and Its Critics

"Do you know that there is a Cultural Revolution happening in Perm?" The upbeat, customer-service-friendly voice belonged to the teenager minding the museum shop at the Perm Museum of Contemporary Art. It was March 2011, and I had come for the opening of an exhibition of avant-garde video art featuring the Moscow-based collective AES+F's Feast of Trimalchio and a number of other installations. Smiling brightly, she directed me to the newest additions: stylized designs featuring Perm's regional symbol—a bear—and a new raft of notebooks, tote bags, and other items featuring "I Love Perm" designs. On the shelves behind these souvenirs stood an impressive array of beautiful and expensive art books for sale, including a growing collection of catalogs from the museum's own exhibits. The shop was tucked into the corner of a vast exhibition space, spread out over two floors in what was once Perm's River Station Hall, originally built in 1940 to serve boat traffic on the Kama River and, in 2008, refurbished and reopened as a contemporary art museum. I had been to the museum several times before, but the practiced jollity of that historically freighted line about cultural revolution brought me up short. It seemed as incongruous as the video art itself. AES+F's reinterpretation of the story of Trimalchio (a former Roman slave turned thrower of lavish feasts who appears in the middle chapters of Petronius's *Satyricon*) was a set of large, digital animations and stills depicting scenes of over-the-top luxury at a twenty-first-century resort hotel. The excesses it depicted were especially striking installed inside the soaring spaces of the Perm River Station's Stalin Empire-style architecture. It is unlikely that the production of such jarring moments was part of that young woman's training, but the idea that contemporary art can produce unsettling moments was at the very heart of the museum's mission to spur individual and collective transformations in Perm. Indeed, the Perm Museum of Contemporary Art, known as PERMM, was the epicenter of a series of high-profile cultural projects and conflicts associated with Governor Chirkunov's campaign to have Perm officially declared a European Capital of Culture in 2016.

Art and Oil

Culturally focused corporate social responsibility projects located at and near sites of oil extraction—of the sort discussed in chapter 6—are a comparatively recent phenomenon worldwide. A much longer-running set of ties between the global oil complex and the production and consumption of culture has lain elsewhere, closer to the urban sites where oil wealth has tended to accumulate. These ties include corporate sponsorship of major urban cultural centers (art museums, theaters, opera houses, and so on); executive philanthropy and trusteeship at this same group of institutions; and the avid participation of both oil companies and individuals with oil-derived fortunes in the collection of works of art. "The major enduring legacy of the wealth of [dynastic families]," write George Marcus and Peter Dobkin Hall in their ethnography of American inheritance and philanthropy, "is Culture itself" (1992, 9). A particularly dense set of such ties, for instance, has long woven together New York City's Museum of Modern Art and several generations of the Rockefeller family (see Rockefeller Archive Center 2011, 28–29), and corporate involvement in the global art world has only increased since the 1980s (Wu 2002).

Relationships of this sort began to emerge rapidly in the mid-2000s boom in Russia, when new fortunes had been largely consolidated, oligarchs were looking to establish long-term legacies, and Russian corporations were responding to critiques from all quarters. Major Russian oil and gas companies sponsored all manner of exhibits, renovations, and performances at Russia's most respected cultural institutions. Big Lukoil, to give but one example, established a standing relationship with the State Tretyakov Gallery in Moscow in late 2007, sponsoring a number of exhibits and restorations and, in 2008, an international traveling exhibit titled "Magical Landscapes." These were also heady days for private collectors on the Moscow art scene, with around $68 million spent at auction for Russian contemporary art over the course of 2007 and 2008 (Hewitt 2010, 77).[1] Some of Russia's new rich went further still, not only expanding their own

1. Degot 2010 and Diaconov 2011 provide useful insiders' snapshots of these years; see also Chukhrov 2011.

collections as a way to reinvest their millions but also helping to build new exhibit spaces, such as the popular art4.ru and Vinzavod galleries (both opened in Moscow in 2007), sponsoring favored artists through personal grants and subsidies, and even curating exhibits and shows (the most-cited example is Dasha Zhukova, partner of Roman Abramovich).

These kinds of practices and relationships began to emerge, on a much more modest scale, as new oil wealth coursed through Perm. In addition to the grant-based competitions for social and cultural project funding in which district-level cultural organizations participated, Lukoil-Perm (together with the Perm Financial-Productive Group [PFPG]) served as general sponsors of both the Perm Opera and Ballet Theater and the Perm Drama Theater. Perm's first independent art gallery, Maris-Art, opened in 2001, and specialized in works by artists from the Perm region, past and present. There was no shortage of art to display or visitors for exhibitions in the gallery's small space in central Perm, across the street from the heavily trafficked Perm State Art Gallery, but sales were another story. Maris-Art initially served a very small number of independent collectors—too few, in the early years, to pay the rent. "Thank God," its director told me, "that in those early years an understanding grew up in Perm that there is such a thing as a corporate gift." The PFPG, Lukoil-Perm, and Permneftorgsintez were her chief clients in the early 2000s, and Maris-Art regularly advised all three on gifts for others in the business world and, just as frequently, filled commissions to provide artwork for corporate offices. A number of Lukoil-Perm executives also took a personal interest in the emerging art world, in at least two cases providing personal contributions to help the gallery publish catalogs and other books. The gallery maintained a warm relationship with Andrei and Nadezhda Agishev, who over the course of the 2000s became perhaps the most widely known independent supporters and collectors of art in the Perm region. (Andrei Agishev, recall, was a PFPG veteran who had also spent time at the head of PermRegionGaz and went from there to a position in the Perm Legislative Assembly; his wife Nadezhda chaired the board of directors at the Ermak Investment Company and, after a period of private collecting, founded the independent New Collection Foundation for the Support of Cultural Projects, which focused specifically on art.)

The global financial crisis and associated plummeting oil prices—from $147 per barrel in July 2008 to a low of just $32 per barrel later that same year—had an immediate impact on Russia's art scene. By 2009, global auction sales of Russian contemporary art dropped to a mere $6.4 million—a small fraction of the previous years' totals (Hewitt 2010, 78). By 2010, the once enormously successful art4.ru gallery in Moscow was open only one day a week, its sponsorship money having nearly dried up and ticket sales an insufficient replacement for a decimated operating budget. It therefore seems counterintuitive that the Perm Museum of Contemporary Art and the cheerful Cultural Revolution it was intended to spark originated in precisely these crisis years. This apparent anomaly, however, was in fact at the very center of ongoing transformations of corporation, state, and cultural politics in the post-Soviet Perm region. Accounting for this anomaly adds a final example to the ways in which the Perm region was constituted as an oil-producing region within Russian state and corporate fields after socialism. Even as Perm's Cultural Revolution was designed to chart a path out of hydrocarbon dependency, I will suggest, the debates and conflicts that surrounded it served largely to extend and entrench the centrality of oil to regional political, economic, social, and cultural sensibilities. The material properties, spatialities, representations, and regional transformations of oil were again highly significant.

Oleg Chirkunov's Team

Vladimir Putin appointed Iurii Trutnev, whose efforts to transform Perm into Russia's Capital of Civil Society helped set the tone for regional state-corporate relationships in the critical early years of the 2000s, to the position of minister of natural resources of the Russian Federation in 2004. Trutnev, along with a number of his key deputies, promptly decamped for Moscow. In Trutnev's place, Putin installed Oleg Chirkunov, first as acting governor in 2004, then as officially appointed governor in the Perm region in 2005. (The post of governor switched from elected official to designee of the president of the Russian Federation in 2004 as part of the Putin administration's efforts to reconsolidate federal power.) Although Chirkunov's initial background was in the Soviet security services (he graduated from the KGB's higher school in 1985 and served, by most accounts, until 1993), he was more well known in the Perm region as a member of Trutnev's former management team at Eks Limited, a sprawling company that had made a fortune as a supplier of consumer goods since the late 1990s—in part based on the European connections Chirkunov had made during a stint as Russia's trade representative in Switzerland. The first years of Chirkunov's term as governor coincided with massive influxes of oil- and gas-based revenues into the Perm region and Russia as a whole, and Chirkunov's former associates at Eks Limited were well positioned to take advantage of the boom in consumption, especially from their new Sem′ia chain of department stores, which rapidly spread through Perm and other major cities in the region. Chirkunov's wife and children remained in Switzerland during his entire term as governor and had dual citizenship; these ties and frequent visits to Western Europe were generally assumed to be motivating, at least in part, his European Capital of Culture campaign and his interest in contemporary art.

Although certainly no overt or persistent critic of the Putin administration, Chirkunov was far from a simple executor of plans hatched in Moscow. He was not, for instance—and never became—a member of Putin's United Russia Party. "He is, basically, a trader," I was told more than once. "What he knows how to do more than anything else," said one acquaintance, "is to buy and sell things." "He's a Thatcherite," said another. In the classification of Russian political/economic styles of the time, Chirkunov was a self-identified and proud *liberal*, a close enough approximation to what is often meant by the term "neoliberal" in the West. He believed fervently in the power of markets and entrepreneurship, and was skeptical of any significant role for government in economic or social spheres. In the introduction to a collection of his own writings and reflections published in 2012 with the support of the Russian Liberal Mission Foundation, Chirkunov wrote that, in the twenty years of reform since 1991,

> there occurred in the public consciousness and in the consciousness of the elite a refusal of a competitive, market model in favor of stability, planned economic development, and, connected to both, the inevitable centralization of power. This book is an attempt to propose an alternative to the strengthening of the state in all spheres of life. (Chirkunov 2012, 24)

As governor, Chirkunov championed tax cuts for regional businesses through the 24–20 Program and did everything he could to slow and, eventually, stop the regional administration's social and cultural project grants that had been so central to his predecessor's administration, believing them to be an unhelpful form of government subsidy. Instead, he traveled the region promoting the virtues of entrepreneurship as a solution to local social problems.

One of the central tensions running along the Moscow–Perm axis in the years of Chirkunov's governorship (2004–12) was, then, a generally increasing centralization of the federal state apparatus in Moscow coupled with an avidly market-oriented and anticentralization (yet federally appointed) governor. The boom years of the mid-2000s masked some of this tension for a time, as recirculated oil wealth coursed through the region and the country and meant that social problems could be temporarily solved with infusions of cash. But, as Chirkunov noted in the same 2009 speech in which he declared Perm's Capital of Culture ambitions, all was no longer well. Between 1998 and 2007, employment in the drilling and mining sector had fallen by 25 percent as Lukoil-Perm and its extractive industry companions restructured their labor forces. The entire sector (including both oil and mining) employed only twenty thousand people in 2009—1 percent of the population of the Perm region. Whatever money Lukoil-Perm and its fellow natural-resource-exploiting corporations were generating, they were not providing jobs for the region even in the boom years of the mid-2000s, and the decline in regional tax revenues beginning in 2008 was exposing problems. Indeed, people were still moving out of both Perm and the Perm region at a steady rate, unemployment was increasing, and many social services continued to be unreliable. In the face of these circumstances, the prospect of further declines in tax receipts from Lukoil-Perm, and increasing demands from the population, what was a market-oriented governor to do?

By his own account, Chirkunov's discovery of culture as a possible answer began in 2007. That was the year in which he named Sergei Gordeev as one of the Perm region's two senators—delegates to the Russian Federation Council (the Upper Chamber of the Russian Legislative Branch). Gordeev, a Moscow-based billionaire who had made his fortune by founding a real estate business in 1995 and then riding it through the astounding boom of the early and mid-2000s, had no previous relationship to the Perm region. His entry into Russian political life had come during a period of service as a senator from the Ust-Orda Buryat Autonomous Republic (also not his home). His appointment from both areas points to a common strategy among Russian provincial regions at the time: recruiting rich and politically connected patrons based in Moscow as senators in the hopes of diverting personal and federal funds to projects in the regions that they represented.

Only in his mid-thirties at the time he began his term as a senator from Perm, Gordeev had developed an interest in urban architecture and, in particular, in rescuing and preserving avant-garde architecture in Moscow—much of which was threatened by decades of neglect and the tendency of most developers, with the support of powerful Moscow mayor Iurii Luzhkov, to level whatever stood in the way of building the next hulking business center or shopping mall. Although professional devotees of architecture greeted Gordeev's interest in the avant-garde and urban design with considerable skepticism at first, he had won over many of his critics through his dogged determination to transform the Melnikov House, near the very center of Moscow, into a museum. In 2007, Gordeev helped to fund and participated in *The Lost Vanguard: Russian Modernist Architecture, 1922–32* at the Museum of Modern Art in New York, and his foundation, Russian Avant-garde, supported collections and the publication of a series of books on major avant-garde artists and movements. Press releases and published retrospective accounts indicate that the idea to establish a Museum of Contemporary Art in Perm emerged fairly early in the collaboration between Governor Chirkunov and Senator Gordeev. Both were interested in making a high-profile move to transform Perm on the model of the European postindustrial cities they were familiar with, Bilbao and Glasgow among them, and they agreed that this long-term project should begin with a contemporary art museum.

The network that set in motion what would eventually be called the Perm Cultural Project thus emerged, on the one hand, out of the new and booming consumption side of the Perm region economy represented by Chirkunov (as opposed to Soviet industry, from which Gennadii Igumenov had hailed, and the Communist Party's Komsomol wing, which had been Iurii Trutnev's primary background and power base) and, on the other hand, out of the Moscow-based nexus of fantastic, hydrocarbon-charged real estate wealth and avant-garde art represented by Gordeev. To help execute their project of transforming Perm, the duo turned to Marat Gel′man, a Moscow-based collector and gallerist at the very center of the contemporary art scene. Gel′man was the founder of the first private art gallery in Moscow (the Marat Gel′man gallery, opened in 1990) and had recently begun dabbling in politics as well, consulting on a number of political campaigns beginning in the mid-1990s through his Foundation for Effective Politics.

The Moscow art world was not without its intrigues and scandals—and Gel′man was rarely far from them—but bigger threats to Gel′man and his colleagues' operations in Moscow were the instabilities that came with their close ties to Russia's hydrocarbon boom. Real estate prices in Moscow, for instance, were becoming prohibitive for gallery and exhibition space, let alone for beginning artists to live. Gel′man related to me in an interview that real estate in Moscow had become a like "gold bars," with everyone "investing their oil money in it and just waiting for the price to go up." It was hard, he said, to maintain a gallery in those conditions, at least without ever-increasing sales and prices for art. Indeed, Gel′man was fond of telling newspaper reporters that 80 percent of his customers at the Marat Gel′man Gallery in 1996–2008 had moved abroad, and that they were replaced not by other wealthy private businesspeople who might dabble in expensive contemporary art but by wealthy members of the Russian state apparatus who were interested in hiding their wealth rather than being conspicuous about their fortunes and sophistication by participating in the Russian corner of the global art market.[2] In these conditions, Governor Chirkunov and Senator Gordeev's invitation to Gel′man presented an opportunity that seemed to be both obvious and insane: a move to the provinces, where, with few exceptions, there had yet to be much interest in contemporary art, where property prices were lower by orders of magnitude, where artists themselves could more easily make a living, and where new regional elites might be interested in following Moscow elites into the art market, both as a form of investment and as an arena of distinction.

2. Masha Charnay, "Marat Guelman: 'Things Can Work Differently,'" *Russia behind the Headlines*, May 17, 2012.

At Chirkunov's urging, Gordeev's ambitions for Perm had never been small. His first thought was apparently to open a branch of the Guggenheim in Perm, and negotiations got as far as site visits by the leadership of the Guggenheim Foundation. Ultimately, however, he and Chirkunov settled on a different plan ("the Guggenheim is the day before yesterday," Gordeev told the magazine *SNOB* in 2010): opening the Perm Museum of Contemporary Art, with Marat Gel′man as director.[3] Gordeev supplied personal funds of around $400,000 to renovate Perm's River Station Hall for this purpose, and the space was reregistered as a state-owned museum in short order.[4] Gordeev described his decision to a local magazine:

> The prevailing idea in the world is that a museum of contemporary art is a multiplex where networked products are rolled out. For example, the exhibition of a big artist's work travels around the world, very stylish, glamorous, and global. People come and look. And then the exhibition is packed up and travels to another city. So it's a circuit, like film distribution. This conception is supported by the Guggenheim Foundation—it's the leader in this movement. And the positions in this approach are only getting stronger. So Vilnius has invited the Guggenheim museum to build a museum there, there's already been a competition to choose the architect. . . . Baku is doing more or less the same thing. And also Astana, Abu-Dhabi. Many oil economies are going according to this script. And this network trend, I'll call it that, will be very influential, and probably dominating in the coming years. But I don't think that's the path that Perm needs. We need to create a museum that will show the world something that is to be found nowhere else . . . something modern, uniquely unusual, and at the same time, ours. "Ours" without the patriotism, with pride in our artists, but without beating our chests.[5]

The Perm Museum of Contemporary Art was, then, another attempt to put Perm and its region on the map—this time on the global map of cities where accumulated oil wealth was being transformed into culture. The something "uniquely unusual" that Gordeev envisioned would carry the same name as PERMM's inaugural exhibit, curated by Marat Gel′man: *Russkoe Bednoe*, or Russian Povera—the Poor Arts of Russia.

---

3. Maksim Kotin, "Permskaia Anomalia," *SNOB*, June 2010, 84.

4. Although Perm was home to a quite well-known classical art museum, the Perm State Art Gallery (Permskaia Khudozhestvennaia Galleria), contemporary and avant-garde art had no previous institutional presence in the city's museums, universities, or artists' unions. As the plans for PERMM moved ahead, Perm also hosted a major summer festival, Perm—A Territory of Culture, that showcased contemporary art and artists in the summer of 2009.

5. Vladislav Gorin, "Prodiuser Gordeev," *Kompan′on Magazine*, November 7–11, 2008.

Russian Povera

The Russian Povera exhibit assembled the work of some three dozen Russian artists and art collectives whose work featured found objects, trash, and ruins. Vladimir Arkhipov contributed photographs and exhibits of socialist and post-socialist objects put to unexpected use: a rusting bed repurposed as a footbridge, a road sign attached to a long pole and used as a shovel. Some contributions were direct, "poor" commentaries on well-known art elsewhere in the world: Vladimir Anzelm's *Skull* recalled Damian Hirst's famous *For the Love of God*—except that it was encrusted in coal rather than diamonds. The Blue Noses' contribution, *Kitchen Supremacism*, commented on Malevich with a photograph of geometrical shapes made of assorted types of salami arranged on the chipped paint of an old table. And there were many others: sculptures of welded wire obtained from fences; a spaceship composed of glued-together cigarettes and cigarette boxes; and a rendering of the familiar *Pravda* masthead in rubber cut from automobile tires. Ol′ga and Aleksandr Florenskii's *Skeletons*, a series of sculptures, cobbled found objects into resemblances of elephants, snakes, fish, and other animals. Their *Map of the Central Part of the City of Perm*, commissioned especially for the exhibit, was made of materials found in Perm's garbage dumps and in the basement of the River Station Hall. Its neatly labeled pieces included a rusted car muffler representing the Kama River and scraps of wood and metal serving as streets and familiar buildings. The goal of reimagining Perm through contemporary art was, perhaps, most explicitly on display in this piece, and it became part of PERMM's permanent collection, frequently exhibited as part of one or another aspect of the larger Perm Cultural Project.

Many of the works exhibited in Russian Povera had previous lives on the glitzy contemporary art scene in Moscow, where they were generally not understood to be in the Povera style at all. It was their grouping together in a provincial city that enabled Marat Gel′man and his collaborators to cast this exhibition as illustrative of a movement in the Russian art world.[6] Indeed, Russian Povera had a distinctly and intentionally grubby aesthetic, one that matched the lightly restored River Station Hall in which it took place. An exhibition catalog featured extensive interviews with many of the artists, asking them questions about their engagements with the material qualities of found objects in light of the exhibition's organizing theme. Aleksandr Brodskii, one of the more well-known artists featured, pointed out that Povera-style art was just about the only art remaining that could be produced without a sponsor and a budget, and that while he himself was now in the position to embark on some of these big-budget art works, it was still

6. I thank Molly Brunson and Bella Grigoriyan for their promptings on this point.

possible and useful to cultivate a poor aesthetic. Such art should be considered every bit the equal of sponsored, big-budget art (PERMM 2008, 26–27).

In his introduction to the catalog, Marat Gel′man expounded on the exhibit's effort to highlight a theme in contemporary Russian art that had, to that point, gone underappreciated. Russian Povera, he wrote,

> has an important subtext—its naturalness. It is much closer to nature than art that is rich, glossy, and built on technologies. The very material of Russian Povera returns us to naturalness . . . wealth has migrated to other spheres; it has been co-opted, consumed, has turned into advertising, design, magazine beauty, what you will, having lost the ability to be art. (PERMM 2008, 24)

Povera art refused the separation of art and life, he went on, as well as any grounding in theories of what art really is, in favor of exploring a wider range of possibilities and processes, with all kinds of materials and experiments. Noted philosopher and critic Boris Groys added his own take, suggesting that "in losing their glamorousness, these things gain their own histories . . . the viewer's attention shifts from the object itself to its genealogy and its practical use" (PERMM 2008, 26). In the version of the catalog that accompanied Russian Povera to the third Moscow Biennial in 2009, Gel′man elaborated further on what he saw as the differences between Russian Povera and its antecedents. The Italian Arte Povera movement of the 1960s, he suggested, was born of a conscious, principled, ascetic rejection of the consumer practices of postwar Italy, of the homogenization of beauty in the commodity form. Russian Povera, by contrast, arose in the late 1980s, 1990s, and early 2000s from the *actual* lack of opportunities, materials, and sponsorship that was enjoyed by Western artists. It was the child, that is, of post-Soviet transformation. Russian Povera's artists, he wrote, made art out of cardboard and found objects because that is what they could afford to work with; only over time had they learned to consciously deploy these materials for a range of purposes, among them the principled critique made by the Italian Arte Povera movement. PERMM's mission, wrote Gel′man, should be to become the world center for the collection of Arte Povera of all kinds—especially but not exclusively the Russian variant (PERMM 2009, 2–4).

Artists and art critics would doubtless have much more to say about the works exhibited at PERMM. I adopt a somewhat different perspective, that of the study of "art worlds," which understands the study of art to extend to lectures and seminars, institutions such as state and corporate agencies, other domains of cultural production such as literature or festivals, and even everyday conversations about art (Becker 1982; Marcus and Myers 1995; Winegar 2006). From this broad perspective, several aspects of Russian Povera are significant. Despite its omnipresence on the Perm region's cultural scene at the time, Lukoil-Perm was *not* a direct sponsor of either PERMM or the Russian Povera exhibit. The project was quite assertively driven by Chirkunov's team and funded through a combination of Moscow-based private donations (chiefly Gordeev's) and the Perm regional budget. It was therefore a break with the practices of Governor Trutnev's administration, which had so closely tied regional oil to regional cultural sponsorship. Indeed, PERMM represented both a new level of intervention in the cultural field by the Perm regional state apparatus and, at the same time, an etatization at the regional level of what had been, in Moscow, largely a private-sector affair—an issue to which I return.

Despite Lukoil-Perm's refusal to participate officially, there were some striking similarities between the cultural projects that the company continued to sponsor in the Perm region's oil-producing districts and Russian Povera. Both, to begin with, relied on powerful notions of authenticity. In Lukoil-Perm's cultural projects, this was an authenticity to be attained through a turn to deep historical, ethnic, and national roots (and, correspondingly, away from contemporary inequalities and exclusions). In Russian Povera, this authenticity was to be attained by leaving behind the world of commodity images. Boris Groys's comments, quoted above, went on to suggest that Russian Povera's artwork had the potential to escape the loss of aura that Walter Benjamin ([1936] 1968) famously argued comes with mechanical reproduction. Both claims to authenticity rested on a similar kind of oil-inflected spatiotemporal shift, accomplished at different scales. Authentic culture could be best found through movement to more provincial areas: from Perm to outlying oil-producing districts of the Perm region in the case of Lukoil-Perm's cultural projects, from glamorous Moscow to provincial yet oil-rich Perm in the case of PERMM. Just as the quest to reclaim national and ethnic identities sponsored by Lukoil-Perm played out at some distance from Perm, that is, it was really only in the poorer and grittier provinces that one could truly locate a museum of Russian Povera as Gel′man and Gordeev imagined it. (A review of the exhibit in the Moscow newspaper *Kommersant* noted obligingly that, in order to get out to see Russian Povera in Perm, one had to "splash through Russian filth.")[7] Russian Povera even traded occasionally in the semiotics of depth and authenticity that were so prominent in Lukoil-Perm's historical and cultural projects, with the curators writing in the exhibit's English-language press release that,

7. "Chtoby popast′ na Russkoe Bednoe, nuzhno proshlepat′ po russkomu griaznomu." The phrases "Russian Povera" and "Russian filth"—*Russkoe bednoe* and *russkoe griaznoe*—form an evocative poetic pair that links movement to the provinces with increasing poverty and dirtiness. Roman Dolzhanskii, "Permi podali bednost," *Kommersant*, September 25, 2008.

> The basis of the exhibition is work by artists who use the simplest, "poor" materials. This approach reveals and demonstrates all of the qualities of contemporary Russian art—an art that is authentic, deep, an art that goes away from surface beauty towards a real miracle.[8]

Whether located at the sites of extraction or accumulation, these cultural projects also took place with international audiences in mind, with the goal of putting both Russia and the Perm region on the map. The frequently heard claim that Lukoil-Perm sponsored local folk arts were well known "in Russia and around the world" (see chapter 6) was not so very different from Gordeev's and Gel'man's ambitions for Russian contemporary art at PERMM. To be sure, Lukoil-Perm's international audience was an investment marketplace increasingly concerned with corporate transparency and citizenship, demanding corporate social responsibility programs as a condition of full participation in capital markets and the oil trade, while PERMM's audience was an international art world actively debating the place and value of Russian art on the global stage. But the questions that were being debated and worked out in the Perm region were quite similar in both the oil and art worlds: What characterized Russian products—whether oil or works of art—in global contexts? Were they distinctive in some way, and who would be the arbiters of that distinctiveness—Western institutions (international capital markets for oil companies, a global museum elite for artists)? If the global circuit of energy specialists was curious about how red directors were meshing with new financiers in the ongoing transformation of the post-Soviet oil industry, then a similar question could be asked of the booming contemporary art world, with its established artists from the Soviet period working in new circumstances and its many young and largely unrecognized artists. PERMM, in sum, brought questions about the global audiences for Russian art and Russian oil into new and close proximity in the provincial Perm region (cf. Adams 2010 on cultural spectacles in Uzbekistan).

There were similarities, finally, between the kinds of relations—among humans and between things and humans—contemplated at Russian Povera and in Lukoil-Perm's sponsorship of folk handicraft production and sale. Both sought to grapple with the involution and impoverishment of the 1990s by putting homemade or found items on display and into monetized circulation in new ways. Making due with materials that were at hand, refashioning them into things that were usable and profitable, and even basing a cultural revitalization program on the results were all features shared by Lukoil-Perm's cultural CSR programs in the early

8. See www.bednoe.ru/eng.

2000s and the Russian Povera exhibit in the late 2000s. "We exist in a society of global castoffs," wrote Petr Belyi, another artist whose work was included,

> They are presupposed by a global economy. . . . We live in a post-Soviet, postindustrial society with a ruined heavy industry, which was so hard to create and fell apart so insanely easily. . . . As an artist, I feel comfortable at a ruined factory, next to the abandoned [excavator] bucket . . . the exploration of materials, the rifling through of cast-offs, trips to trash heaps, are a constant need. (PERMM 2008, 79)

His words echo those of the Lukoil-Perm's Connections with Society employee who summarized the company's proposal to the impoverished and unemployed residents of oil-producing districts: "Sit home, sew, make some pottery, and maybe you can make a living of some sort" (see chapter 6).

This was, however, a similarity that supporters and artists connected with Russian Povera were only willing to take so far, and they responded to my queries on this topic by deploying a version of the commonly made hierarchical distinction between handicrafts and art (compare, for instance, Winegar 2006, 63). Folk handicrafts and Povera art, one interlocutor told me, might look superficially similar in their transformations of everyday items in the contexts of post-Soviet economic involution, but folk handicrafts, it was important to remember, "have no conception of themselves as art." One of the contributors to the Russian Povera exhibit, Vladimir Arkhipov, addressed this question in his catalog interview in a similar way, speaking of his works featuring everyday items:

> [In my art] I preserve the authorship of people who make sincere things. They create while paying no mind to aesthetics or quality; their things exist in an everyday environment and are not perceived as works of art. . . . For its maker, a handcrafted thing is, on the one hand, a solution of a domestic problem; on the other, it's creativity, *but he does not understand that.* (PERMM 2008, 28; emphasis added)

This last point is debatable as a matter of principle: inasmuch as many folk artisans (and even nonartisan everyday improvisers, as I learned in my previous fieldwork on a former Soviet state farm) are highly attuned to their own creativity. But the perception that the artists exhibited at PERMM knew and projected something about creativity that others did not was absolutely central to the entire Perm Cultural Project. It was precisely this point—the possibilities of unrecognized, untapped creativity in the ways in which the Perm region's residents transformed their material surroundings—that the backers and sponsors of PERMM hoped to inculcate through a cultural revolution spreading outward from a museum of contemporary art. It was here, in this cultural creativity, that new entrepreneurs would be born and new paths to postindustrial, post-hydrocarbon economic development opened. In his own introduction to the exhibit catalog, Senator Gordeev elaborated on his vision of contemporary art as the "new engine" of Perm's economy. "The factories that served as the motors of Permian civilization have not disappeared," he wrote, but

> they have also grown "tired" of pulling the city along all by themselves. They need help. Perm's new contemporary art museum, he said, is called upon to turn a giant layer of Permian subconscious, of hidden ambitions, into a project, a point on the map, a destination. (PERMM 2008, 22)

In 2009, Oleg Chirkunov began a high-profile effort to ensure that the transformative project he had set in motion would not stop at the walls of the River Station Hall. He announced that the defining project of the remainder of his governorship would be an effort to have Perm displace Saint Petersburg as Russia's unofficial cultural capital. Less than a year later, on March 1, 2010, he revised this goal upward to having Perm officially designated a European Capital of Culture in 2016. In his speech announcing this target, Chirkunov argued that Perm would become a place where "creative people will live—and that means that many different sectors will develop differently—not just culture." There would be a new "economy of the intellect, where we will create not with our hands but with our heads." Every problem that needed to be solved in the region was in large or small part ("and probably in large part") a factor in regional cultural development. Anticipating one line of critique, he concluded that: "It's not cities like Paris, or London, or Moscow that aim for these projects. It is cities that need a breakthrough, that, thanks to that project and the gathering of resources at a specific point in a specific place, break through and become great cities." Perm, he contended, needed a breakthrough as much as any small European city. If Istanbul, not even located in a European Union member state, could be a European Capital of Culture, then what was stopping Perm?

Chirkunov's subsequent budgets in the Perm region reflected these priorities. Spending on cultural projects rose nearly tenfold between 2008 and 2012, and of the 567 million rubles allocated to cultural programming in 2012, 540 million, or 95 percent, was dedicated in one way or another to the European Capital of Culture campaign.[9] In an earlier era of global political economy, political leaders often sought to reinvest soaring oil revenues in the development of local industry. Fernando Coronil's treatment of the political economy of oil in mid-twentieth century Venezuela, for instance, follows attempts to "sow the oil" by building a

9. "95% kraeovogo biudzheta Minkulta poidet na proekt Perm—Kul'turnaia Stolitsa," *Argumenty i Fakty*, February 14, 2012. See www.perm.aif.ru/society/details/603031 (accessed August 17, 2014).

domestic auto industry to compete with Detroit and foster domestic Venezuelan agricultural production (1997, 237–85). Chirkunov, Gordeev, and their associates picked a development project for a different era of global capitalism, one that that privileged creativity, culture, consumption, and entrepreneurship over industry, labor, and social sector spending. Among other things, they said, it was much cheaper and more realistic to develop culture than to attempt to rehabilitate industry.

The Perm Project: Creative Life in a Cultural Capital

Post-Soviet art worlds have been instructively theorized in a variety of ways: as a domain in which new understandings of—and borders between—religious and secular are hashed out (Bernstein 2014); as vehicles for the legitimation of new kinds of power or authority (Bazylevych 2010); and as sites of specifically postsocialist combinations of state power and commercial value (Nauruzbayeva 2011). PERMM and its broader cultural revolution point us in still another direction: to contemporary art as a technique of transformation, an arena where it comingled with other domains of culture—from theater to festivals to film—and with the transformative agendas of state agencies, corporations, and others. In her masterful ethnography of Egyptian artists and their interlocutors in the late 1990s and 2000s, for instance, Jessica Winegar explores ways in which "cultural policy . . . is part of modern strategies of governance and the move to neoliberalism" (2006, 142). She goes on to argue that Egyptian "national arts policy was the cultural accompaniment to neoliberalism, stressing international progress, flexibility, and democratic openness as made manifest in 'avant-garde' art" (2006; see also Miller and Yúdice 2002).[10] In the Perm region as in Egypt, state-led cultural policy played out in apparently contradictory ways: increased state participation in the cultural sphere, for instance, proceeded in tandem with ongoing privatization in other spheres (even in the cultural sphere itself in oil-producing districts, as Lukoil-Perm's cultural CSR projects demonstrated).

As it began to extend outside the walls of PERMM, the Perm Cultural Project was led by another addition to Governor Chirkunov's team: Boris Mil′gram. Mil′gram was a theater director who, although born in Odessa, had studied in Perm in the 1970s, eventually receiving an advanced degree in chemical sciences

10. As Winegar is well aware, links between the creativity associated with avant-garde art and national and international transformations of political economy are not limited to the neoliberal era; see, for instance, Guilbaut's (1983) study of American abstract impressionism in the early Cold War. I thank Brinton Ahlin for his prompts and references on this point.

and participating in an active and popular student theater. (Oleg Chirkunov was a student in Perm at the same time, several people pointed out to me.) Mil'gram trained as a theater director in Moscow in the late 1980s, and went on to gain significant renown both in Moscow at the Mossovet Theater (in the 1990s) and subsequently back in Perm, beginning in 2004, as head of the Perm Drama Theater. Chirkunov appointed him minister of culture and mass communications in 2008, as plans for PERMM were getting under way, and promoted him again in 2010 to the post of vice-premier of the Perm regional government, a position he held until Chirkunov's resignation in the spring of 2012.

"I came to the Ministry [of Culture] to put on a play," Mil'gram told the newspaper Gazeta.ru.[11] If there was an overall script for that play, it was a 122-page manifesto titled *The Perm Project: A Conception of Cultural Policy for the Perm Region*, released in 2010 (Zelentsova 2010).[12] The lengthy document, which Mil'gram had a central role in composing, presented itself as the views and work of a collection of project managers, experts, and economists in the field of culture and cultural politics. *The Perm Project* elaborates in much greater detail the view of culture as engine of postindustrial development that Gordeev and Chirkunov had set out. Culture, in the understanding of the authors, is primarily a zone for the "creative self-expression and self-realization of every person"; this "creative potential," in their view, "will become the main factor in the development of all economic subjects" (Zelentsova 2010, 1). The basic lexicon of *The Perm Project* will be familiar to those well versed in the language of creative cities. Culture in the Perm region was to be "open and dynamic," "self-organized," and would be based not on sector-wide programs but on "clusters" bringing together "innovative companies, firms, and creative groups" to serve social and cultural needs. Projects in the field of culture would unfold quickly, creating jobs and attracting tourists far faster than any attempt to rebuild the region's industrial or defense sector. *The Perm Project* was conceived of as a way of making new kinds of people—creative, free, independent—and a new kind of society. "We rely on the resources of culture, art, and heritage," the coauthors wrote. "But that is not even the most important. The primary resource of the project is the creativity and creative potential of society itself" (Zelentsova 2010, 45). In the straightforward terms in which Mil'gram put it to me in an interview, the project as a whole aimed for "the formation of a mechanism for the self-realization of the person." Or, as Marat Gel'man told me, culture should be understood as impacting everything having to do with the life of a city, and to invest in culture was

---

11. Vadim Nesterov, "Ia prishel v ministry sdelat' spektakl'," *Gazeta.ru*, September 22, 2010.

12. See also Abashev 2009 for an earlier collaborative effort between Chirkunov's team and members of the local intelligentsia.

therefore to drive all other sectors, from large to small business. "An entrepreneur and an artist," he said,

> it's the same mentality . . . [one that] makes you want something. It's just that for the artist, you want to express something, for the entrepreneur, that's not as important . . . and so, we are beginning to produce, for now in small quantities, these new kinds of people. . . . This is how I am considering art, as an instrument for solving problems in the territory as a whole.

Such were the transformations contemplated under the rubric of Cultural Revolution in Perm—new entrants into the crowded field of technologies of the self in the second postsocialist decade (see, for instance, Matza 2009 and Zigon 2011).

Although frequent aspirational comparisons for Perm as cultural hub were postindustrial cities of Europe and the United States (Glasgow, Pittsburgh, Bilbao, and others), at least as common were comparative references to Russia's own hydrocarbon sector. In a public address in 2009, for instance, Mil′gram sought to shake the default assumption that cultural potential paled in comparison to the revenues, prestige, and influence of the oil and gas industry:

> The expense for the development of culture is nothing. The financial valuation of the heritage of Picasso is higher than the capitalization of Gazprom. The biggest taxpayer in England is not the president of BP but the author of *Harry Potter*. Financial turnover in the film industry is comparable to that of oil, and the cultural heritage of Italy, France, or Spain makes them richer than the biggest hydrocarbon deposits would.

In *The Perm Project* and other visions for the future of Perm, oil clearly belongs to the industrial past, not the postindustrial present and future. Whereas natural resources like oil are by definition scarce, culture is portrayed as endlessly creative, everywhere, and inexhaustible. Or, as Mil′gram put it to me in one of our conversations, "There is no cultural sector, because culture can't be controlled. There is the infrastructure of culture: departments, organizations, creative unions. But that is only a part of culture, because culture is a whole, it is everything. It is everything, especially in a cultural capital." Indeed, as a resource with endless potential—after initial private and state investment—culture was the ideal fix for the declining oil reserves in the region and the specter of the resource curse that had begun to worry Russian economists and policymakers. What better sector to diversify into than one that could be framed as not beset by problems of scarcity at all? In the 1990s Perm region, dreams of unimaginable wealth, possibilities of unfettered growth, and hopes of putting Perm on the map gathered around pyramid schemes and rags-to-riches privatization tales. In the early and mid-2000s, these dreams shifted to the realm of oil wealth. By the end of the decade, the plan was for culture to take up that role: dreams of ever-expanding oil wealth would be replaced by the endless potential of human cultural creativity. Chirkunov's signature line on this topic was, simply, "A City Should Have a Dream."

As I have noted, Lukoil-Perm was not a sponsor of the Russian Povera exhibit or of PERMM, although, of course, the steady stream of tax revenue from the company underwrote a significant part of the state budget allocations for these and all other state cultural initiatives. The company continued to distance itself from Chirkunov's Perm Cultural Project in official and general terms, although it did chip in some significant sponsorship funding when the Perm Cultural Project began to include city-wide public festivals.[13] "They have their projects," Mil′gram told me,

> that take place out in the districts, where they invest in cultural projects, where they work on historical memory, historical cities. They do this on their own level, and we don't interfere much. . . . It's called parallel processes, and we only welcome that. We never want to tell them "Do this, or don't do that." We will find other resources to do our work here [on the Perm Cultural Project].

Indeed, despite its advocacy of flexible and mobile state-corporate-culture projects as central to life in a cultural capital, *The Perm Project* does not mention Lukoil-Perm at all (although big Lukoil in Moscow does come up as a suggested funder of cultural projects of national significance). For its part, Lukoil-Perm kept its primary focus on oil-producing districts and the depths of cultural and national traditions.

The Perm Project and those behind it remained quiet on their opinions about the ways in which Lukoil-Perm's social and cultural projects were clearly connected to boosting the company's own image, but they were often direct and unsparing in their criticism of those in Russian government on this score, including the federal Ministry of Culture—who, they believed, continued to see culture as a domain of mere heritage preservation and/or state ideological work (Zelentsova 2010, 21). Importantly, in their view, the Perm Cultural Project was *not* an instance of a state using the cultural sphere as a domain of legitimation in the guise of state tutelage (*gosudarstvennaia opeka*), or shoring up the legitimacy of the state, as classic Soviet cultural work attempted to. "Culture in the Soviet Union was a very strong tool," said Mil′gram in one of his many public addresses on the matter,

13. In 2006, however, the company was an ardent support of Perm's status as Cultural Capital of the Volga Federal District—a project that unfolded under the auspices of the social and cultural project competitions central to regional development plans at the time; see Nikolai Trukhonin, "Skoro Gorod Izmenitsia!" *Permskaia Neft′* 21 (195), October 2006.

> it was an ideological instrument for the restraint (*uderzhanie*) of the country. An ideological influence (*vozdeistve*). And inasmuch as it was a tool, it was well financed and it was a territory of status, that is, people who worked in culture had some status.

It was true that the state picked what kind of art or theater you could and could not engage in, and that artists suffered for this, he went on, but the status of culture in the Soviet Union was no doubt higher than it was in the post-Soviet period. Culture could regain that high status again, he said, *not* by returning to being the ideological partner of a powerful state, but by becoming its own resource, its own locomotive for economic growth. Indeed, it is clear from the first pages of *The Perm Project* that Chirkunov and his collaborators' goal was to make culture a topic of conversation among economists, businesspeople, and politicians in addition to its traditional practitioners and managers.[14]

The Perm Cultural Project was, then, envisioned just as much as a transformation of the state as of individuals and communities, and its architects were anxious to restructure the state cultural bureaucracy, which they saw as built around outdated expectations and understandings of culture. Much of *The Perm Project* described changes that the organizers deemed necessary to create the larger-scale transformations they envisioned: the legal status of cultural institutions needed to be changed in order to accommodate a range of flexible and evolving collaborations among state agencies, businesses, and artists of various sorts; officials in significant positions in the cultural affairs hierarchy needed to be concerned with fundraising, marketing, and branding; budget forecasts needed ways to calculate the effectiveness of cultural work; and the entire process needed to be streamlined, made more flexible and adaptive, and professionalized. The language of business plans, in short, needed to find its way more firmly into the Ministry of Culture itself. Culture as a budget item, moreover, needed to be moved out of its current location in the social sphere of the state budget. Funding for the arts and culture, Boris Mil′gram emphasized in a 2009 address, had long been connected with Soviet ideology; when that ideology collapsed, culture took a place next to other categories and other groups that expected help from the state. Artists, museums, and theaters were lumped into the same overall category as pensioners, hospitals, and disabled people in the state's "social territory." Gel′man told me that, "Tradi-

14. One of the ways in which they accomplished this was by making the cultural capital project a regular theme of the annual Perm Economic Forum, at which regional elites across all sectors gathered and met with outside guests. The opening of the Russian Povera exhibit was timed to coincide with the 2008 Perm Economic Forum, and culture played a prominent role in the 2009 and 2010 iterations. The Perm Culture Project was then made into the overarching theme of the 2011 Forum, at which a parade of featured experts from around the world advised Perm's leading politicians and corporate executives on matters of city branding, cultural revival, urban renewal, and the creative industries.

tionally, in Russia, and not only in Russia, culture was in the same place as beggars at the church gate, hospitals, elderly people receiving pensions . . . in other words, in the formation of any budget, it was in the social sector."

And there, he said, it always loses, because, of course, a hospital will be repaired before a museum is built. So long as culture was funded in accordance with the remainder principle (*ostatochnyi printsip*), in which it received only what was left over after all other budget items were funded, the Perm Cultural Project would founder. "We are saying," Marat Gel′man went on, "that culture in the contemporary world plays a different role—an infrastructural role."

The Perm Project was, finally, a deeply regional document, insistent on the ways in which cultural projects of the sort it envisioned must come from the provinces and not from the federal center. The broader Perm Cultural Project, in other words, set out to challenge existing patterns of state cultural production in Russia by rethinking their spatiality. "For three hundred years," Gel′man told me, "culture has been concentrated in the center"—in Moscow and Saint Petersburg. He and his collaborators were actively revising that organization of space, he went on, allowing the provinces to develop their own initiatives and advertise their own cultural distinctiveness, and to do so on their own terms. This was not, he hastened to add, a critique of Putin-era Russia. It was a critique of much of the sweep of Russian history: whatever the period, whatever the politics, whatever the regime, the field of culture had been dominated by the centrality of Moscow and Saint Petersburg. "We're breaking prejudices [against the provinces], said Gel′man, "Everywhere I go, in other [provincial] cities, people feel themselves to be revolutionaries. They tell me, 'we want to do what Perm has done.' "

In Governor Chirkunov's estimation, this remaking of cultural space was not only a challenge to the federal center but also, in fact, a promising strategy in the never-ending battle to beat out other Russian regions for federal funds. By staking out the terrain of cultural capital, he hoped to position Perm to receive federal investments of the same sort—if perhaps not at the same scale—that had been earmarked for Saint Petersburg's three-hundredth anniversary in 2003; Kazan's thousandth anniversary in 2005, and the reconstruction of Sochi for the 2014 Olympics. If, he argued persistently, Perm were to become the best place to make artistic careers, both out of city residents and new arrivals, and if it had the best new art museum, the best experimental theater, the most renowned festival life in the summer months, then federal aid for infrastructural projects such as renovating the train station, airport, or local roads would follow in short order.[15]

---

15. On links between region or identity branding and the quest for investment, see also Comaroff and Comaroff 2009.

• • •

At the center of the cultural field at the end of the 2000s in the Perm region was, then, a not uncommon paradox: massive state-funded cultural production, recognized by all as such, that was nonetheless cast as a way to diminish the power of the regional state apparatus and to direct the source of creativity, transformation, and entrepreneurship away from the state itself. In the 1990s, the active work of creating selves and state agencies in the wake of socialism was quintessentially that of Western aid agencies, nongovernmental organizations, and other sorts of advisers (see, in a very large literature, Mandel 2002b; Rivkin-Fish 2005; and Hemment 2007). In the early 2000s in the Perm region, as I showed in earlier chapters, much of this task was taken up by the oil industry and state agencies working in close concert, in good part on the terrain of culture, through state-corporate grant competitions and the spread of the project movement. By the late 2000s, as Lukoil-Perm and the regional state apparatus grew further apart, both continued to pursue this cultural effort on parallel, in many ways similar, and highly spatialized tracks—the oil company through its ongoing cultural CSR work in the region's oil-producing districts, and the state apparatus through the increasingly omnipresent Perm Cultural Project focused largely in Perm itself.

Miracles and Heroes of Culture

For Chirkunov's team, Cultural Revolution was less a domain of conceptions and plans than a domain of resolute action: any attempt to work through the existing, outmoded state cultural and bureaucratic channels was, they believed, doomed to fail. The Perm Cultural Project would thus need to rely in significant part on outside, contracted cultural managers and on seizing opportunities as they arose, without waiting for broad public consultation or the buy-in of the regional cultural intelligentsia. "Now is the time," Marat Gel'man was fond of saying, "for miracles and heroes" in the cultural sphere. Training a new generation of Perm-based cultural managers and producers was, to be sure, part of the plan, but that task would have to wait for later stages.

Indeed, the Perm Cultural Project team was acutely aware how rare and unusual was the vein of regional state funding that they had tapped. As Boris Mil'gram put it in one of his first major speeches on the issue, "I said earlier that we don't have any specific competitive advantage with other regions. That was a mistake! We do! We have a liberal-thinking governor, and that is already a resource." One of the lower-level members of his team voiced agreement in one of my conversations with her: "In addition to the stable financial situation [provided by regional budget allocations], we had a terrific administrative resource—the governor himself, who opened up an umbrella over all our cultural projects." But Oleg Chirkunov served at the pleasure of the president of the Russian Federation, and it was unlikely that the Perm Cultural Project would extend past his governorship; in fact, because the project was so closely identified with Chirkunov's office, his successor was likely to pursue a very different agenda for the region, just as Chirkunov had quickly distanced himself from Governor Trutnev's signature Capital of Civil Society campaign.

One primary hallmark of the Perm Cultural Project was thus the breakneck speed at which it unfolded. Public art installations sprang up overnight in central city spaces, sometimes entirely without warning—such as the *Power* installation outside the Legislative Assembly building discussed in the introduction.[16] One of the project managers brought in from outside of Perm recalled to me her shock at the pace:

> Marat [Gel′man] said that we had very little time and would not work according to the schedules to which we were accustomed. I said that in the beginning we need to do research, plans, presentations, and so on. He said, "No, no, no, we are doing things differently: break ground first, then carry out the project, then dream."

It was, she reported, exhilarating to be freed of the normally sluggish bureaucracy. Although she and nearly everyone I spoke with agreed with Gel′man that this kind of pace was necessary in order to get anything accomplished at all, they also agreed that, in retrospect, this speed was an important reason that the Perm Cultural Project became so controversial.

Culture Everywhere

After its start at PERMM, the Perm Cultural Project played out across a great many cultural domains simultaneously. Among the most visible of its dimensions was a large, state-funded public art campaign, run out of the Perm Museum of Contemporary Art, with the explicit goal of moving contemporary art outside of the walls of the museum and directly into the squares and streets of Perm. In the introduction to the definitive catalog of the dozens of public art projects realized in Perm, Gel′man wrote of this effort in the context of Governor Chirkunov's long-term plans to develop the city of Perm into a modern, European city. Those urban redesign plans would take a long time to realize, "but people needed to see

16. On the reception of some of the Perm Cultural Project's initiatives in the city, see especially Kruglova 2013.

something concrete right now, an improvement in the present, not in some distant future. Real visible changes needed to start right away" (PERMM 2011, 6–7). Through high profile installations such as *Power, Red People*, and *Apple*, as well as dozens of other smaller projects ranging from commissioned graffiti on crumbling Soviet-era facades to a series of painted angels that began to appear on balconies scattered throughout the city center, residents of Perm would immediately feel themselves connected to the transformation of urban culture and space in a new way. At the very least, they would be drawn into debates and discussions about what constituted art, the appropriate use of government funds and city spaces, or the meaning of the headless wooden red figure with a raised arm—was it voting?—that appeared perched atop the regional administration building. On this enthusiasm (or at least engagement) a new era of economic development could ride.

Projects like PERMM, the new Hammer Stage experimental theater, and public art installations were often cast in a language of urban rather than regional development, and so one of the questions that followed the Perm Cultural Project throughout its life was whether it was focused exclusively on the city of Perm or on the Perm region as a whole. The primary way in which the Perm Cultural Project did extend throughout the region was through increased regional funding for and coordination among festivals. According to Russia's official region numbering system—most commonly encountered on license plates—the Perm region was Russia's fifty-ninth. This number provided Mil′gram's Ministry of Culture with the inspiration for a new, region-wide program: Fifty-Nine Festivals for the Fifty-Ninth Region. All manner of festivals—from film festivals in Perm itself to new and revived district-level and local festivals up and down the Perm region—were eligible to be included. In 2010, this program featured twenty festivals billed as regional, an additional twenty-seven classified as intermunicipality collaborations, ten as interdistrict, and two as international. Only fifteen took place primarily in Perm, with the remainder spread throughout the smaller cities and districts of the Perm region.

These festivals were not only allocated new and ample funds for their own celebrations, but also were periodically invited to exhibitions of the entire Fifty-Nine Festivals project and other celebrations in Perm. Delegations from many of these festivals appeared onstage in succession, for instance, at the annual celebrations of City Day in Perm—held on the same June Day as the national Russia Day holiday. Under the auspices of the Perm Cultural Project, arrangements like this began to reconfigure the spatial organization of culture that had, since the beginning of Lukoil-Perm's Historical Cities of the Kama River Region festivals in 2001, been closely keyed to the geography of the oil industry. The entire region, that is, rather than primarily its constituent oil-producing districts, was transformed in the festival space. "The territory of creativity is expanding," proclaimed the Perm Cultural Project's promotional materials.

Public art installations and the Fifty-Nine Festivals movement—and many more dimensions of the Perm Cultural Project as well—came to a collective, spectacular crescendo in June 2012, when the Ministry of Culture sponsored a month-long "festival of festivals" called White Nights in Perm. An open poke at the legendary White Nights celebrations held during the longest summer days of the year in Saint Petersburg, White Nights in Perm was a month-long marathon of morning-to-evening cultural events designed to "create favorable conditions for living in and visiting Perm and the Perm region, and for the self-realization of individuals."[17] White Nights in Perm was overwhelming by design. It boasted a total of 750 events; just one of its numerous constituent festivals, the Perm Alive festival of contemporary art, included 165 events in five days. The vast majority of the considerable bill for White Nights in Perm was picked up by the regional administration, with a substantial contribution from the Perm city budget as well. With the festival taking place on the doorstep of its headquarters, and with many of its features in line with the company's own goals, Lukoil-Perm contributed around a fifth of the White Nights in Perm budget—30 million rubles in 2012—and many other local businesses signed up as sponsors as well.[18]

At the center of White Nights in Perm was a fenced-in Festival Village erected in front of the Regional Administration building on Perm's esplanade. Just over three hectares in size, the Festival Village included two small and one large outdoor stages for concerts and other performances; numerous alleys for small shops and displays; two restaurants and two cafés; and a Festival Club for nearly fifty planned discussions and presentations. In order to cope with inevitable summer muddiness, boardwalk-style walkways were constructed to funnel crowds from space to space; they were repainted white nearly every night. Booths arrayed alongside these walkways provided spaces where folk artisans and other culture producers could display and sell their wares, and the grassy spaces between the walkways hosted small-scale performances and exhibitions, from clowns to blacksmiths. Everywhere there were nooks and crannies—many of them in two massive towers at one end of the Festival Village—where little exhibits or performances sprang up. Most stunningly to many observers, there was even a "festival beach": a large circular pool, suitable for dozens of children at a time, erected within a raised

17. "Polozhenie o Provedenii Festivalia 'Belye Nochi v Permi,'" January 23, 2012, www.permfest.com/about/polozjenie.

18. Precise state budget figures for the White Nights in Perm festival are hard to come by. Official regional expenses were in the neighborhood of 150 million rubles, but that number is likely quite low as a total budget figure, given the numerous other sources and lines of funding that underwrote the festival's many elements.

platform that could accommodate hundreds of sunbathers. Showers and changing rooms were located in a sandy area beneath.

For nearly a month, events of all sorts were scheduled for all hours of the day in the Festival Village. One day was devoted to folkloric performances by all the national and cultural groups represented in the Perm region; with each folklore troupe permitted only a few songs, the benches and walkways of the Festival Village were crowded with groups of mostly elderly women in costume awaiting their turns on stage. In a variation on the same take on the location of culture, another day featured a five-hour stretch in which the wedding celebrations of the national groups of the Perm region—from Chuvash to Russian to Tatar to Udmurt to Jewish—were showcased one after another. Many of these performers had experience with, and even owed their initial funding to, Lukoil-Perm's sponsorship of festivals and handicrafts earlier in the decade. In this context, however, those forms of corporation-sponsored cultural production were subsumed into the goals of the Perm Cultural Project. In the evenings, as buses carrying elderly folklore performers made their ways back to the region's rural districts, disc jockeys and pop groups from all over Russia took over the main stages. A delegation of more than 250 performers arrived from Mexico for a festival of Mexican Days in Perm, and dance and musical troupes hailing from places from Moscow and Saint Petersburg to Italy and Zimbabwe performed on those same stages on other days. Less than a hundred yards from the folklore performance stage hung exhibits from the Russian Povera art collection in a temporary exhibit space for PERMM.

Every hour or so when the Festival Village was open, teams of two or four young men and women would appear at the tops of the two tall towers that flanked the main entranceway from the Perm esplanade into the Festival Village. Clad all in white and standing in formation, they would raise red semaphore-style flags, one in each hand, and go through series of signals in unison (see fig. 13). The friends and acquaintances with whom I walked through the Festival Village often commented on the flag signals, but, as far as we could tell, there did not seem to be a regular coded message to be deciphered. Less important than decoding precisely what they were signaling, it seems to me now, was the direction in which they signaled. From the highest point in the Festival Village, they looked down Perm's central esplanade, out beyond the Perm-2 railway station, in the direction of Moscow and, beyond that, to the cultural capitals of Europe. Perm, they signaled repeatedly, is here.

On one of the festival days in early June 2012, I attended a public interview with Boris Mil′gram in the Festival Village's club, where he was presented by the interviewer as "one of the main creators of the whole idea of festival life in Perm in the format we see today." In answer to a series of questions, Mil′gram presented the White Nights in Perm Festival as clearly emerging from his life in the "zone

FIGURE 13 One of the three-story towers marking the corners of the Festival Village at the White Nights in Perm Festival. The periodic semaphore signals attracted a good deal of puzzled attention.

Photo by author, 2012.

of unexpectedness" and creativity that was the world of stage theater. "I didn't always want to change Perm, but it seems to me that theater can change life. . . . It's possible to look at the [Festival] Village and believe that we live in an excellent city and that *therefore* the roads will be repaired and the apartment buildings will change—in *that* order, and not the other way around." White Nights in Perm was, he said, a "model for life" that, if one believed in it, "could be found all the time"—not just in the summer festival months. But that model for life did not feature much in the way of Russian political figures or political parties—the hundreds of events in the Festival Village were directed, rather, at the creative transformations of individual lives and urban space.

By the organizers' reckoning, over half a million people visited the White Nights in Perm Festival Village in June 2012. Many more attended other associated events and productions nearby, such as a sold out run of *Scarlet Sails*, a romantic musical that was a White Nights staple in Saint Petersburg. Among my friends and acquaintances, it was clear that a number of skeptics had finally been won over, at least to the point where they—and especially their children—were enjoying the nightly possibilities for strolling and entertainment. "We've gotten used to [the Perm Cultural Project] now . . . even to the crazy public art," said one friend. "Five or six years ago," said another, "there was nowhere to go in Perm . . . and now you can't possibly make it to everything interesting."

Against State Culture

I returned to the Festival Club tent ten days later for the opening of a workshop for cultural managers from other regions of Russia who had come to network and learn about how to stage large-scale cultural projects. Boris Mil′gram was again one of those welcoming the guests, and, as a way of orienting them, he drew an explicit contrast between the Festival Village and the celebrations of City Day in Perm that had taken place at the other end of the esplanade a couple of days earlier, on Russia Day, June 12. Perm's city administration, he said, had decided that it needed its own separate celebration, one that included more standard and familiar elements: welcoming words from the mayor, a parade of representatives from different city districts, each presenting their district's distinctive characteristics, a subsequent parade of companies and other groups (with Lukoil-Perm's large contingent taking the lead), participation from regional representatives of the United Russia Party, and a long set of musical numbers. Mil′gram pronounced the entire affair "a monstrosity" and continued:

> It's just that state power (*vlast′*) wants to present itself, wants to create something around itself. If you just want to present yourself, then why

> are the people even there? [If that's what you want to do, then] the best way is to do it like Putin. Empty roads, just a camera. Present yourself to the whole country and move on. It's great. Everyone understands. There's the tsar. And even if you do want to add the people in, why would you contaminate the whole thing with such an awful concert when here [in the Festival Village] we have good ones?

This comparison resonates well with similar kinds of statements in the official vision of the Perm Cultural Project. In the understanding of its architects, that is, the Perm Cultural Project, in all its various incarnations and especially in their intersection in the White Nights in Perm Festival, had moved some significant distance from the modes of cultural production that had focused on the legitimation or sacralization of the federal state, whether Soviet-era or Putin-era.[19] It is certainly useful to understand, as many have, the grand Russian cultural spectacles of these years—up to and including the Opening Ceremonies at the Winter Olympics in Sochi—as contributing to the fetishization of the Russian state and its embodiment in the person of president Vladimir Putin, but we must also acknowledge that these Russia-wide cultural spectacles unfolded in dynamic relationship with many other possibilities, and even cross-currents, at regional and local levels. In the same way that I showed for regional state agencies and regional corporate subsidiaries in previous chapters, the central Russian state apparatus was far from in control when it came to the regional cultural field in Perm and the Perm region. Indeed, many of governor Chirkunov's most vociferous critics wished, at least at times, that the cultural field *was* more centralized, for they regularly appealed, without success, to the federal Ministry of Culture to shut the Perm Cultural Project down and send Gel′man, Mil′gram, and their colleagues packing. But even these critics, when asked about what they would prefer to see in the sphere of culture, did not wish for more federal projects or more centralization. They sought an alternative sort of region-based cultural production, one that marked its own careful distance from the federal center.

The Cultural Opposition

In 2009, I ran into an old acquaintance who had spent most of her career working in a Perm museum and asked about what were then the early stages of the Perm

19. In their focus on creativity, entrepreneurship, and economic development, the architects of the Perm Cultural Project also avoided talk of *kul′turnost′*, "culturedness," despite its centrality to Soviet (Kelly 1999) and post-Soviet (Patico 2008) discourse. Many people in Perm did talk about the Perm Cultural Project in these terms, but these conversations lie outside the scope of my present argument.

Cultural Project. "We have a new minister of culture," she said, "who says that it's his job to make the people of Perm smile." She added acidly: "They'll be laughing soon enough." The most vocal source of discontent with the Perm Cultural Project was what was usually referred to as the "local intelligentsia"—a mixed and shifting set of cultural producers that is more internally diverse and contentious than the label implies (see esp. Fadeeva 2011). (Not surprisingly, perhaps, those affiliated with Perm's vibrant theater scene were less critical of Boris Mil′gram than, for instance, either the museum workers or music school instructors with whom I spoke.) Although the wide array of artists, writers, and other cultural producers that took up the cause against the Perm Cultural Project had not objected much to the entry of Lukoil-Perm into the cultural field or to the independent Kamwa festivals, the opening of the Perm Museum of Contemporary Art and what they viewed as the takeover of the regional Ministry of Culture by the Perm Cultural Project provoked a sustained response that was vigorously reported by regional newspapers and endlessly debated on blogs and in public forums. Like the European Capital of Culture campaign in Skopje, Macedonia, studied by Andrew Graan, the Perm Cultural Project also "authorize[d] citizens to demand responsible representation and to leverage the branding process to agitate for alternative conceptions of the social order" (2013, 165). Unlike the Macedonian case, however, critiques of the Perm Cultural Project were largely about center–region relationships. They also continued to proceed through institutions, languages, and expectations shaped by the regional oil complex.

One of the first and most acrimonious battles came in 2009, and it drew in a number of Perm's most prominent post-Soviet businesspeople, politicians, and cultural producers in ways that illustrate the continuing centrality of various aspects of the regional oil industry to cultural production. Recall that, by the mid-2000s, Andrei Kuziaev (of the Perm Commodity Exchange, Perm Financial-Productive Group, and Lukoil-Perm), Evgenii Sapiro (Perm Commodity Exchange "godfather," former deputy governor, and Lukoil-Perm adviser), and Iurii Trutnev (former governor) had all left their 1990s and early 2000s positions in Perm regional political and economic circles and moved up to the federal level in Moscow—Kuziaev at Lukoil Overseas Holding, Sapiro as his freelance consultant, and Trutnev as Russia's minister of natural resources. In addition to their day jobs in Moscow, the three were among the key members of the Moscow-based Perm Fraternity (Permskoe Zemliachestvo), a network of expatriates from the Perm region who were dedicated to improving the region's image in the capital (and, at the same time, maintaining a hand in regional affairs from afar). In addition to its ongoing center–region networking possibilities, the Perm Fraternity launched a new initiative in 2006: the annual Stroganov Prizes, meant to honor contributions to the life and image of the Perm region in a number of categories, ranging from sport to economic life to culture. Nominations were accepted each year in all categories, with a panel of experts making a recommendation to the leadership of the fraternity. Winners were decided in secret, by an unnamed committee, and awarded the ruble equivalent of $10,000 at a gala celebration.

The earliest Stroganov Prizes for achievement in culture and art were uncontroversial. The 2006 prize went to the novelist Aleksei Ivanov for his bestselling and acclaimed *The Rebellion's Gold, Or Down the Craggy River*, a historical novel based in late eighteenth-century Perm province (and about which more below). In 2007, the prize went to Georgii Isaakian, director of the Perm Opera and Ballet Theater. No prize in the culture category was awarded in 2008. However, following the opening of the Perm Museum of Contemporary Art, the Perm Fraternity announced that the 2009 Stroganov Prize for achievement in the field of culture would be given to Marat Gel′man. The announcement provoked an immediate and intense backlash. Novelist Aleksei Ivanov quickly declared that, if Gel′man were actually awarded the Stroganov Prize, he would return his inaugural 2006 prize in protest. Newspaper editorial pages, local academics, cultural producers, and politicians of all stripes lined up on either Gel′man's or Ivanov's side—the first in a set of running battles that lasted through the end of the Perm Cultural Project years later. In the narrow matter of the Stroganov Prize, Andrei Kuziaev personally brokered a compromise. Ivanov agreed to keep his award in name, but donated his prize money to the Stroganov museum complex in Usol′e. Kuziaev matched his donation with a personal contribution of the same amount. For his part, Gel′man agreed to donate his prize monies to opening a House of Photography in Perm, a project that would be led by Andrei Bezkladnikov, a Perm photographer who was at that time living in Moscow and who Gel′man hoped to woo back to Perm as another element of the Perm Cultural Project.

The Stroganov Prizes constituted a new arena of cultural competition and distinction in the Perm region. The world of cultural prizes and competitions and their intersection with capitalist sponsors has long been noted in studies of cultural politics inspired by Pierre Bourdieu (e.g., 1993) and other cultural sociologists, but it was as new to the Russian cultural scene as were Lukoil-Perm's grant competitions in the mid-2000s. It brought with it a new field for cultural debate and competition and a new way in which capitalist corporate as well as state interests intersected in the sphere of regional cultural production, consumption, and evaluation. It is notable, then, that even as Chirkunov's Perm Cultural Project sought new directions for regional culture and economy, some of the key final arbiters of how distinction was allocated hailed, in fact, from the state-corporate field of the previous governor, despite the fact that its key players had been transplanted to Moscow. Even without direct sponsorship, Lukoil-Perm remained a powerful arbiter as the regional cultural field stretched out over the Moscow–Perm axis, and it did so, once again, under the aegis of the Stroganov name.

Echoes of the Russian oil complex were also to be found in the substance of critics' objections to the Perm Cultural Project. One of the chief complaints registered by Aleksei Ivanov and his fellow critics about Marat Gel′man's Stroganov Prize, for instance, was that neither PERMM nor the growing Perm Cultural Project was designed to benefit Perm at all. The evidence for this was to be found in the very nature of Chirkunov's team and the times and places in which they operated. In this view, the Perm Cultural Project was primarily a conduit for Moscow-based culture producers like Marat Gel′man to siphon money out of the regional budget at a time when Moscow was no longer as lucrative a place as it had been in the early 2000s. As Igor Averkiev put it in the opening pages of a lengthy dissection of the Perm Cultural Project published in a regional newspaper and on the Perm Citizens' Chamber's website in late 2009, "Perm has become an attraction for Moscow and international creative types who are bored and broke in the [global] economic crisis. . . . [They want] to ride out these crisis years under the administrative and financial wing of Permian power structures."[20]

These critics pointed out that the entire project seemed to emerge from a very specific personal network and to come precisely as the mid-2000s Russian oil boom was stalling. Moreover, many of the management contracts for the Perm Cultural Project were going to Moscow-based firms, and most of the artists and performers were not from Perm either. Moreover, after the widely acclaimed and admired Russian Povera exhibit, many residents of Perm told me that subsequent shows at PERMM were second hand (*sekond khend*) or factory seconds (*shtok*)—kinds of art that were no longer wanted or appreciated in more prestigious locales like Moscow and were therefore being dumped on an unsuspecting Perm. A traditional focus of the regional Ministry of Culture—the development of new musicians, artists, performers, and cultural managers through a network of local schools and institutes—was suddenly neglected in favor of large, spectacular, expensive, and short-term projects that never seemed to include artists from Perm. The Perm Union of Artists and the Professional Union of Culture Workers were thus in the forefront of the opposition, and they demanded attention to Perm-based artists, artwork, and the schools and institutions dedicated to producing more of them. Gel′man had little patience in response: "Look at the Pinault Museum in Venice or the Guggenheim in New York—are there lots of Venetians and New Yorkers [on display] there?"[21]

20. Igor Averkiev, "Permskii Kul′turnyi Puzyr′," part 1, *Novyi Kompan′on*, September 29, 2009.

21. "Marat Gel′man Protiv Alekseia Ivanova," *Russkii Portret*, July 19, 2009. http://rupo.ru/m/1743/marat_gelyman_protiw_alekseya_iwanowa.html.

Given all this, and the fact that Chirkunov, Gel′man, Mil′gram, and Gordeev all had preexisting friendships and other ties, it could therefore be deduced that the Perm Cultural Project was the coordinated effort of a network interested in seizing power and profit for its own gain, innovative only in that culture was an unexpected corner of the state budget in which to attempt this. In this view, the Perm Cultural Project was just a new twist on the familiar practice of Muscovite raiding: This time, instead of wealthy Muscovites coming to the provinces in search of privatization deals and companies to acquire when possibilities had diminished in the capital, the raid was taking place on the terrain of culture, which had gone through a similar, somewhat delayed, trajectory of boom and bust in Moscow and whose leading practitioners were looking for new arenas of profit making. Some version of this accusation was quite commonly mentioned to me in conversations with all sorts of friends and acquaintances. Anthropologist Anna Kruglova (personal communication) reported a succinct version to me from one of her own interviews with a Perm-based culture worker:

> [The Perm Cultural Project team] has created a product that has no relationship to reality, but sucks perfectly real money out of the budget. Why is that happening now? Because people have appeared who aren't able to steal metallurgy, but know how to steal culture.

As I learned more about the 1990s raids and deals in the oil sector discussed in previous chapters, I began to try out my own twist on this understanding of events with my interlocutors. This might be a similarity, I would agree, but the difference between the periods was that in the 1990s, a powerful coalition of local businessmen and state administration officials stood up to the corporate raiders from big Lukoil, leading to the compromise that produced ZAO Lukoil-Perm and OOO Lukoil-Permneft and retained control over important circuits of regional exchange and their ties to senses of Permianness. This time around, I went on, it seemed as though Perm's cultural producers had no analogous champions in regional business or state administration. It was, in fact, precisely those local officials—Chirkunov and Mil′gram—who allied themselves with the powerful outsiders. Among opponents of the Perm Cultural Project, this was a narrative that entirely fit with their understanding of what was happening; "Yes," one of my interlocutors replied immediately to my framing, "[Chirkunov and Mil′gram] were just kidnapped by those Muscovites." I have noted that that the Perm Cultural Project's architects assiduously avoided what they saw as cultural production that served the legitimating ideological purposes of the state on the model of either the Soviet period or the Putin era. It is interesting that, for the most part, their critics conceded this point: there were few objections that the cultural project was glorifying or sacralizing the regional state apparatus or Oleg Chirkunov personally. Instead of seeing state abstractions accomplished by state-led cultural production, critics saw a specific Moscow–Perm coalition pursuing its own interests at the expense of the Perm region as a whole (see also Leibovich and Shushkova 2011).

To be sure, those on the inside of the Perm Cultural Project disputed that significant money from Perm tax receipts was going to Moscow, emphasizing that the permanent collection at PERMM was being assembled by an array of businesspeople and that even the renovation of the River Station Hall was financed in good part by Senator Gordeev. This was private money serving a public good, they maintained, and, in any case, the long-term economic benefits to the Perm region—whether from other subsidies and sponsorships or a coming culturally focused economic boom—would far outweigh any budget outlays. Culture, they said, remained far and away the cheapest part of the budget in which to encourage economic growth. I have no reliable budget figures to cite that could prove the accuracy of either side; the point, rather, is that many of the debates and conflicts in the cultural field borrowed language, concepts, and expectations that stuck closely to patterns familiar from an earlier epoch of struggles over regional values in a federal system—one that configured oil, regional cultural distinctiveness, and state-corporate alliances in ways both similar and different.

Much the same could be said of another critique of the Perm Cultural Project, an essay called "The Perm Cultural Bubble," by Igor Averkiev, a noted human rights activist and organizer in the Perm region whose writings were widely read by local intellectuals. In this essay, Averkiev argued that the rise of interest in contemporary art—around the world but especially in Russia—was an outcome of the commodity and speculative booms of the 2000s:

> The virtual-speculative economy of financial bubbles and pyramids, which has driven the global economy to today's sorry state, must find its reflection in culture. . . . Contemporary art, like the bubble economy, lives in the airless space of synthetic symbols and thoughts, where real "consumer prices" are determined not by objects, images, and feelings, but by opinions about them, or more accurately, "the status of opinions in the system of opinions." "The virtual economy" and "contemporary art" are, in fact, both speculative constructs. . . . In any speculative model, the consumer is an insubstantial figure, because the "market operators" have found a way to manipulate the continuum of supply and demand to their own ends (remember the pre-crisis history of oil prices). . . . The price of culture (in the sense of fine arts) in Perm is now overstated and must collapse, just like oil prices, and with the same consequences.[22]

---

22. Igor Averkiev, "Permskii Kul′turnyi Puzyr′," part 2, *Novyi Kompan′on*, October 6, 2009. Averkiev uses the term *aktual′noe isskustvoe* rather than *sovremennoe isskustvo*; although specialists

Although the article is without footnotes or other references, this element of Averkiev's critique belongs to a family of approaches to culture and finance capital taken by a number of Marx-influenced scholars, among them Frederic Jameson, who writes that "the problem of abstraction—of which this one of finance capital is a part—must also be grasped in its cultural expressions" (1997, 252). The Retort collective, drawing on Guy Debord's famous dictum that "the spectacle is capital accumulated to the point where it becomes an image" ([1967] 1977, 34), likewise posits a newly salient relationship between accumulation strategies in and around the global oil industry and a newly intense, highly abstracted, "battle for the control of appearances" (2005, 31) that was central to the attacks of September 11, 2001, and the ensuing U.S. War on Terror.

I am less concerned with whether these ways of linking oil and images of various sorts are, in general germs, analytically accurate than with drawing attention to the ways in which this brand of critique was deployed in the context of Perm's ongoing cultural upheavals. Averkiev's critiques of conceptual art as a speculative system and his invocations of pyramid schemes and fluctuating oil prices, like criticisms of PERMM as an instance of culture raiding, are notable because they extended into a new era the kinds of regional debate over the relationship between oil and Permianness that began in the early 1990s with petrobarter chains and fuel veksels and moved to Lukoil-Perm's CSR projects and state-run grant competitions in the early 2000s. In all these cases, materiality and abstraction, region and center, and Permianness and non-Permianness formed the grid on which claims and counterclaims were made about what it meant, and what it should mean, to live in the Perm region. Moreover, various elements of the regional oil complex continued to inflect regional sensibilities—it was on this grid that the Perm region continued to be fashioned as an oil region, even by those who sought other paths for it.

Landscapes, Matrices, Spines: Aleksei Ivanov's Urals

Averkiev's "Perm Cultural Bubble" begins by noting that even if some of the projects on which Chirkunov's team had embarked were attractive and welcome, the Permian land (*Permskaia zemlia*) was "not native [*rodnoi*] to them. . . . They just use it, and sometimes trample it."[23] The essay winds down on a similar point,

might draw distinctions, the two terms were largely interchangeable in Russia in the 1990s. Inasmuch as Averkiev is clearly speaking about all contemporary art here, I have used this term in my translation.

23. Averkiev, "Permskii Kul'turnyi Puzyr'," part 1.

noting that if Oleg Chirkunov had been governor of any other region, that region would have become Russia's next aspiring cultural capital. That is, if *The Perm Project* was a region-based challenge to the center, it had nothing to do with the Perm region in particular, and that was the problem: "They don't care where—they are indifferent to the place on the map. But *we* are not indifferent." Averkiev went on to argue for a kind of contemporary change that *was* place-based: "The world is changing in different ways—not only in globalization but also in the rapidly growing regionalization of national territories. New and mighty regional centers are forming everywhere."[24]

In fact, the "Perm Cultural Bubble" opens not only with critique, but with a grudging, "Thanks for the Summons"—a summons to help imagine a place for culture in the Perm region different from that proposed by Chirkunov and his team. Nearly all opponents of the Perm Cultural Project had a long list of counterproposals for the development of culture in the region, and, indeed, claimed that regional cultural producers had long been at work on them only to have their efforts interrupted and diverted by the Perm Cultural Project (see, e.g., Karzarinova and Abashev 2000; Nikitina, Ustiugova, and Chernysheva 2001). Perm Animal Style as a brand, the independent Kamwa festivals, and the field of "Permistika" as explored by V. V. Abashev and collaborators based at Perm State University (e.g., Abashev 2008) were all cited to me as region-based alternatives. The most popular alternative, however, was a set of cultural projects proposed by novelist and culturologist Aleksei Ivanov. Academics I met referred me to Ivanov's culturological essays (often enough with warnings about their historical accuracy), librarians slipped his work into my hands when I asked about contemporary regional studies, and he also edited a series of books and collections called *Perm as Text* that brought together region-based fiction and nonfiction by a range of authors. Ivanov's numerous novels, which had gained local and national fame even before the Perm Cultural Project began, were always prominently on display in Perm's bookstores. I focus on projects associated with Ivanov because they were perhaps the most widespread, in part due to his sheer productivity and in part due to his willingness to engage in frequent, public, and contentious disputes with Chirkunov's team.

Regional Magical Historicism

Aleksei Ivanov was born in 1969 in Nizhnii Novgorod, but his family moved to the Perm region in 1971, and he eventually studied journalism at Urals State University. He worked as a tour guide and local historian on the Chusovaia River

24. Igor Averkiev, "Permskii Kul′turnyi Puzyr′," part 3, *Novyi Kompan′on*, October 13, 2009.

in the Perm region between 1992 and 1998, and many of his later projects featured careful, detailed descriptions of rivers, mountains, and landscapes of the Perm region that he had come to know intimately. Ivanov's first novel, released to significant critical acclaim in 2002, was *Heart of the Uplands* (Ivanov [2002] 2012), set in and around fifteenth-century Cherdyn, capital of Perm Velikaia—the Principality of Great Perm, a Komi-Permiak feudal state in what is now the north of the Perm region. The novel traces the bloody and ultimately unsuccessful effort by the Permian Prince Mikhail to resist annexation by Ivan III, grand prince of all the Rus (known to historians as "gatherer of Russian lands" and founder of the Russian state).[25] *Heart of the Uplands* tells the historically familiar story of annexation and early Russian state formation largely from the perspective of the annexed, including Komi and Mansi tribes, their warriors and shamans, and indeed the Permian lands themselves. We learn in the book's very first pages, in a scene-setting conversation between a warrior prince and a shaman, that the Rus-Novgorodians have come "because they want our riches . . . and in addition to our treasures they also want all our land" (Ivanov [2002] 2012, 16). Ivanov's *The Rebellion's Gold, or Down the Craggy River* (Ivanov [2006] 2012), published in 2006, takes place almost entirely on and around the Chusovaia River in the immediate aftermath of the chaos and confusion that followed the Pugachev Rebellion of 1773–74; its protagonist is a river barge operator searching for gold stolen by Pugachev. This novel, too, is about the reestablishment of central Russian state power in the wake of disorder and challenges to state sovereignty.

Ivanov's work can certainly be read as a commentary on Russian life after the parcelization of sovereignty and chaos of the 1990s (Kukulin 2007). The main character in *The Rebellion's Gold* is, after all, nicknamed Perekhod—"transition." As Ivanov frequently recounted in media interviews at the time, *Heart of the Uplands*, too, could be read as a critique of the reemergence of central Russian state power in the late 1990s and early 2000s. This was not necessarily what he set out to write, Ivanov claimed, but, given the centuries-long historical pattern in which Muscovites come to the Perm region in search of land and riches for themselves, it could hardly have been otherwise. With these novels already popular in the region, Ivanov had a ready-made critique of Chirkunov's team that he did not hesitate to deploy far and wide: the precedents for Muscovite raiding of Perm and Permian cultural distinctiveness went back centuries, and were foundational to the establishment of the Russian state. The Gordeev-Gel′man raid was a story as old as Russia itself.

---

25. *Heart of the Uplands* was actually the second of Ivanov's major works, but his first, *The Geographer Drank Away His Globe* (2005) (later made into a major movie filmed in the Perm region), did not find a publisher until after the success of *Uplands*.

But Ivanov's novels captured regional imaginations for reasons other than historical analogy to present-day patterns of state formation. His books are filled with lengthy, detailed descriptions of the landscape, especially of the rivers and mountains of the Perm region that Ivanov himself came to know so well through his own travels. His characters often reflect on the entwined nature of humans and their surroundings, and he borrows liberally from the mythologies of Finno-Ugric tribes to create and animate otherworldly actors. Spirits and witches, gods and demons, and enchanted objects are everywhere in his pages—participants in the battles, tempters and lovers of the protagonists, and casualties of Russian Orthodox missionaries. Perm-based literature scholars and semioticians V. V. Abashev and M. P. Abashev (2010), in fact, see Ivanov primarily as a poet of the landscape, drawing on Roman Jakobson, Rudolph Otto, and a long Russian literary tradition to show numerous ways in which the materiality of the Perm region itself participates in the action of the novels.

Ivanov's novels, wrote one of his popular press interviewers, "turned Perm into a magical land with a living history."[26] The mystical inhabitants of Ivanov's novels are not alone in contemporary Russian literature, where they find good company with the supernatural Others and vampires of Sergei Lukianenko's *Nightwatch* series (e.g., 1998; 2000), the time-traveling werewolves and werefoxes of Viktor Pelevin's *The Sacred Book of the Werewolf* (2004), and numerous other tales of the undead, the resurrected, or the quasi-human.[27] Alexander Etkind (2013, 220–42) has appropriately termed this writing magical historicism to distinguish it from the magical realism of postcolonial literature and situate it in the specificities of the post-Soviet moment. Against this backdrop, Ivanov's work is remarkable in part because its magical historicism is relentlessly regional, laboriously carved into the landscape of the Perm region. The world conjured in Ivanov's texts is, moreover, to be encountered not only between the covers of a bestseller, but also through river tours, festivals, new museums, and extreme sports—all projects in the Perm region in which Ivanov was personally engaged, all projects that unfolded, at least in 2009–12, as parts of an explicit challenge to the abstractions of Chirkunov's Perm Cultural Project.[28]

After the success of *Heart of the Uplands*, for instance, Ivanov worked with the district administration in the Cherdyn district of the northern Perm region to host a large and successful series of outdoor festivals by the same name in 2006–9. The

26. Konstantin Mil'chin, "Rossia: sposob sushchestvovaniia," *Russkii Reporter* 39 (167), October 10, 2010.

27. The Perm region is not the only oil region visited by such quasi-human beings and forces on the field of cultural production; see Atkinson 2013 on vampires and the United States Gulf Coast in the wake of BP's *Deepwater Horizon* oil spill.

28. On Ivanov's novels and cultural projects as a seamless whole and part of a larger effort to seek out the "cultural-symbolic self-determination of Russian territories," see Abashev and Firsova 2013.

festivals attracted many thousands of guests, who were treated to all manner of games and events drawn from the era described in the book. The festival's name was later changed to The Call of the Uplands: An Ethno-Landscape Festival, and began to attract many of the same young people as the Kamwa festivals—nearly twenty thousand of them in 2011. Indeed, Ivanov himself was an occasional contributor to Kamwa projects (see Ivanov 2008) and some of the Kamwa organizers I met expressed their hope that the new festivals with historical themes picked up from Ivanov's novels would replace the Kamwa festival as it wound down.

"The Urals Matrix" and Ivanov's Cultural Projects

In 2008, Ivanov began supplementing his fiction with a series of culturological essays, the most widely known of which was called "The Urals Matrix" (Ivanov 2009). These essays lay the groundwork for *The Spine of Russia*, a major television series and accompanying large-format book of photographs and text (Ivanov 2010a; 2010b). The overarching claim of these culturological projects—the closest there was to an explicit countermanifesto to the Chirkunov team's *The Perm Project*—was that the Urals region presents a sui generis configuration of many elements, both natural and human—a matrix. This matrix has structured and informed life in the Urals, including in the Perm region, for centuries, and it can be traced explicitly from to the present day back to the time of Ermak, whose famous sixteenth-century expeditions "opening" Siberia were launched from Perm. The term "matrix" itself is indicative of the kind of history that Ivanov wants to project. In the first place, this history is, in keeping with the original Latin meaning of matrix, generative: it gives birth repeatedly across the centuries to a particular and identifiable lineage of complex forms—political, economic, social, and cultural. Matrix is also appropriate because of the term's long history of use in paleography and geology, where it indicates the rocks onto which fossils attach themselves, and metalworks of various sorts—from industrial stamping to the casting of figurines. The long geological history of the Perm region, along with its deposits of salt, potash, coal, copper, oil, and the mines, factories, and factory towns that have grown up to exploit them, are at the heart of Ivanov's vision of the Urals as a mining civilization.

In Ivanov's view, several elements and characteristics of this matrix recur throughout history in new configurations—a result of enduring landscapes and, to no small extent, enduring mentalities. The Urals have always been, and continue to be, a meeting place of difference: between Europe and Asia, among Christianity and Islam and paganism. The "genetic code" (Ivanov 2010a, 201) of this matrix is the landscape of the Urals itself: its rivers, mountains, mineral deposits, and the ways that all these elements channel and enable the movement and organization of human settlements. The meetings that take place on this landscape, Ivanov argues, lead to personal and historical transformations that have recurred again and again in the history of the Urals. The mixing of Finno-Ugric paganism and Christianity, for instance, yielded the Permian gods, hauntingly beautiful wooden sculptures of Christ, now housed in the Perm State Art Gallery, that are quite specific to the Perm region and exceptional because sculpted images of the divine are generally not permitted in Eastern and Russian Orthodoxy. Another recurrent theme in the history of the Urals is expansive local sovereignty: the tendency for industrial leaders to strike deals with central power that create domains of power within domains of power (*derzhava v derzhave*). The Stroganov and Demidov families, both of which had special concessions from the tsar for their factories and other operations in the Urals, are the archetypal cases of this expansive local sovereignty. The 1990s–2000s privatization schemes and oligarchs of the post-Soviet period, in Ivanov's view, are their most recent incarnations.

Especially in the popularizations of Ivanov's perspectives on the Urals that make up *The Spine of Russia*, the point of attending to the Urals matrix in this way is to convince the viewer/reader to understand not just the Urals, but all of Russia differently. Ivanov argues that it was in the Urals that Rus became Russia—it was through Ermak's encounters in the Urals, he notes, that the famous explorer turned from raiding brigand into settler in the service of the tsar. We find out, in fact, that many firsts were accomplished in the Urals: the samovar was actually invented in the Urals; the first steam engine was produced in the Urals (although its importance and potential were not recognized by its builders); Russian business started in a small (now abandoned) accounting office; Russian fairy tales owe many of their characters and plots to the myths of the old Perm region, including those associated with Perm Animal Style. The first island in the Gulag Archipelago, too, is to be found in the Urals. All these firsts are the children of the Urals matrix, the precipitate of centuries' worth of encounters and entanglements on the landscapes of the Urals. They are the contributions, good and bad, of the Urals region to larger Russian history and identity. This combination of identity, distinctiveness, and generativity is, Ivanov insists, premised on the provinciality, the noncentrality, of the Urals within Russia. It is, in his view, because the Urals are a meeting place, and have been for centuries, that so many innovations and departures in Russian history can emerge here. To declare that a capital of anything might be found or constructed in the Urals is, therefore, *precisely* to miss the importance of the Urals as noncentral and peripheral. This was the point that Ivanov again and again drove home in his public and vocal opposition to Chirkunov's European Capital of Culture campaign.

Academic historians of the Perm region have not always agreed with the details and particulars of Ivanov's historical and culturological analyses; in my own view, they surely attribute too much to mentalities and abstract, underspecified historical and personal forces that can explain any and all things. As I noted in the introduction, they clearly neglect local historical consciousness. It is, nevertheless, in good part because of its sweep and ability to tell a new and coherent story about the Urals, one discoverable in the landscapes of the Perm region, that Ivanov's work gained such a following and spread across so many platforms. Survey research by Abashev and Firsova (2013) revealed that 25 percent of the residents of Cherdyn had read Ivanov's *Heart of the Uplands*, set five centuries earlier in their hometown, and 81 percent had participated in one way or another in the eponymous festival that Cherdyn began hosting in 2006. Nearly 35 percent of surveyed travel agents in the Perm region agreed that Ivanov's work had significantly and positively influenced the image of the Perm region among tourists.

For all his disagreement with the Perm Cultural Project, though, Ivanov's corpus of works ended in a similar place: for the postindustrial Perm region, he believed, marketing cultural brands and historical distinctiveness was the way forward.[29] Ivanov was not, however, much of a supporter of folk handicrafts production and fairs in the manner of Lukoil-Perm. His proposed answers also include tourism, but tourism of the sort that includes river tours, outdoor sports, and long hikes that pass through old factory towns. One of the three main hosts for the television-series version of *The Spine of Russia* is Iulia Zaitseva, an extreme sports enthusiast. The historical excursions and commentaries provided by Ivanov and journalist host Leonid Parfenov are interspersed with Zaitseva demonstrating the ways in which tourists might experience the Perm region and the larger Urals. After Parfenov and Ivanov discuss and trudge along one of the imperial-era roads running through the Perm region, for instance, Zaitseva notes that it is easiest to traverse these roads these days in a four-wheeler, and we see her churning through the mud. Zaitseva shoots rapids in a catamaran, climbs towering factory smokestacks in the latest mountaineering gear, views archaeological sites from a hot-air balloon, and drives a tunneling machine through salt mines. The series concludes with the trio passing an old factory town, which Ivanov informs viewers is in the process of becoming a dacha settlement—a place used for leisure rather than work. Leisure, he says, is the latest trend for the Urals as a whole.

Ivanov himself had been engaged in a multiyear effort to turn the old town of Kyn, in the Lys′va district, into a new tourist destination. He helped win the designation of Perm Regional Cultural Center for Kyn in 2010, and, in the same

29. On other nation- and culture-branding efforts in the postsocialist world, see Dzenovska 2005; Jansen 2008; and Marat 2009. My analysis here has its closest kinship with Manning and Uplisashvili, who write that, "Brands, like the commodities they are attached to, are material semiotic forms whose circulation defines a broader social imaginary, . . . whether it is the market, the nation, or the empire that in part gives them meaning" (2007, 628; internal reference omitted).

year, opened the tourist program Kyn-Reality that combined lessons in the history of Kyn as an exemplary Urals factory town with rafting excursions on the Chusovaia River in the summer or snowmobile excursions in the winter. In his public disputes with the Perm regional administration, Ivanov often held up Kyn as an alternate model of regional cultural development for tourism, one that would benefit the region rather than the federal center and would provide tangible infrastructural benefits right away—especially a new road linking Perm and Sverdlovsk—to handle increased tourist traffic.

Ivanov's Oil

The production and refining of oil occupies a minor place in Ivanov's conceptualization of the Urals: among the many items emerging from the subsoil charted in "The Urals Matrix" and the television version of *The Spine of Russia*, oil is not mentioned directly—either in its discovery in the twentieth century or its contemporary significance. There is a single page devoted to oil production in the book version of *The Spine of Russia*, and it deals with the Republic of Bashkiria, further south of the Perm region in the Volgo-Urals oil basin. Indeed, it is not especially clear where oil fits into the model of "mining-metallurgical civilization" that Ivanov outlines; the section on Bashkiria only situates oil in relationship to the broader dynamics of the Urals matrix in one way—by noting that, in Bashkiria, oil became the resource around which the proposal for a sovereign Bashkortostan was floated in the 1990s.

Strikingly, however, oil does feature in another way—in its circulation through the Urals and the Urals matrix *as oil*, rather than as money. Early on in the television version of *The Spine of Russia*, at the end of a discussion of the sixteenth-century trading routes that formed one element of the Urals as a meeting place at the time, the following exchange takes place between Parfenov and Ivanov:

> PARFENOV: It turns out that sable coat was our barrel [i.e., unit of measure] of the sixteenth century.
>
> IVANOV: Yes, it was a natural currency of that epoch. There were four in all.
>
> PARFENOV: So, Russian furs, American gold . . .
>
> IVANOV: . . . Chinese silk, Indian spices. That is, furs were the oil of the sixteenth century.

The illustration on the screen at this point is as significant as the dialogue. It is a colorful ink drawing by Moscow-based artist Il′ia Viktorov, a frequent collaborator of Parfenov's, that depicts Russian traders posing proudly next to a wooden barrel overflowing with sable coats. Above the pair and their barrel, in block letters, is the word "fur"; in stylized lettering and ornamentation on a gold banner below, the words "oil of the sixteenth century" appear.

A later segment of the film employs a similar strategy. In a discussion of the ironworks of the imperial era, a time when, Ivanov argues, the Urals became the blast furnace of world civilization, the film discusses the intricate system by which cannons and other products produced in the factories of the Stroganovs and Demidovs made their way to the rest of Russia. In the winter, iron products were stacked at docks along the Chusovaia River, waiting for the spring thaw, when they would be loaded onto wooden barges. After the ice had cleared from the river, and spring melting had raised water levels, a series of artificial ponds near factory towns was released into the river in a carefully orchestrated sequence. The combination of melting snow and released pond water created a rolling flood that sent scores of heavily laden barges careening along the Chusovaia to central Russia.

Parfenov explains this system to the viewer via analogy to the circulation of oil in the present day: "If you compare a factory to an oil well, then the Chusovaia River is a pipeline, and the dock is a valve. Whoever sits on the pipe and turns the valve rules over all." The visual accompanying this comment is another Viktorov ink drawing, this one showing the Chusovaia River snaking through the Perm region. The river, though, is encased in a pipeline (see fig. 14). In the foreground, the top half of the pipeline is cut away, revealing barges stacked with iron products passing through a large valve, with a massive horizontal on–off wheel atop it. At the valve sits an unidentified Russian noble in full fur and boots and hat, Gulliver-sized by comparison to the barges and people, slowly cranking the wheel as the barges float through on their way west to Moscow. Once again, the Russian nobles of the old Perm region are reincarnated as oil magnates.

Two things are significant about these images of oil in the television version of Ivanov's *The Spine of Russia*. First, the exchange and circulation of oil is the background assumption, the terms in which other aspects of history—the fur trade or barge transport—can be best explained to a wide television audience. Fur is the oil of the sixteenth century; river transport by barge is the oil pipeline of the Stroganov era. There could scarcely be a better indication, I think, of the ways in which images and vocabularies from Russia's oil complex have worked their way into everyday discourse about history and culture. Second, circulating oil is presented *not* as abstracted circulating money, but as a natural currency and as a specific, material, regionally produced object under the control of a local notable and moving through the landscape of the Urals. The 1990s era of pervasive petrobarter exchanges crisscrossing the Perm region may have faded by 2010, but when *The Spine of Russia* reached for a way to showcase the distinctiveness of the Perm region in the context of the Russian federal state over the long term,

FIGURE 14 "Pipeline." Illustration by Il'ia Viktorov for the television series *The Spine of Russia*, conceived by Aleksei Ivanov, directed by Leonid Parfenov, and produced by Iulia Zaitseva. Studio Namednyi, 2009.

Image courtesy of Aleksei Ivanov, Iulia Zaitseva, and Il'ia Viktorov.

it arrived in a familiar place: the nonmonetized exchange and specifically Permian characteristics of oil.[30]

• • •

The chapters of part III have traced a cultural field shaped in significant part by the political economy of the Russian oil complex at a regional level, a field in which the low-labor, high-profit, highly spatialized nature of the oil industry, along with the material properties and transformational potentials of that complex—and of oil itself—again and again had crucial implications for cultural projects and possibilities. Lukoil-Perm's Connections with Society Division, other energy sector and independent cultural projects, Governor Chirkunov's Perm Cultural Project team, and the many projects of Aleksei Ivanov and his fellow travelers all sought to put the Perm region on the map of Russia and the world. They disagreed vehemently about how to do so, but they were all inhabitants of an institutional, discursive, and material field that was powerfully shaped by the oil complex of twenty-first-century Russia.

Oleg Chirkunov resigned his post as governor of the Perm region in April 2012, and even as I walked through the White Nights in Perm Festival Village that summer, word was circulating that the Perm Cultural Project would quickly wind down without its chief cheerleader. Speaking with an employee of the Perm Museum of Contemporary Art one day in 2012, shortly after Chirkunov's resignation, I asked whether Lukoil-Perm would ever sponsor an exhibit at the museum, perhaps stepping in to help if state funding dried up under the next governor. "No," she replied, using the familiar lexicon of depth, oil, space, and cultural production:

> Lukoil completely distances itself from us. Lukoil supports only that which relates to classical [art]. It's a resource economy, you understand . . . it's to their advantage to support Ivanov, who mythologizes the subsoil and says: what a magical subsoil (*volshebnyi nedr*) we have, how sacred it is, how deep and beautiful our roots are. We [at PERMM] say nothing about the subsoil, [but for Lukoil-Perm] it's all roots, roots, especially when they rot through and turn to oil.

At the recent inauguration of Oleg Chirkunov's successor, she went on to say, rolling her eyes, that, she had heard, Aleksandr Leifrid, head of Lukoil's operations in the Perm region, had presented the new governor with a painting—a classical

30. Alexander Etkind (2011) argues that Russia's ancient fur-based and more recent oil-based economies are similar in that they both led to a resource curse. Ivanov and Parfenov's analogy, I believe, operates very differently, running as it does through the materiality of oil/fur in their regional locations rather than through their accumulation as money in a federal center.

landscape. But I found out later that her information was wrong. The gift was, in fact, a handcrafted item made by one of the company's sponsored folk artisans from an oil-producing district. It was a wood carving of several items that, together, were meant to represent the Perm region—a container labeled "Permian salt," a cannon (for the defense sector), and a barrel (for all other products). Standing behind these three items as if to present them, and nearly twice their size, was a rotund, smiling, bearded, big-eared figure in a peasant shirt—the folk archetype of the jolly, welcoming, simple Russian peasant man. He wore a Lukoil hard hat.

As nearly all my interlocutors predicted that summer, the Perm Cultural Project did indeed end quickly. The 2013 White Nights in Perm Festival (already planned and budgeted in 2012) was scaled back considerably and, by the summer of 2014, it had disappeared from the central esplanade entirely. Mil'gram himself returned to more traditional theater directing in 2012, and, a year later, Marat Gel'man resigned his post as head of the Perm Museum of Contemporary Art. In 2014, PERMM moved to smaller and less central quarters on Gagarin Boulevard, where it continued to host exhibits and, on a less grand scale, sponsor and maintain public art projects. As was the case with Governor Trutnev's Capital of Civil Society initiative, then, the effort to fashion Perm into a European Capital of Culture did not outlast the political network that championed it. Chirkunov's critics initially rejoiced at the fracturing of his network and its signature project, and more homegrown cultural projects like the Kamwa festival began to plan and stage a new series of cultural events and spectacles. However, at least some of Chirkunov's critics soon found that they had received more attention—in the media and for their own preferred styles of cultural production—when they were in the midst of high-stakes debates rather than when culture was relegated to its more usual subordinate station.

Chirkunov's replacement as governor, Viktor Basargin, was new to Perm. A native of the neighboring Sverdlovsk region, he had risen through the Party ranks, spent time in the central offices of the Urals Federal District (which, recall, did not include the Perm region) and, from 2008 to 2012, served in Russian president Medvedev's administration as minister of regional development. Basargin's leadership team in Perm was a mix of his own loyal deputies brought in from Moscow and Perm regional powerbrokers from Chirkunov's and Trutnev's teams. In the early years of his term, local elites and outside observers debated frequently whether a new, coherent regional agenda—an effort on the order of those that defined his predecessors' terms to put this provincial region on the map—would emerge out of this three-way hybrid. What was not in dispute, however, as that inaugural gift bespoke so clearly, was that Lukoil-Perm stood ready to serve as general partner of the governor's office, through the same structure of agreements, projects, and initiatives that it had begun nearly two decades earlier. One of Basargin's first initiatives in the social sphere, for instance, was to build and refurbish kindergartens in Perm and throughout the region. Lukoil-Perm was an early and enthusiastic collaborator, and pictures of Lukoil-Perm general director Leifrid and Governor Basargin surrounded by happy schoolchildren quickly became standard fare in the regional media. The company's 2013 signed agreement with the Perm region promised 820 million rubles for development projects in oil-producing districts.

The beginning of Basargin's term also coincided with the return of Vladimir Putin to the Russian presidency in 2012 and, in the wake of the street protests that accompanied Putin's election bid, another round of efforts to centralize power through federal mandates and projects. But there was little sign that the center-region dynamics and frameworks I have traced in this book were set to change dramatically. Responding to President Putin's designation of 2014 as a Year of Culture, for instance, Lukoil-Perm announced Routes of Culture, a special category of social and cultural project grants that would celebrate the eighty-fifth anniversary of Permian oil and showcase the distinctiveness of the Perm region's many cultures and traditions. Central initiatives were once again being routed through regional political and economic networks, landscapes, and sensibilities—as they had been in the postsocialist 1990s, in the socialist Second Baku, in the Stroganov era, and, at least according to widely circulating versions of regional history, long before that.

Appendix

GOVERNORS OF THE PERM REGION IN THE POST-SOVIET PERIOD

Boris Kuznetsov	December 1991–January 1996
Gennadii Igumenov	January 1996–December 2000
Iurii Trutnev	December 2000–March 2004
Oleg Chirkunov	March 2004–April 2012
Viktor Basargin	April 2012–

Glossary

OIL SECTOR ENTERPRISES, FIRMS, COMPANIES, AND SUBSIDIARIES

Lukoil or big Lukoil: Moscow-based holding company, founded by former deputy minister of oil and gas Vagit Alekperov as the Soviet oil complex unraveled. Big Lukoil was the parent company for Lukoil Overseas Holding (1996–present), OOO Lukoil-Permneft (1995–2004), ZAO Lukoil Perm (1995–2004), Lukoil-Perm (2004–present), PNOS (1991–present), and Permnefteprodukt (1993–present).

Lukoil Overseas Holding: Big Lukoil subsidiary founded in 1997 to manage company operations outside of Russia, both in the former Soviet Union and further abroad. Lukoil Overseas Holding was based in Perm—and financed largely by ZAO Lukoil-Perm—from 2001 until its move to Moscow in 2006.

Lukoil-Perm: Since 2004, big Lukoil's sole production subsidiary in the Perm region, formed by combining OOO Lukoil-Permneft and ZAO Lukoil-Perm.

NGDU (neftegazodobyvaiushchaia upravleniia): Oil and gas production administration, the lowest level of organization in the socialist oil complex. NGDUs were based in oil towns and responsible for oil production in a given geographical territory; in the Soviet period, they also administered the local social sphere of housing, kindergartens, clubs, and so on. Permneft, the Perm region's oil production association, was composed of five NGDUs in the Soviet period, which were divided between OOO Lukoil-Permneft and ZAO Lukoil-Perm at the time of big Lukoil's takeover in 1995.

OOO Lukoil-Permneft: Big Lukoil production subsidiary based in the southern Perm region from 1995 until unification with ZAO Lukoil-Perm in 2004. OOO Lukoil-Permneft was a full subsidiary of big Lukoil and included most of the Perm region's heritage oil industry, labor force, and production infrastructure from the Soviet period.

Permneft: Soviet-era oil production association in the Perm region, eventually including NGDUs based in Krasnokamsk, Polazna, Chernushka, Osa, and Kungur.

Permnefteprodukt: Perm regional oil products sales and marketing company, controlled by big Lukoil since 1991.

Permneftorgsintez (PNOS): The major oil refinery and petrochemicals plant based in the industrial district of Perm, completed in 1959. PNOS entered State Concern Lukoil in 1991.

Soiuznefteksport: The single, federal-level Soviet entity that coordinated international oil export from the Soviet Union.

State Concern Lukoil: The first incarnation of Lukoil as a holding company, founded in 1991 and including the West Siberian production fields Langepas, Urai, and Kogalym and Perm's refinery PNOS. State Concern Lukoil became big Lukoil as privatization proceeded.

ZAO Lukoil-Perm: Big Lukoil production subsidiary based in the northern Perm region from 1995 until unification with OOO Lukoil-Permneft in 2004. ZAO Lukoil-Perm's ownership was split fifty–fifty between big Lukoil and the PFPG, and the company earmarked a set amount of oil each year to be refined for the uses of the Perm regional administration. It also specialized in financial operations—including fuel veksels—and served as a shareholding consolidation center for big Lukoil.

OTHER COMPANIES, ORGANIZATIONS, AND TERMS

24–20 Program: Tax incentive program in which Perm regional corporate taxes were lowered from 24 percent to 20 percent, beginning in 2006; Lukoil-Perm agreed to reinvest its 4 percent savings in the region, divided evenly between social and cultural development projects and investment in its own operations and infrastructure.

Connections with Society Office: Corporate division of Lukoil-Perm tasked with government and public relations, including running the annual social and cultural projects grant competitions.

Cooperation (Sodeistvie): Perm-based noncommercial organization engaged in social and cultural project sponsorship and design. Cooperation was contracted to run PermRegionGaz's corporate social and cultural projects in the mid-2000s.

FIG: Financial-Industrial Group, a brand of conglomerate including several companies (often but not always including banks) that emerged in the early post-Soviet period. FIGs were influential in the privatization of large state-owned enterprises and often served as vehicles for the enrichment of oligarchs. The PFPG was the Perm region's largest and most influential regional FIG.

finansisty: Bankers and financial specialists who rose to influence across Russian sectors in the 1990s, and whose new expertise often sat uneasily with that of career oilmen and engineers (the neftianiki). In the Perm region, oil-sector finansisty were concentrated in ZAO Lukoil-Perm and the northern and eastern districts of the region.

Glavsnab (Glavnoe Snabzhenie): The official Soviet-era supply and distribution organization. The Perm region's decision to replace Glavsnab offices (and personnel) with a commodity exchange was instrumental in charting the Perm region's course as Soviet central planning disappeared.

Kamwa: Independent ethnofuturist festival, sponsor of the yearly Kamwa festivals, a range of other activities from fashion shows to raves, and a key booster of Perm Animal Style in the Perm region.

Komsomol: The Communist Youth League, an important training ground for many of the post-Soviet regional elite in both business and government, and especially in the administration of social and cultural projects.

neftianiki: Oil workers understood as a profession or a collective, usually with an emphasis on long careers working in the oilfields or in oil engineering. At the executive level, neftianiki were the red directors—the Soviet-era managers—of the oil industry, and they often clashed with new financiers (finansisty). In the Perm region, oil-sector neftianiki were concentrated in OOO Lukoil-Permneft and the southern and western districts of the region.

Oil Products Marketing (Neftsintezmarket): A company controlled by the PTB (and then PFPG) and specializing in petrobarter transactions in the early 1990s. Although it was never formally part of big Lukoil, Oil Products Marketing was instrumental in bringing together the production, refining, and distribution of oil at the regional level after the end of the Soviet Union.

Perm Animal Style: Generic name for copper or bronze castings in the shape of humans, animals, and human-animal hybrids fashioned by Komi-Permiak peoples in the seventh through twelfth centuries and uncovered in archaeological digs. Perm Animal Style was an increasingly popular brand for Perm and the Perm region in the mid- and late 2000s, in part due to the activities of the Kamwa festival.

Perm Commodity Exchange, or PTB (Permskaia Tovarnaia Birzha): The PTB was established in 1991 as a mechanism for regional enterprises to exchange goods in the absence of Soviet central planning. Initial profits from the PTB launched the careers of many of the Perm region's new oil elite, especially as oil began to trade on the exchange in 1992 and the PTB subsidiary Oil Products Marketing initiated wide-ranging petrobarter operations.

Perm Cultural Project: Official name for Governor Oleg Chirkunov's effort to have Perm designated a European Capital of Culture by 2016 and to make cultural creativity into the new driver of regional development across all sectors.

Perm Financial-Productive Group (PFPG): founded in 1993 out of the remains of the PTB and active across a wide range of financial, industrial, and consumer industries in the Perm region, from privatization auctions to the sale of household goods. Originally incorporated as a regional FIG in 1993 and named the Perm Financial-Industrial (*Promyshlennaia*) Group, the holding company was quickly forced to switch its name to Perm Financial-Productive (*Proizvodstvennaia*) Group due to changes in federal regulations. The PFPG's accumulated shares in Permneft helped block big Lukoil's raid on the Perm region in the mid-1990s and helped turn many of its executives into oilmen. It also controlled the crucial subsidiary Oil Products Marketing and helped to manage the fuel veksel circuits of the mid-1990s.

Perm Fraternity (Permskoe Zemliachestvo): Club of notable executives and state officials based in Moscow, but hailing from the Perm region. The Perm Fraternity awarded the annual and sometimes controversial Stroganov Prizes recognizing special achievement of various sorts in the Perm region.

PERMM: The Perm Museum of Contemporary Art, the first step in (and eventual centerpiece of) the Perm Cultural Project. PERMM opened in Perm's River Station Hall in 2009 with the *Russian Povera* exhibit.

PermRegionGaz: Beginning in 2001 (and largely replacing UralGazServis), Gazprom's distribution subsidiary in the Perm region. Through short-term contracts with Cooperation, PermRegionGaz ran social and cultural project competitions in the electoral district of its chief executive, Igor Shubin, for a number of years.

UralGazServis: The Perm region's primary natural gas distribution company from the end of the Soviet Union through the establishment of PermRegionGaz in 2001.

Uralkalii: One of the world's largest and most profitable potash mining companies, based in Berezniki in the north of the Perm region. Uralkalii was the only industrial company that could match Lukoil's regional oil subsidiaries when it came to income and political clout, although both were tightly concentrated in and around Uralkalii's areas of operation, as opposed to spread out through much of the region (as was the case for oil production).

veksel: A promissory note. Fuel-backed veksels, issued by Oil Products Marketing and then ZAO Lukoil-Perm, became a surrogate currency in the Perm region in the mid-1990s, paying down billions of rubles in debt and helping to establishing the centrality of oil to the region after privatization.

VKMKS: The Upper Kama Potassium Salts Deposit, home to the mining operations of Uralkalii and Silvinit. The fact that significant untapped reserves of oil were located in geological strata below the salt mines made the VKMKS an area of high tension and negotiation between Lukoil subsidiaries and Uralkalii, especially in the late 1990s.

References

Abashev, V. V. 2008. *Perm' kak tekst: Perm' v russkoi kul'ture i literature XX veka*. Perm: Zvezda.

——, ed. 2009. *Gorod Perm': Smyslovye struktury i kul'turnye praktiki*. Perm: Ministerstvo Kul'tury i Massovykh Kommunikatsii Permskogo Kraia.

Abashev, V. V., and M. P. Abasheva. 2010. "Poeziia prostranstva v proze Alekseia Ivanova." *Sibirskii Filologicheskii Zhurnal* 2: 81–90.

Abashev, V. V., and A. V. Firsova. 2013. "Tvorchestvo Alekseia Ivanova kak faktor razvitiia vnutrennego turisma v permskom krae." *Vestnik Permskogo Universiteta* 3 (23): 182–90.

Abaturova, O. A. 2003. *Pervenets bol'shoi ural'skoi nefti Krasnokamsk*. Perm: OOO Studiia Zebra.

Abrams, Philip. 1988. "Notes on the Difficulty of Studying the State." *Journal of Historical Sociology* 1 (1): 58–89.

Adams, Laura. 2010. *The Spectacular State: Culture and National Identity in Uzbekistan*. Durham, NC: Duke University Press.

Adams, Laura, and Assel Rustemova. 2009. "Mass Spectacle and Styles of Governmentality in Kazakhstan and Uzbekistan." *Europe-Asia Studies* 61 (7): 1249–76.

Adunbi, Omolade. 2013. "Mythic Oil: Resources, Belonging and the Politics of Claim Making Among the Ìlàjẹ Yorùbá of Nigeria." *Africa* 83 (2): 293–313.

Aksartova, Sada. 2009. "Promoting Civil Society or Diffusing NGOs? U.S. Donors in the Former Soviet Union." In *Globalization, Philanthropy, and Civil Society*, edited by David C. Hammack and Steven Heydemann, 160–91. Bloomington: Indiana University Press.

Alekperov, Vagit. 2001. *Neft' Rossii: vzgliad top-menedzhera*. Moscow: OAO Vneshtorgizdat.

——. 2011. *Neft' Rossii: proshloe, nastoiashchee i budushchee*. Moscow: Kreativnaia ekonomika.

Alexander, Jeffrey. 2008. "Iconic Experience in Art and Life: Surface/Depth Beginning with Giacometti's *Standing Woman*." *Theory, Culture, and Society* 25 (5): 1–19.

Alexeev, Michael. 1987. "The Underground Market for Gasoline in the USSR." Berkeley Duke Occasional Papers on the Second Economy in the USSR 9.

Alexeev, Michael, and Robert Conrad. 2009. "The Russian Oil Tax Regime: A Comparative Perspective." *Eurasian Geography and Economics* 50 (1): 93–114.

Allina-Pisano, Jessica. 2010. "Social Contracts and Authoritarian Projects in Post-Soviet Space: The Use of Administrative Resource." *Communist and Post-Communist Studies* 43 (4): 373–82.

Alonso, José, and Ralph Galliano. 1999. "Russian Oil-for-Sugar Barter Deals 1989–1999." *Cuba in Transition* 9: 335–41.

Anand, Nikhil. 2011. "Pressure: The PoliTechnics of Water Supply in Mumbai." *Cultural Anthropology* 26 (4): 542–62.

Appel, Hannah. 2012a. "Offshore Work: Oil, Modularity, and the How of Capitalism in Equatorial Guinea." *American Ethnologist* 39 (4): 692–709.

——. 2012b. "Walls and White Elephants: Oil Extraction, Responsibility, and Infrastructural Violence in Equatorial Guinea." *Ethnography* 13 (4): 439–65.

Appel, Hilary. 2008. "Is It Putin or Is It Oil? Explaining Russia's Fiscal Recovery." *Post-Soviet Affairs* 24 (4): 301–23.

Applegate, Celia. 1999. "A Europe of Regions: Reflections on the Historiography of Sub-National Places in Modern Times." *American Historical Review* 104 (4): 1157–82.

Apter, Andrew. 2005. *The Pan-African Nation: Oil and the Spectacle of Culture in Nigeria.* Chicago: University of Chicago Press.

Åslund, Anders. 1995. *How Russia Became a Market Economy.* Washington, DC: Brookings Institution Press.

Atkinson, Ted. 2013. "'Blood Petroleum': *True Blood*, the BP Oil Spill, and Fictions of Energy/Culture." *Journal of American Studies* 47 (1): 213–29.

Auld, Graeme, Steven Bernstein, and Benjamin Cashore. 2008. "The New Corporate Social Responsibility." *Annual Review of Environment and Resources* 33 (1): 413–35.

Austin, Diane E., Thomas R. McGuire, and Rylan Higgins. 2006. "Work and Change in the Gulf of Mexico Offshore Petroleum Industry." In *Markets and Market Liberalization: Ethnographic Reflections*, edited by Norbert Dannhaeuser and Cynthia Werner, 89–122. Amsterdam: Elsevier.

Auyero, Javier, and Debora Swistun. 2009. *Flammable: Environmental Suffering in an Argentine Shantytown.* New York: Oxford University Press.

Ballard, Chris, and Glenn Banks. 2003. "Resource Wars: The Anthropology of Mining." *Annual Review of Anthropology* 32 (1): 287–313.

Balmaceda, Margarita. 2008. "Corruption, Intermediary Companies, and Energy Security: Lithuania's Lessons for Central and Eastern Europe." *Problems of Post-Communism* 55 (4): 16–28.

——. 2013. *Politics of Energy Dependency: Ukraine, Belarus, and Lithuania between Domestic Oligarchs and Russian Pressure.* Toronto: University of Toronto Press.

Balzer, Harley. 2005. "The Putin Thesis and Russian Energy Policy." *Post-Soviet Affairs* 21 (3): 210–25.

——. 2006. "Vladimir Putin's Academic Writings and Russian Natural Resource Policy." *Problems of Post-Communism* 53 (1): 48–54.

Balzer, Marjorie. 2006. "The Tension between Might and Rights: Siberians and Energy Developers in Post-Socialist Binds." *Europe-Asia Studies* 58 (4): 567–88.

Barnes, Andrew. 2006. *Owning Russia: The Struggle Over Factories, Farms, and Power.* Ithaca, NY: Cornell University Press.

Barrett, Ross. 2012. "Picturing a Crude Past: Primitivism, Public Art, and Corporate Oil Promotion in the United States." *Journal of American Studies* 46 (2): 395–422.

Barry, Andrew. 2004. "Ethical Capitalism." In *Global Governmentality: Governing International Spaces*, edited by Wendy Larner and William Walters, 195–211. London: Routledge.

——. 2013. *Material Politics: Disputes along the Pipeline.* Chichester: Wiley.

Baudrillard, Jean. 1981. *For a Critique of the Political Economy of the Sign.* St. Louis, MO: Telos.

Bazylevych, Maryna. 2010. "Public Images, Political Art, and Gendered Spaces: Construction of Gendered Space in Socialist and Post-Socialist Ukraine." *Journal of Contemporary Anthropology* 1 (1): 2–19.

Becker, Howard S. 1982. *Art Worlds.* Berkeley: University of California Press.

Behrends, Andrea, Stephen P. Reyna, and Günther Schlee, eds. 2011. *Crude Domination: An Anthropology of Oil.* New York: Berghahn Books.

Benjamin, Walter. [1936] 1968. "The Work of Art in the Age of Mechanical Reproduction." In *Illuminations: Essays and Reflections*, 217–52. New York: Schocken.

Benson, Peter. 2012. *Tobacco Capitalism: Growers, Migrant Workers, and the Changing Face of a Global Industry.* Princeton, NJ: Princeton University Press.

Benson, Peter, and Stuart Kirsch. 2010. "Capitalism and the Politics of Resignation." *Current Anthropology* 51 (4): 459–86.

Berliner, Joseph. 1957. *Factory and Manager in the USSR*. Cambridge, MA: Harvard University Press.

Bernstein, Anya. 2013. *Religious Bodies Politic: Rituals of Sovereignty in Buryat Buddhism*. Chicago: University of Chicago Press.

——. 2014. "Caution, Religion!: Iconoclasm, Secularism, and Ways of Seeing in Post-Soviet Art Wars." *Public Culture* 26 (3): 419–48.

Bird, Robert. 2011. "The Poetics of Peat in Soviet Literary and Visual Culture, 1918–1959." *Slavic Review* 70 (3): 591–614.

Bitušková, Alexandra. 2009. "Regions and Regionalism in Social Anthropology." *Anthropological Journal of European Cultures* 18 (2): 28–49.

Black, Brian. 2003. *Petrolia: The Landscape of America's First Oil Boom*. Baltimore: Johns Hopkins University Press.

Blus', P. I. 2004. *Sotsial'nye proekty v Permskoi oblasti: metodiki i tekhnologii*. Perm: Permskoe Knizhnoe Izdatel'stvo.

Bolotova, A. 2007. "Habitual Risk Taking in Dzerzhinsk: Daily Life in the Capital of Soviet Chemistry." In *Cultures of Contamination: Legacies of Pollution in Russia and the U.S*, edited by M. Edelstein, M. Tysiachniouk, and L.V. Smirnova, 223–52. Amsterdam: JAI Press.

Bondarenko, N. 2003. *Golosa vremeni*. Perm: Permskoe Knizhnoe Izdatel'stvo.

Borenstein, Eliot. 2014. "Conspiracy as Information: The Afterlife of Bad Ideas." Plenary Address, Council on European Studies Conference, Washington, DC, March 15.

Borisova, N. V. 2002. "Perm': kvaziavtonomiia." In Gel'man et al., eds, *Avtonomiia ili Kontrol'?*, 189–227.

——. 2008. "Permskie pravozashchitniki: mezhdu obshchestvom i vlast'iu." *Vestnik Permskogo Universiteta Seriia Sotsiologiia* 2 (18): 198–213.

——. 2009. "Institutsional'naia sreda i uchastniki mezhsektornogo vzaimodeistviia v Permskom krae." *Vestnik Permskogo Univesiteta Seriia Politologiia* 4 (8): 40–47.

Borisova N. V., E. E. Reneva et al. 2003. "Strategii vzaimodeistviia negosudarstvennogo sektora i vlastei v Prikam'e," *Politicheskii Al'manakh Prikam'ia* 4: 238–48.

Botkin, O., M. Kozlov, and B. Collins. 1997. The Ural Plants Financial-Industrial Group: First Year of Operation." *Russian and East European Finance and Trade* 33 (2): 59–71.

Bourdieu, Pierre. 1993. *The Field of Cultural Production*. New York: Columbia University Press.

——. 1999. "The Genesis and Structure of the Bureaucratic Field." In *State/Culture: State Formation After the Cultural Turn*, edited by George Steinmetz, 53–75. Ithaca, NY: Cornell University Press.

Bouzarovski, Stefan. 2010. "Post-Socialist Energy Reforms in Critical Perspective: Entangled Boundaries, Scales and Trajectories of Change." *European Urban and Regional Studies* 17 (2): 167–82.

Bouzarovski, Stefan, and Mark Bassin. 2011. "Energy and Identity: Imagining Russia as a Hydrocarbon Superpower." *Annals of the Association of American Geographers* 101 (4): 1–12.

Bowden, Gary. 1985. "The Social Construction of Validity in Estimates of US Crude Oil Reserves." *Social Studies of Science* 15 (2): 207–40.

Boyer, Dominic. 2006. "Conspiracy, History, and Therapy at a Berlin *Stammtisch*." *American Ethnologist* 33 (3): 327–39.

——. 2014. "Energopower: An Introduction." *Anthropological Quarterly* 87 (2): 309–33.

Boyer, Dominic, and Claudio Lomnitz. 2005. "Intellectuals and Nationalism: Anthropological Engagements." *Annual Review of Anthropology* 34: 105–20.

Breglia, Lisa. 2013. *Living with Oil: Promises, Peaks, and Declines on Mexico's Gulf Coast.* Austin: University of Texas Press.

Bridge, Gavin, and Andrew Wood. 2010. "Less Is More: Spectres of Scarcity and the Politics of Resource Access in the Upstream Oil Sector." *Geoforum* 41 (4): 565–76.

Brinegar, Sara. 2014. "Baku at All Costs: The Politics of Oil in the New Soviet State." PhD diss., University of Wisconsin-Madison.

Brown, Kate. 2013. *Plutopia: Nuclear Families, Atomic Cities, and the Great Soviet and American Plutonium Disasters.* New York: Oxford University Press.

Burawoy, Michael, and Janos Lukacs. 1992. *The Radiant Past: Ideology and Reality in Hungary's Road to Capitalism.* Chicago: University of Chicago Press.

Burawoy, Michael, and Katherine Verdery, eds. 1999. *Uncertain Transition: Ethnographies of Change in the Postsocialist World.* Lanham, MD: Rowman and Littlefield.

Büscher, Bram, and Veronica Davidov, eds. 2013. *The Ecotourism-Extraction Nexus: Political Economies and Rural Realities of (un)Comfortable Bedfellows.* London: Routledge.

Campbell, Robert. 1968. *The Economics of Soviet Oil and Gas.* Baltimore: Johns Hopkins University Press.

——. 1976. *Trends in the Soviet Oil and Gas Industry.* Baltimore: Johns Hopkins University Press.

——. 1980. *Soviet Energy Technologies: Planning, Policy, Research, and Development.* Bloomington: Indiana University Press.

Carmin, JoAnn, and Adam Fagan. 2010. "Environmental Mobilisation and Organisations in Post-Socialist Europe and the Former Soviet Union." *Environmental Politics* 19 (5): 689–707.

Cepek, Michael. 2014. "There Might Be Blood: Oil, Cosmo-Politics, and the Composition of a Cofán Petro-Being." Paper Presented at the Annual Meeting of the American Anthropological Association, Washington, DC, December 6.

Chadwick, Margaret, David Long, and Machiko Nissanke. 1987. *Soviet Oil Exports: Trade Adjustments, Refining Constraints, and Market Behavior.* Oxford: Oxford University Press.

Chagin, Georgii. 2008. "Etnokul'turnyi krai." In Shostina, *Kamwa: Al'manakh*, 6–13. Perm: Master-Flok.

Chantsev, A. 2009. "The Antiutopia Factory: The Dystopian Discourse in Russian Literature of the Mid-2000s." *Russian Studies in Literature* 45 (2): 6–41.

Chari, Sharad, and Katherine Verdery. 2009. "Thinking between the Posts: Postsocialism, Postcolonialism, and Ethnography after the Cold War." *Comparative Studies in Society and History* 51 (1): 6–34.

Chebankova, Elena. 2010. "Business and Politics in the Russian Regions." *Journal of Communist Studies and Transition Politics* 26 (1): 25–53.

Chernikov, Nikolai. 2000. *Vertikal' Alekperova.* Moscow: Izdatel'skii Dom Zhurnala Smena.

——. 2003. *Neft' i rozy.* Moscow: Izdatel'skii Dom Zhurnala Smena.

Chirkunov, Oleg. 2012. *Liberal: Gosudarstvo i konkurentsiia.* Moscow: Novoe literaturnoe obozrenie.

Chukrov, Keti. 2011. "Art after Primitive Accumulation: Or, on the Putin-Medvedev Cultural Politics." *Afterall: A Journal of Art, Context, and Enquiry* 26: 127–36.

Chung, Han-Ku. 1987. *Interest Representation in Soviet Policymaking: A Case Study of a West Siberian Energy Coalition.* Boulder, CO: Westview Press.

Clay, Catherine. 1995. "Russian Ethnographers in the Service of Empire, 1856–1862." *Slavic Review* 54 (1): 45–61.

Collier, Stephen. 2011. *Post-Soviet Social: Neoliberalism, Social Modernity, Biopolitics.* Princeton, NJ: Princeton University Press.

Collins, John. 2011. "Melted Gold and National Bodies: The Hermeneutics of Depth and the Value of History in Brazilian Racial Politics." *American Ethnologist* 38 (4): 683–700.

Comaroff, John L., and Jean Comaroff. 2009. *Ethnicity, Inc.* Chicago: University of Chicago Press.

Confino, Alon, and Ajay Skaria. 2002. "The Local Life of Nationhood." *National Identities* 4 (1): 7–24.

Cook, Linda. 2007. *Postcommunist Welfare States: Reform Politics in Russia and Eastern Europe.* Ithaca, NY: Cornell University Press.

Coronil, Fernando. 1997. *The Magical State: Nature, Money, and Modernity in Venezuela.* Chicago: University of Chicago Press.

——. 2011. "Oilpacity: Secrets of History in the Coup against Hugo Chávez." *Anthropology News* (May): 6.

Costlow, Jane. 2004. "*Nedra*, Nature, and the Depths of Despair: Journeying with Turgenev into Poles′e." Unpublished manuscript, Bates College.

Coumans, Catherine. 2011. "Occupying Spaces Created by Conflict: Anthropologists, Development NGOS, Responsible Investment, and Mining." *Current Anthropology* 52 (S3): S29–43.

Creed, Gerald. 2002. "Economic Crisis and Ritual Decline in Eastern Europe." In *Postsocialism: Ideals, Ideologies, and Practices in Eurasia*, edited by C. M. Hann, 57–73. London: Routledge.

——. 2010. *Masquerade and Postsocialism: Ritual and Cultural Dispossession in Bulgaria.* Bloomington: Indiana University Press.

Davidov, Veronica. 2013. *Ecotourism and Cultural Production: An Anthropology of Indigenous Spaces in Ecuador.* New York: Palgrave Macmillan.

Davis, Jerome. 1998. "Russian Commodity Exchanges: A Case Study of Organized Markets in the Transition Process, 1990–96." *Economics of Transition* 6 (1): 183–96.

Davis, John. 1987. *Libyan Politics: Tribe and Revolution.* Berkeley: University of California Press.

Debord, Guy. [1967] 1977. *Society of the Spectacle.* Detroit, MI: Black and Red.

Dedov, G. I. 1959. *Kizelovskii ugol′nyi bassein v gody velikoi otechestvennoi voiny.* Perm: Permskoe Knizhnoe Izdatel′stvo.

Degot, E. 2010. "A New Order." *Artforum International* 49 (3): 107–10.

Dement′ev, L. 1967. *Povest′ o Prikamskoi nefti.* Perm: Permskoe Knizhnoe Izdatel′stvo.

Denison, M. 2009. "The Art of the Impossible: Political Symbolism, and the Creation of National Identity and Collective Memory in Post-Soviet Turkmenistan." *Europe-Asia Studies* 61 (7): 1167–87.

Diaconov, Valentin. 2011. "The Russian Question." *Artforum International* 49 (6): 52–54.

Dickinson, J. A. 2005. "Post-Soviet Identities in Formation: Looking Back to See Where We Are Now." *Nationalities Papers* 33 (3): 291–301.

Dienes, Leslie. 1985. "The Energy System and Economic Imbalances in the USSR." *Soviet Economy* 1 (4): 340–72.

——. 2004. "Observations on the Problematic Potential of Russian Oil and the Complexities of Siberia." *Eurasian Geography and Economics* 45 (5): 319–45.

Dienes, Leslie, and Theodore Shabad. 1979. *The Soviet Energy System: Resource Use and Policies.* Washington, DC: V. H. Winston.

DiMaggio, Paul, ed. 2001. *The Twenty-First-Century Firm: Changing Economic Organization in International Perspective.* Princeton, NJ: Princeton University Press.

Dolan, Catherine, and Dinah Rajak. 2011. "Introduction: Ethnographies of Corporate Ethicizing." *Focaal* 60: 3–8.

Dominiak, A. V. 2010. *Svidetel'stva utrachennykh vremen: chelovek i mir v permskom zverinom stile*. Perm: Knizhnyi Mir.

Donahoe, Brian, and Joachim Otto Habeck, eds. 2011. *Reconstructing the House of Culture: Community, Self, and the Makings of Culture in Russia and Beyond*. New York: Berghahn Books.

Dunn, Elizabeth C. 2004. *Privatizing Poland: Baby Food, Big Business, and the Remaking of Labor*. Ithaca, NY: Cornell University Press.

Dzenovska, Dace. 2005. "Remaking the Nation of Latvia: Anthropological Perspectives on Nation Branding." *Place Branding* 1 (2): 173–86.

Ekhlakov A. V., and P. V. Panov. 2002. "NK Lukoil i organy vlasti v Permskoi oblasti: teoreticheskie modeli i politicheskaia praktika." *Politicheskii al'manakh Prikam'ia* 3: 138–75.

Erenburg, B. 2011. *Tsivilizatsiia khranitelei: Iazik zverinogo stilia*. Perm: Senator.

Etkind, Alexander. 2011. "Barrels of Fur: Natural Resources and the State in the Long History of Russia." *Journal of Eurasian Studies* 2 (2): 164–71.

——. 2013. *Warped Mourning: Stories of the Undead in the Land of the Unburied*. Stanford, CA: Stanford University Press.

Evtuhov, Catherine. 2012. "Voices from the Regions." *Kritika: Explorations in Russian and Eurasion History* 13 (4): 877–87.

Fadeeva, L. A. 2006. *Skvoz' prizmu politicheskoi kul'tury: natsiia, klass, region*. Perm: Pushka.

——. 2008. "Permskoe politologicheskoe soobshchestvo." In *Politicheskaia nauka v rossii: problemy, napravleniia, shkoly (1990–2007)*, 347–65. Moscow: ROSSPEN.

——. 2011. "Identity Politics and Interaction between Authorities and Society: The Case of Perm." Paper Presented at Association for Slavic, East European, and Eurasian Studies Conference, Washington, DC, November 17–20.

Fedotova, S.L. 2006. *Vol'nyi putevoditel'*. Perm: Kompan'on.

——. 2009. *Molotovskii kokteil'*. Perm: Zvezda.

Fehérváry, Krisztina. 2013. *Politics in Color and Concrete: Socialist Materialities and the Middle Class in Hungary*. Bloomington: Indiana University Press.

Ferguson, James. 1999. *Expectations of Modernity: Myths and Meanings of Urban Life on the Zambian Copperbelt*. Berkeley: University of California Press.

——. 2005. "Seeing Like an Oil Company: Space, Security, and Global Capital in Neoliberal Africa." *American Anthropologist* 107 (3): 377–82.

Ferry, Elizabeth Emma. 2005. "Geologies of Power: Value Transformations of Mineral Specimens from Guanajuato, Mexico." *American Ethnologist* 32 (3): 420–36.

Ferry, Elizabeth Emma, and Mandana Limbert, eds. 2008. *Timely Assets: The Politics of Resources and Their Temporalities*. Santa Fe: School for Advanced Research Press.

Fish, M. Steven. 2005. *Democracy Derailed in Russia: The Failure of Open Politics*. New York: Cambridge University Press.

Foster, Robert J. 2014. "Corporations as Partners: 'Connected Capitalism' and the Coca-Cola Company." *PoLAR* 37 (2): 246–58.

Freinkman, Lev. 1995. "Financial-Industrial Groups in Russia: Emergence of Large Diversified Private Companies." *Communist Economies and Economic Transformation* 7 (1): 51–66.

Frye, Timothy. 1995. "Caveat Emptor: Institutions, Contracts and Commodity Exchanges in Russia." In *Institutional Design*, edited by David Weimer, 37–62. Boston: Kluwer Academic.

——. 2000. *Brokers and Bureaucrats: Building Market Institutions in Russia*. Ann Arbor: University of Michigan Press.

——. 2006. "Original Sin, Good Works, and Property Rights in Russia." *World Politics* 58 (4): 479–504.

Frynas, Jedrzej George. 2005. "The False Developmental Promise of Corporate Social Responsibility: Evidence From Multinational Oil Companies." *International Affairs* 81 (3): 581–98.

Gaddy, Clifford. 1996. *The Price of the Past: Russia's Struggle with the Legacy of a Militarized Economy*. Washington, DC: Brookings Institution Press.

——. 2004. "Perspectives on the Potential of Russian Oil." *Eurasian Geography and Economics* 45 (5): 346–51.

Gaddy, Clifford, and Barry Ickes. 2002. *Russia's Virtual Economy*. Washington, DC: Brookings Institution Press.

Gaidar, Yegor. 2007. *Collapse of an Empire: Lessons for Modern Russia*. Washington DC: Brookings Institution Press.

Gal, Susan, and Gail Kligman. 2000. *The Politics of Gender after Socialism*. Princeton, NJ: Princeton University Press.

Ganev, Venelin. 2005. "Post-Communism as an Episode of State Building: A Reversed Tillyan Perspective." *Communist and Post-Communist Studies* 38: 425–45.

Gasheva, N. 2007. *Osa—Istoricheskii gorod Rossii*. Kirov: Kirovskaia Oblastnaia Tipografiia.

Gasheva, N., and V. Mikhailiuk. 1999. *Slovo o permskoi nefti*. Perm: OOO Lukoil-Permneft.

Gasratian, K. 2004. "Problems of Cultural Development in Russia." *Russian Social Science Review* 45 (5): 28–43.

Geertz, Clifford. 1973. *The Interpretation of Cultures*. New York: Basic Books.

Gel'man, Vladimir. 1999. "Regime Transition, Uncertainty and Prospects for Democratisation: The Politics of Russia's Regions in a Comparative Perspective." *Europe-Asia Studies* 51 (6): 939–56.

——. 2002. "The Politics of Local Government in Russia: The Neglected Side of the Story." *Perspectives on European Politics and Society* 3 (3): 496–508.

Gel'man, Vladimir, and Otag Marganiya, eds. 2011. *Resource Curse and Post-Soviet Eurasia*. Lanham, MD: Rowman and Littlefield.

Gel'man, Vladimir, and Sergei Ryzhenkov. 2011. "Local Regimes, Sub-National Governance and the 'Power Vertical' in Contemporary Russia." *Europe-Asia Studies* 63 (3): 449–65.

Gel'man, Vladimir, Sergei Ryzhenkov, Elena Belokurova, and Nadezhda Borisova, eds. 2002. *Avtonomiia ili kontrol'? Reforma mestnoi vlasti v gorodakh Rossii, 1991–2001*. Saint Petersburg: European University Press.

Ghodsee, Kristen. 2004. "Feminism-By-Design: Emerging Capitalisms, Cultural Feminism, and Women's Nongovernmental Organizations in Postsocialist Eastern Europe." *Signs* 29 (3): 727–53.

Ghosh, Amitav. 1992. "Petrofiction." *New Republic* 206 (9): 29–34.

Gille, Zsuzsa. 2007. *From the Cult of Waste to the Trash Heap of History: The Politics of Waste in Socialist and Postsocialist Hungary*. Bloomington: Indiana University Press.

Glatter, Peter. 1999. "Federalization, Fragmentation, and the West Siberian Oil and Gas Province." In Lane, *The Political Economy of Russian Oil*, 143–60. Lanham, MD: Rowman and Littlefield.

——. 2003. "Continuity and Change in the Tyumen' Regional Elite 1991–2001." *Europe-Asia Studies* 55 (3): 401–35.

Goldman, Marshall. 1980. *The Enigma of Soviet Petroleum: Half Empty or Half Full?* London: Allen & Unwin.

——. 2008. *Petrostate: Putin, Power, and the New Russia*. Oxford: Oxford University Press.

Golub, Alex. 2014. *Leviathans at the Gold Mine: Creating Indigenous and Corporate Actors in Papua New Guinea*. Durham, NC: Duke University Press.

Golub, Alex, and Mooweon Rhee. 2013. "Traction: the Role of Executives in Localising Global Mining and Petroleum Industries in Papua New Guinea." *Paideuma* 59: 215–36.

Gond, Jean-Pascal, Nahee Kang, and Jeremy Moon. 2011. "The Government of Self-Regulation: On the Comparative Dynamics of Corporate Social Responsibility." *Economy and Society* 40 (4): 640–71.

Gorbatova, Larisa. 1995. "Formation of Connections Between Finance and Industry in Russia: Basic Stages and Forms." *Communist Economies and Economic Transformation* 7 (1): 21–34.

Graan, A. 2013. "Counterfeiting the Nation? Skopje 2014 and the Politics of Nation Branding in Macedonia." *Cultural Anthropology* 28 (1): 161–79.

Graeber, David. 2011. *Debt: The First Five Thousand Years.* Brooklyn: Melville House.

Grant, Bruce. 1995. *In the Soviet House of Culture: A Century of Perestroikas.* Princeton, NJ: Princeton University Press.

——. 2009. *The Captive and the Gift: Cultural Histories of Sovereignty in Russia and the Caucasus.* Ithaca, NY: Cornell University Press.

——. 2011. "Epilogue: Recognizing Soviet Culture." In *Reconstructing the House of Culture,* edited by Donahoe and Habeck, 263–76. New York: Berghahn Books.

Greenberg, Jessica. 2006. "Noć Reklamoždera: Democracy, Consumption, and the Contradictions of Representation in Post-Socialist Serbia." *PoLAR: Political and Legal Anthropology Review* 29 (2): 181–207.

Greenfeld, Liah. 1996. "The Bitter Taste of Success: Reflections on the Intelligentsia in Post-Soviet Russia." *Social Research* 63 (2): 417–38.

Gubkin, I. M. 1940. *Uralo-Volzhskaia neftenosnaia oblast': Vtoroe Baku.* Moscow: Izdatel'stvo akademii nauk SSSR.

Guilbaut, Serge. 1983. *How New York Stole the Idea of Modern Art.* Chicago: University of Chicago Press.

Gulbrandsen, Lars H., and Arild Moe. 2005. "Oil Company CSR Collaboration in 'New' Petro-States." *Journal of Corporate Citizenship* 20: 53–64.

——. 2007. "BP in Azerbaijan: A Test Case of the Potential and Limits of the CSR Agenda?" *Third World Quarterly* 28 (4): 813–30.

Guss, David. 2000. *The Festive State: Race, Ethnicity, and Nationalism as Cultural Performance.* Berkeley: University of California Press.

Gustafson, Bret. 2011. "Flashpoints of Sovereignty: Territorial Conflict and Natural Gas in Bolivia." In Behrends, Reyna, and Schlee, *Crude Domination,* 220–41. New York: Berghahn Books.

Gustafson, Thane. 1981–82. "Energy and the Soviet Bloc." *International Security* 6 (3): 65–89.

——. 1989. *Crisis Amid Plenty: The Politics of Soviet Energy.* Princeton, NJ: Princeton University Press.

——. 2012. *Wheel of Fortune: The Battle for Oil and Power in Russia.* Cambridge, MA: Harvard University Press.

Guyer, Jane. 2015. "Price Fluctuation and the Production of Confusion in Two West African Oil Producing Economies." In *Subterranean Estates: Lifeworlds of Oil and Gas,* edited by Hannah Chadeayne Appel, Michael Watts, and Arthur Mason. Ithaca, NY: Cornell University Press.

Hale, Henry. 2003. "Explaining Machine Politics in Russia's Regions: Economy, Ethnicity, and Legacy." *Post-Soviet Affairs* 19 (3): 228–63.

——. 2005. *Why Not Parties in Russia? Democracy, Federalism, and the State.* Cambridge: Cambridge University Press.

Handler, Richard. 1988. *Nationalism and the Politics of Culture in Quebec.* Madison: University of Wisconsin Press.

Hann, Chris. 1996. "Introduction: Political Society and Civil Anthropology." In *Civil Society: Challenging Western Models,* edited by Chris Hann and Elizabeth Dunn, 1–24. London: Routledge.

——, ed. 1998. *Property Relations: Renewing the Anthropological Tradition.* New York: Cambridge University Press.

Hann, Chris, and Keith Hart. 2011. *Economic Anthropology: History, Ethnography, Critique.* Cambridge: Polity.

Hann, Chris et al. 2007. "Anthropology's Multiple Temporalities and Its Future in Central and Eastern Europe: A Debate." Max Planck Institute for Social Anthropology Working Paper No. 90. Halle/Salle, Germany.

Harper, Krista. 2006. *Wild Capitalism: Environmental Activists and Post-Socialist Political Ecology in Hungary.* New York: Columbia University Press.

Harris, James. 1999. *The Great Urals: Regionalism and the Evolution of the Soviet System.* Ithaca, NY: Cornell University Press.

Healey, Nigel, Vladimir Leksin, and Aleksandr Svetsov. 1999. "The Municipalization of Enterprise-Owned 'Social Assets' in Russia." *Post-Soviet Affairs* 15 (3): 262–80.

Hellebust, Rolf. 2003. *Flesh into Metal: Soviet Literature and the Alchemy of Revolution.* Ithaca, NY: Cornell University Press.

Hemment, Julie. 2004. "Global Civil Society and the Local Costs of Belonging: Defining Violence against Women in Russia." *Signs* 29 (3): 815–40.

——. 2007. *Empowering Women in Russia: Activism, Aid, and NGOs.* Bloomington: Indiana University Press.

——. 2012a. "Redefining Need, Reconfiguring Expectations: The Rise of State-Run Youth Voluntarism Programs in Russia." *Anthropological Quarterly* 85 (2): 519–54.

——. 2012b. "Nashi, Youth Voluntarism and Potemkin NGOs: Making Sense of Civil Society in Post-Soviet Russia." *Slavic Review* 71 (2): 234–60.

——. 2015. *Youth Politics in Putin's Russia: Producing Patriots and Entrepreneurs.* Bloomington: Indiana University Press.

Henry, Laura. 2010. *Red to Green: Environmental Activism in Post-Soviet Russia.* Ithaca, NY: Cornell University Press.

Herrera, Yoshiko M. 2004. *Imagined Economies: The Sources of Russian Regionalism.* New York: Cambridge University Press.

Hertog, Steffen. 2008. "Petromin: the Slow Death of Statist Oil Development in Saudi Arabia." *Business History* 50 (5): 645–67.

——. 2010. *Princes, Brokers, and Bureaucrats: Oil and the State in Saudi Arabia.* Ithaca, NY: Cornell University Press.

Hewett, Ed. 1984. *Energy, Economics, and Foreign Policy in the Soviet Union.* Washington, DC: Brookings Institution Press.

Hewitt, Simon. 2010. "What Makes a Market? The State of Russian Contemporary Art." *Art and Auction* 34 (1): 74–79.

Hinton, Diana Davids. 2008. "Creating Company Culture: Oil Company Camps in the Southwest, 1920–1960." *Southwestern Historical Quarterly* 111 (4): 369–87.

Hirsch, Francine. 2005. *Empire of Nations: Ethnographic Knowledge and the Making of the Soviet Union.* Ithaca, NY: Cornell University Press.

Ho, Karen. 2009. *Liquidated: An Ethnography of Wall Street.* Durham, NC: Duke University Press.

Hoffman, David. 2002. *The Oligarchs: Wealth and Power in the New Russia.* New York: Public Affairs.

Hough, Jerry. 1969. *The Soviet Prefects: The Local Party Organs in Industrial Decision-Making.* Cambridge, MA: Harvard University Press.

Huber, Matthew. 2009. "Energizing Historical Materialism: Fossil Fuels, Space and the Capitalist Mode of Production." *Geoforum* 40 (1): 105–15.

——. 2011. "Enforcing Scarcity: Oil, Violence, and the Making of the Market." *Annals of the Association of American Geographers* 101 (4): 816–26.

——. 2013. *Lifeblood: Oil, Freedom, and the Forces of Capital.* Minneapolis: University of Minnesota Press.

Humphrey, Caroline. 1985. "Barter and Economic Disintegration." *Man* 20 (1): 48–72.

——. 1999. *Marx Went Away—But Karl Stayed Behind.* Ann Arbor: University of Michigan Press.

——. 2001 "Inequality and Exclusion: A Russian Case Study of Emotion in Politics." *Anthropological Theory* 1 (3): 331–53.

——. 2002. *The Unmaking of Soviet Life: Everyday Economies after Socialism.* Ithaca, NY: Cornell University Press.

——. 2003. "Rethinking Infrastructure: Siberian Cities and Great Freeze of January 2001." In *Wounded Cities: Destruction and Reconstruction in a Globalized World*, edited by Jane Schneider and Ida Susser, 91–107. New York: Berg.

——. 2005. "Ideology in Infrastructure: Architecture and the Soviet Imagination." *Journal of the Royal Anthropological Institute* 11 (1): 39–58.

——. 2009. "Historical Analogies and the Commune: The Case of Putin/Stolypin." In *Enduring Socialism: Explorations of Revolution and Transformation, Restoration and Continuation*, edited by Harry G. West and Parvathi Raman, 231–49. New York: Berghahn Books.

Humphrey, Caroline, and Stephen Hugh-Jones. 1992. *Barter, Exchange, and Value: An Anthropological Approach.* Cambridge: Cambridge University Press.

Humphrey, Caroline, and Ruth Mandel. 2002. "The Market in Everyday Life: Ethnographies of Postsocialism." In *Markets and Moralities: Ethnographies of Postsocialism*, edited by Humphrey and Mandel, 1–16. Oxford: Berg.

Humphreys, Macartan, Jeffrey Sachs, and Joseph Stiglitz. 2007. *Escaping the Resource Curse.* New York: Columbia University Press.

Ignat′eva, O. V. 2009. *Permskii zverinyi stil′: Istoriia kollektsii i ikh izucheniia.* Perm: PGPU.

Igolkin, A. A. 1999a. *Otechestvennaia neftianaia promyshlennost′ v 1917–1920 godakh.* Moscow: RGGU.

——. 1999b. *Sovetskaia neftianaia promyshlennost′ v 1921–1928 godakh.* Moscow: RGGU.

——. 2005. *Neftianaia politika SSSR v 1928–1940 godakh.* Moscow: Institut rossiskoi istorii RAN.

——. 2009. *Sovetskaia neftianaia politika v 1940–1950 godakh.* Moscow: Institut rossiskoi istorii RAN.

Igumenov, Gennadii. 2008. *Linii sud′by: ot slesaria do gubernatora.* Perm: Aster.

Iuri, Kuchyran. 2008. "Etnofuturizm v Udmurtii." In Shostina, *Kamwa: Al′manakh*, 114–17.

Iuzifovich, O. 2008. *Zavodchane.* Moscow: Maktsentr.

Ivanov, Aleksei. [2002] 2012. *Serdtse parmy, ili Cherdyn′—kniaginia gor.* Saint Petersburg: Azbuka.

——. 2005. *Geograf globus propil.* Saint Petersburg: Azbuka-Klassika.

——. [2006] 2012. *Zoloto bunta, ili Vniz po reke tesnin.* Saint Petersburg: Azbuka.

——. 2008. "Tochka otscheta—Zemlia." In Shostina, *Kamwa: Al′manakh*, 86–88.

——. 2009. "Uralskaia matritsa: Kul′turologicheskie ocherki." In *Permistika*, edited by Ivanov, 241–349. Perm: Zvezda.

——. 2010a. *Khrebet Rossii: Geroi, zavody, mastera, matritsa.* Saint Petersburg: Azbuka-Klassika.

——. 2010b. *Khrebet Rossii.* Documentary film, Parts 1–4. Produced by Aleksei Ivanov, Leonid Parfenev, and Iulia Zaitseva. Studiia "Namednyi" and Prodiuserskii Tsentr "Iul′."

Jameson, Frederic. 1997. "Culture and Finance Capital." *Critical Inquiry* 24 (1): 246–65.

Jansen, Sue Curry. 2008. "Designer Nations: Neo-Liberal Nation Branding–Brand Estonia." *Social Identities* 14 (1): 121–42.

Johnson, Juliet. 2000. *A Fistful of Rubles: The Rise and Fall of the Russian Banking System.* Ithaca, NY: Cornell University Press.

Jones, Toby Craig. 2010. *Desert Kingdom: How Oil and Water Forged Modern Saudi Arabia.* Cambridge, MA: Harvard University Press.

Jones Luong, Pauline, and Erika Weinthal. 2010. *Oil Is Not a Curse: Ownership Structure and Institutions in Soviet Successor States.* New York: Cambridge University Press.

Jørgensen, Dolly. 2012. "Mixing Oil and Water: Naturalizing Offshore Oil Platforms in Gulf Coast Aquariums." *Journal of American Studies* 46 (2): 461–80.

Kalimullin, R. M. 1993. *Letopis′ osinskoi nefti 1963–1993.* Perm: Permneft′.

Kalinin, Il′ia. 2013. "Proshloe kak ogranichennyi resurs: Istoricheskaia politika i ekonomika renty." *Neprikosnovennyi Zapas* 88 (2): 200–214.

Karl, Terry Lynn. 1997. *The Paradox of Plenty: Oil Booms and Petro-States.* Berkeley: University of California Press.

Karzarinova, N. V., and V. V. Abashev, eds. 2000. *Iskusstvo Permi v kul′turnom prostranstve Rossii. Vek XX.* Vereshchagino: Pechatnik.

Keane, Webb. 2003. "Semiotics and the Social Analysis of Material Things." *Language and Communication* 23: 409–25.

——. 2007. *Christian Moderns: Freedom and Fetish in the Mission Encounter.* Berkeley: University of California Press.

——. 2008. "Market, Materiality and Moral Metalanguage." *Anthropological Theory* 8 (1): 27–42.

Kellison, Bruce. 1999. "Tiumen, Decentralization, and Center-Periphery Tension." In Lane, *The Political Economy of Russian Oil,* 127–42.

Kelly, Catriona. 1999. "*Kul′turnost′* in the Soviet Union: Ideal and Reality." In *Reinterpreting Russia,* edited by Geoffrey Hosking and Richard Service, 198–213. London: Arnold.

——. 2004. "Byt: Identity and Everyday Life." In *National Identity in Russian Culture: An Introduction,* edited by Simon Franklin and Emma Widdis, 147–67. Cambridge: Cambridge University Press.

Kennedy, Michael D. 1992. "The Intelligentsia in the Constitution of Civil Societies and Postcommunist Regimes in Hungary and Poland." *Theory and Society* 21 (1): 29–76.

——. 2008. "From Transition to Hegemony: Extending the Cultural Politics of Military Alliances and Energy Security." In *Transnational Actors in Central and East European Transitions,* edited by Mitchell Orenstein, Steven Bloom, and Nicole Lindstrom, 188–212. Pittsburgh: University of Pittsburgh Press.

——. 2014. *Globalizing Knowledge: Intellectuals, Universities, and Publics in Transformation.* Palo Alto, CA: Stanford University Press.

Kirsch, Stuart. 2007. "Indigenous Movements and the Risks of Counterglobalization: Tracking the Campaign Against Papua New Guinea's Ok Tedi Mine." *American Ethnologist* 34 (2): 303–21.

——. 2014. *Mining Capitalism: The Relationship Between Corporations and Their Critics.* Berkeley: University of California Press.

Kligman, Gail. 1988. *The Wedding of the Dead: Ritual, Poetics, and Popular Culture in Transylvania.* Berkeley: University of California Press.

Klumbyte, Neringa. 2010. "The Soviet Sausage Renaissance." *American Anthropologist* 112 (1): 22–37.

Knight, Nathaniel. 1998. "Science, Empire, and Nationality: Ethnography in the Russian Geographical Society, 1845–1855." In *Imperial Russia: New Histories for the Empire,* edited by Jane Burbank and David Ransel, 108–42. Bloomington: University of Indiana Press.

Kolbas, V. S. 2002. *Industrial′nyi raion: Istoriia prodolzhaetsia.* Perm: Zvezda.

Konrád, George, and Ivan Szelényi. 1979. *The Intellectuals on the Road to Class Power: A Sociological Study of the Intelligentsia in Socialism.* New York: Harcourt Brace Jovanovich.

Kornai, János. 1980. *Economics of Shortage.* Amsterdam: North Holland.

Kotkin, Stephen. 1995. *Magnetic Mountain: Stalinism as a Civilization.* Berkeley: University of California Press.

——. 2001. *Armageddon Averted: The Soviet Collapse, 1970–2000.* Oxford: Oxford University Press.

Krasil'nikova, Tat'iana Grigor'evna. 2004. *Rol' sviazei s obshchestvennost'iu v sotsial'nykh protsessakh trudovykh kollektivov sovremennogo rossiiskogo obshchestva.* Kandidat dissertation, Perm State Technical University.

Kruglova, Anna. 2013. "Sensory Utopia in the Times of 'Cultural Revolution': On Art, Public Space, and the Moral Ontology of Class." *Laboratorium: Russian Review of Social Research* 5 (1): 25–51.

Kryukov, Valeriy, and Arild Moe. 1991. "The Changing Structure of the Russian Oil Industry." *Oil and Gas Law and Taxation Review* 9: 367–72.

——. 1994. "Observations on the Reorganization of the Russian Oil Industry." *Post-Soviet Geography* 35 (2): 89–101.

Kukulin, Il'ia. 2007. "Geroizatsiia vyzhivaniia." *Novoe Literaturnoe Obozrenie* 86: 302–30.

Kurbatova, T. V. 2006. *Istoriia promyshlennosti Permskogo kraia XX vek.* Perm: Liter-A.

Kut'ev, O., ed. 2004. *Il'inskii: Stranitsy istorii. K 425-Letiiu poselka.* Perm: Pushka.

Labban, Mazen. 2008. *Space, Oil and Capital.* New York: Routledge.

Lane, David, ed. 1999. *The Political Economy of Russian Oil.* Lanham, MD: Rowman and Littlefield.

Larkin, Brian. 2013. "The Politics and Poetics of Infrastructure." *Annual Review of Anthropology* 42 (1): 327–43.

Laruelle, Marlene. 2012. "Larger, Higher, Farther North . . . Geographical Metanarratives of the Nation in Russia." *Eurasian Geography and Economics* 53 (5): 557–74.

Ledeneva, Alena. 2006. *How Russia Really Works: the Informal Practices that Shaped Post-Soviet Politics and Business.* Ithaca, NY: Cornell University Press.

Ledeneva, Alena, and Paul Seabright. 2000. "Barter in Post-Soviet Societies: What Does It Look Like and Why Does It Matter?" In Seabright, *The Vanishing Rouble*, 93–113.

Leibovich, Oleg. 2009. *V Gorode M: Ocherki politicheskoi povsednevnosti sovetskoi provintsii v 40–50-kh godakh XX veka.* Perm: Zvezda.

Leibovich, O. L., and N. V. Shushkova. 2011. "Chuzhye igry . . . sotsial'nyi analiz permskogo kul'turnogo proekta." *Vestnik PGIIK* (12): 80–90.

LeMenager, Stephanie. 2014. *Living Oil: Petroleum Culture in the American Century.* New York: Oxford University Press.

Lemon, Alaina. 1998. " 'Your Eyes Are Green Like Dollars': Counterfeit Cash, National Substance, and Currency Apartheid in 1990s Russia." *Cultural Anthropology* 13 (1): 22–55.

Limbert, Mandana. 2010. *In the Time of Oil: Piety, Memory, and Social Life in an Omani Town.* Palo Alto, CA: Stanford University Press.

Lindquist, Galina. 2005. *Conjuring Hope: Magic and Healing in Contemporary Russia.* New York: Berghahn Books.

Liu, Morgan Y. 2005. "Hierarchies of Place, Hierarchies of Empowerment: Geographies of Talk about Postsocialist Change in Uzbekistan." *Nationalities Papers* 33 (3): 423–38.

Liubanovskaia, S. 1998. "Long-Term Directions of the [Perm] Oblast Economy: Interview with E. S. Sapiro." *Problems of Economic Transition* 40 (9): 72–81.

Luehrmann, Sonja. 2011a. "The Modernity of Manual Reproduction: Soviet Propaganda and the Creative Life of Ideology." *Cultural Anthropology* 26 (3): 363–88.

——. 2011b. *Secularism Soviet Style: Teaching Atheism and Religion in a Volga Republic.* Bloomington: Indiana University Press.

Lukianenko, Sergei. 1998. *Nochnoi dozor.* Moscow: AST.

——. 2000. *Dnevnoi dozor.* Moscow: AST.

Lukoil-Perm. 2005a. *Sotsial'naia politika OAO "Lukoil" v Prikam'e: Praktika primenenia sotsial'nogo kodeksa v OOO "Lukoil-Perm'" v 2003–2004 gg.* Perm: Avgust.

——. 2005b. *IV korporativnyi konkurs sotsial'nykh i kul'turnykh proektov.* Perm.

——. 2007. *VI konkurs sotsial'nykh i kul'turnykh proektov.* Perm.

——. 2008. *VII konkurs sotsial'nykh i kul'turnykh proektov.* Perm.

——. 2010. *IX konkurs sotsial'nykh i kul'turnykh proektov.* Perm.

Lukoil-Permnefteprodukt. 2003. *Produkt veka—Produkt na veka 1903–2003: 100-Letie obrazovanie sistemy nefteproduktoobespechenaia Prikam'ia.* Perm.

Lutz, Catherine. 2014. "The US Car Colossus and the Production of Inequality." *American Ethnologist* 41 (2): 232–45.

Makarychev, Andrey S. 2000. "Russian Financial-Industrial Groups and the Regions: Toward New Patterns of Horizontal Networking." *International Journal of Political Economy* 30 (3): 44–57.

Mandel, Ruth. 2002a. "Seeding Civil Society." In *Postsocialism: Ideals, Ideologies, and Practices in Eurasia*, edited by C. M. Hann, 279–96. New York: Routledge.

——. 2002b. "A Marshall Plan of the Mind: The Political Economy of a Kazakh Soap Opera." In *Media Worlds: Anthropology on New Terrain*, edited by Faye Ginsburg, Lila Abu-Lughod, and Brian Larkin, 211–28. Berkeley: University of California Press.

——. 2012. "Introduction: Transition to Where? Developing Post-Soviet Space." *Slavic Review* 71 (2): 223–33.

Mann, Michael. 1986. *The Sources of Social Power, Vol. 2: The Rise of Classes and Nation-States, 1760–1914.* Cambridge: Cambridge University Press.

Manning, Paul, and Ann Uplisashvili. 2007. " 'Our Beer': Ethnographic Brands in Postsocialist Georgia." *American Anthropologist* 109 (4): 626–41.

Marasanova, I. V. 2010. "Korporativnaia sotsial'naia otvetstvennost': Teoriia i praktika." In *Ural'skaia Gornozavodskaia Tsivilizatsiia Monogoroda: Al'manakh*, edited by Marasanova, 53–60. Perm: Aster.

Marat, Erika. 2009. "Nation Branding in Central Asia: A New Campaign to Present Ideas about the State and the Nation." *Europe-Asia Studies* 61 (7): 1123–36.

Marcus, George, with Peter Dobkin Hall. 1992. *Lives in Trust: The Fortunes of Dynastic Families in Late Twentieth-Century America.* Boulder, CO: Westview Press.

Marcus, George, and Fred Myers. 1995. "The Traffic in Art and Culture: An Introduction." In *The Traffic in Culture: Refiguring Art and Anthropology*, edited by Marcus and Myers, 1–51. Berkeley: University of California Press.

Markelova, O. A. 2004. *Istoriia dobychi nefti v Permskoi oblasti, 1928–2004 gody: Khronika.* Perm: GOPAPO.

Matza, Tomas. 2009. "Moscow's Echo: Technologies of the Self, Publics, and Politics on the Russian Talk Show." *Cultural Anthropology* 24 (3): 489–522.

Maurer, Bill. 2005. *Mutual Life, Limited: Islamic Banking, Alternative Currencies, Lateral Reason.* Princeton, NJ: Princeton University Press.

Mazzarella, William. 2003. *Shoveling Smoke: Advertising and Globalization in Contemporary India.* Durham, NC: Duke University Press.

Meneley, Anne. 2008. "Oleo-Signs and Quali-Signs: The Qualities of Olive Oil." *Ethnos* 73 (3): 303–26.

Merrill, Karen R. 2012. "Texas Metropole: Oil, the American West, and US Power in the Postwar Years." *Journal of American History* 99 (1): 197–207.

Miller, Toby, and George Yúdice. 2002. *Cultural Policy.* Thousand Oaks, CA: Sage.

Mitchell, Timothy. 1999. "Society, Economy, and the State Effect." In *State/Culture: State-Formation after the Cultural Turn*, edited by George Steinmetz, 76–97. Ithaca, NY: Cornell University Press.

——. 2011. *Carbon Democracy: Political Power in the Age of Oil.* London: Verso.

Mitchneck, Beth. 2005. "Geography Matters: Discerning the Importance of Local Context." *Slavic Review* 64 (3): 491–516.

Mitrofankin, Evgenii. 2006. "Oil Companies' Relations with Reindeer Herders and Hunters in Noglitskii District, Northeastern Sakhalin Island." *Sibirica* 5 (2): 139–52.

Moser, N. R. 2009. *Russian Industrial Development 1861–2008: A Case Study of the Oil Industry.* PhD diss., University of Manchester.

Munn, Nancy. 1986. *The Fame of Gawa: A Symbolic Study of Value Transformation in a Massim (Papua New Guinea) Society.* New York: Cambridge University Press.

Nauruzbayeva, Zhanara. 2011. "Portraiture and Proximity: 'Official' Artists and the State-ization of the Market in Post-Soviet Kazakhstan." *Ethnos* 76 (3): 375–97.

Neroslov, A. M., ed. 2009. *Permskii period: Vagit Alekperov i ego kommanda.* Perm: Aster.

Nikitina, O. V., V. V. Ustiugova, and N. M. Chernysheva. 2001. *Periferiinost′ v kul′ture XX veka.* Perm: Perm State University.

Nitzan, Jonathan, and Shimshon Bichler. 2002. *The Global Political Economy of Israel.* London: Pluto Press.

OFPI Bastion. 2000. *Elita Permskoi oblasti.* Online report, www.bastion.ru/files/elit.doc (accessed February 13, 2015).

Ong, Aiwah, and Stephen Collier. 2004. *Global Assemblages: Technology, Politics, and Ethics as Anthropological Problems.* London: Wiley-Blackwell.

OOO Lukoil-Permneft. 2003. *Permskaia neft′: Iskusstvo byt′ vyshe obstoiatel′stv.* Moscow: Delo.

Opitz, Petra, Hella Engerer, and Christian von Hirschausen. 2002. "The Globalisation of Russian Energy Companies: A Way Out of the Financial Crisis?" *International Journal of Global Energy Issues* 17 (4): 292–310.

Orttung, Robert W. 2004. "Business and Politics in the Russian Regions." *Problems of Post-Communism* 51 (2): 48–60.

Oushakine, Serguei Alex. 2000. "In the State of Post-Soviet Aphasia: Symbolic Development in Contemporary Russia." *Europe-Asia Studies* 52 (6): 991–1016.

——. 2009a. *The Patriotism of Despair: Nation, War, and Loss in Russia.* Ithaca, NY: Cornell University Press.

——. 2009b. "Introduction: Wither the Intelligentsia: The End of the Moral Elite in Eastern Europe." *Studies in East European Thought* 61 (4): 243–48.

——. 2014. "'Against the Cult of Things'": On Soviet Productivism, Storage Economy, and Commodities with No Destination." *Russian Review* 73 (2): 198–236.

Panov, P. V. 2009. "Elektoral′nye praktiki i partikularistikaia distributsiia resursov v Permskom krae." *Vestnik Permskogo Universiteta, Seriia Politologiia* 4 (8): 65–71.

Panov, P. V., K. A. Sulimov, and L. A. Fadeeva, eds. 2009. *Soobshchestva kak politicheskii fenomen.* Moscow: Rosspen.

Pappe, Ia. Sh. 2000. *Oligarkhi: Ekonomicheskaia khronika: 1992–2000.* Moscow: Vysshaia Shkola Ekonomiki.

Patico, Jennifer. 2008. *Consumption and Social Change in a Post-Soviet Middle Class.* Palo Alto, CA: Stanford University Press.

Paxson, Margaret. 2005. *Solovyovo: The Story of Memory in a Russian Village.* Bloomington: Indiana University Press.

Peebles, Gustav. 2008. "Inverting the Panopticon: Money and the Nationalization of the Future." *Public Culture* 20 (2): 233–65.

Pelevin, Viktor. 2004. *Sviashchennaia kniga oborotnia.* Moscow: Eksmo.

Peregudov, S. P. 2003. *Korporatsii, obshchestvo, gosudarstvo: Evolutsiia otnoshenii*. Moscow: Nauka.

Peregudov, S. P., and I. S. Semenenko. 2008. *Korporativnoe grazhdanstvo: Kontsepsiia, mirovaia praktika, i rossiiskie realii*. Moscow: Progress-Traditsiia.

Perm Museum of Contemporary Art (PERMM). 2008. *Russkoe bednoe. Katalog*. Perm: PERMM.

——. 2009. *Russkoe bednoe. Moskovskaia biennale sovremennogo iskusstva*. Moscow: Nemetskaia Fabrika Pechati.

——. 2011. *Iskusstvo v gorode: Pablik-Art programma, 2008–2011. Katalog*. Perm: PERMM.

Perm Regional Administration. 2002. *Konkurs sotsial'nykh i kul'turnykh proektov*. Perm: Lukoil-Perm.

Perm Regional Department of Culture. 1997. *Informatsial'no-analiticheskii sbornik*. Perm: Perm Regional Administration.

Pesmen, Dale. 2000. *Russia and Soul: An Exploration*. Ithaca, NY: Cornell University Press.

Petrone, Karen. 2000. *Life Has Become More Joyous, Comrades: Celebrations in the Time of Stalin*. Bloomington: Indiana University Press.

Platz, Stephanie. 1996. "Pasts and Futures: Space, History, and Armenian Identity, 1988–1994." PhD diss., University of Chicago.

Podvintsev, O. B. 2004. *Kto est' kto v politicheskoi zhizni Prikam'ia—2004*. Perm: Permskoe Knizhnoe Izdatel'stvo.

Poussenkova, Nina. 2004. "From Rigs to Riches: Oilmen vs. Financiers in the Russian Oil Sector." *The Energy Dimension in Russian Global Strategy*. Houston, TX: Rice University Baker Institute.

Prokop, Jane. 1995. "Industrial Conglomerates, Risk Spreading and the Transition in Russia." *Communist Economies and Economic Transformation* 7 (1): 35–50.

Rachkov, B. 1965. "The Soviet Union: A Major Exporter of Petroleum and Petroleum Products." *American Review of Soviet and Eastern European Trade* 1 (1): 26–38.

Rajak, Dinah. 2011. *In Good Company: An Anatomy of Corporate Social Responsibility*. Palo Alto, CA: Stanford University Press.

——. 2014. "Corporate Memory: Historical Revisionism, Legitimation, and the Invention of Tradition in a Multinational Mining Company." *PoLAR* 37 (2): 259–80.

Reddaway, Peter, and Robert W. Orttung, eds. 2004. *The Dynamics of Russian Politics: Putin's Reform of Federal-Regional Relations*. 2 vols. Lanham, MD: Rowman and Littlefield.

Reed, Kristin. 2009. *Crude Existence: Environment and the Politics of Oil in Northern Angola*. Berkeley: University of California Press.

Reeves, Madeleine. 2014. *Border Work: Spatial Lives of the State in Rural Central Asia*. Ithaca, NY: Cornell University Press.

Retort. 2005. *Afflicted Powers: Capital and Spectacle in a New Age of War*. London: Verso.

Richardson, Tanya, and Gisa Weszkalnys. 2014. "Resource Materialities." *Anthropological Quarterly* 87 (1): 5–30.

Ries, Nancy. 1997. *Russian Talk: Culture and Conversation during Perestroika*. Ithaca, NY: Cornell University Press.

——. 2002. "'Honest Bandits' and 'Warped People': Russian Narratives about Money, Corruption, and Moral Decay." In *Ethnography in Unstable Places: Everyday Lives in Contexts of Dramatic Political Change*, edited by Carol J. Greenhouse et al., 276–315. Durham, NC: Duke University Press.

——. 2009. "Potato Ontology: Surviving Postsocialism in Russia." *Cultural Anthropology* 24 (2): 181–212.

Rivkin-Fish, Michele. 2005. *Women's Health in Post-Soviet Russia: The Politics of Intervention*. Bloomington: Indiana University Press.

Rockefeller Archive Center. 2011. *Rockefeller Philanthropy: A Selected Guide*. Sleepy Hollow, NY: Rockefeller Archive Center.

Rogers, Douglas. 2006. "How to Be a *Khoziain* in a Transforming State: State Formation and the Ethics of Governance in Post-Soviet Russia." *Comparative Studies in Society and History* 48 (4): 915–45.

——. 2009. *The Old Faith and the Russian Land: A Historical Ethnography of Ethics in the Urals*. Ithaca, NY: Cornell University Press.

——. 2011. "Certainty and Vulnerability in Oil Boom Russia." *Anthropology Now* 3 (2): 10–16.

——. 2012. "The Materiality of the Corporation: Oil, Gas, and Corporate Social Technologies in the Remaking of a Russian Region." *American Ethnologist* 39 (2): 284–96.

——. 2014a. "Petrobarter: Oil, Inequality, and the Political Imagination in and after the Cold War." *Current Anthropology* 55 (2): 131–53.

——. 2014b. "Energopolitical Russia: Corporation, State, and the Rise of Social and Cultural Projects." *Anthropological Quarterly* 87 (2): 431–51.

——. 2014c. "The Oil Company and the Crafts Fair: From *Povsednevnost'* to *Byt* in Post-socialist Russia." In *Everyday Life in Russia: Subjectivities, Perspectives, and Lived Experience*, edited by Mary Cavender, Choi Chatterjee, Karen Petrone, and David L. Ransel, 71–93. Bloomington: Indiana University Press.

——. 2015. "Oil and Anthropology." *Annual Review of Anthropology* 44.

Rogers, Douglas, and Katherine Verdery. 2013. "Postsocialist Societies: Eastern Europe and the Former Soviet Union." In *The Handbook of Sociocultural Anthropology*, edited by James G. Carrier and Deborah B. Gewertz, 439–55. London: Bloomsbury.

Rolston, Jessica Smith. 2013. "The Politics of Pits and the Materiality of Mine Labor: Making Natural Resources in the American West." *American Anthropologist* 115 (4): 582–94.

——. 2014. *Mining Coal and Undermining Gender: Rhythms of Work and Family in the American West*. New Brunswick, NJ: Rutgers University Press.

Ross, Cameron. 2009. *Local Politics and Democratization in Russia*. London: Routledge.

Ross, Cameron, and Adrian Campbell, eds. 2009. *Federalism and Local Politics in Russia*. London: Routledge.

Ross, Michael. 2012. *The Oil Curse: How Petroleum Wealth Shapes the Development of Nations*. Princeton, NJ: Princeton University Press.

Rupp, Stephanie. 2013. "Considering Energy: $E = mc^2 = (\text{Magic} \cdot \text{Culture})^2$." In *Cultures of Energy: Power, Practices, Technologies*, edited by Sarah Strauss, Stephanie Rupp, and Thomas Love, 79–98. Walnut Creek, CA: Left Coast Press.

Rutland, Peter. 1985. *The Myth of the Plan: Lessons of Soviet Planning Experience*. London: Hutchinson.

——. 1993. *The Politics of Economic Stagnation in the Soviet Union: The Role of Local Party Organs in Economic Management*. Cambridge: Cambridge University Press.

——. 2008. "Putin's Economic Record: Is the Oil Boom Sustainable?" *Europe-Asia Studies* 60 (6): 1051–72.

——. 2015. "'Petronation?' Oil, Gas, and National Identity in Russia." *Post-Soviet Affairs* 31 (1): 66–89.

Sabin, Paul. 2004. *Crude Politics: The California Oil Market, 1900–1940*. Berkeley: University of California Press.

Sagafi-Nejad, Tagi, and John H. Dunning. 2008. *The UN and Transnational Corporations: From Code of Conduct to Global Compact*. Bloomington: Indiana University Press.

Sagers, Matthew J. 1984. Refinery Throughput in the U.S.S.R. *CIR Staff Paper*, no. 2. Washington, DC.

——. 1993. "The Energy Industries of the Former USSR: A Mid-Year Survey." *Post-Soviet Geography* 34 (6): 341–418.

Sagers, Matthew J., and John D. Grace. 1993. "Observations on the Russian Oil Sector in 1992 and 1993." *International Geology Review* 35 (9): 855–77.

Sagers, Matthew J., and Albina Tretyakova. 1985. "Restructuring the Soviet Petroleum Refining Industry." CIR *Staff Paper*, no. 4. Washington, DC.

Sakwa, Richard. 2014. *Putin and the Oligarchs: The Khodorkovsky-Yukos Affair*. London: I. B. Tauris.

Sampson, Steven. 1996. "The Social Life of Projects: Importing Civil Society to Albania." In *Civil Society: Challenging Western Models*, edited by C. M. Hann and Elizabeth Dunn, 121–42. London: Routledge.

Samuelson, Lennart. 2011. *Tankograd: The Formation of a Soviet Company Town: Cheliabinsk*. New York: Palgrave Macmillan.

Sanchez-Sibony, Oscar. 2010. "Soviet Industry in the World Spotlight: The Domestic Dilemmas of Soviet Foreign Economic Relations, 1955–1965." *Europe-Asia Studies* 62 (9): 1555–78.

Sapiro, Evgenii. 2003. *Striptiz s iumorom*. Perm: Izdatel′stvo Kompan′on.

——. 2009. *Traktat ob udache (vospominaniia i razmyshleniia)*. St. Petersburg: Mamatov.

Sawyer, Suzana. 2004. *Crude Chronicles: Indigenous Politics, Multinational Oil, and Neoliberalism in Ecuador*. Durham, NC: Duke University Press.

——. 2006. "Disabling Corporate Sovereignty in a Transnational Lawsuit." *PoLAR: Political and Legal Anthropology Review* 29 (1): 23–43.

——. 2012. "Commentary: the Corporation, Oil, and the Financialization of Risk." *American Ethnologist* 39 (4): 710–15.

Schönle, Andreas. 2011. *Architecture of Oblivion: Ruins and Historical Consciousness in Modern Russia*. Dekalb: Northern Illinois University Press.

Seabright, Paul, ed. 2000. *The Vanishing Rouble: Barter Networks and Non-Monetary Transactions in Post-Soviet Societies*. Cambridge: Cambridge University Press.

Shabanova, L. N. 2004. *Put′ k sotsial′nomu proektu: Prakticheskie sovety nachinaiushchim*. Perm: Administratsiia permskoi oblasti i tsentr sotsial′nykh initsiativ.

——, ed. 2002. *Konkurs sotsial′nykh i kul′turnykh proektov (Rekomendatsii po organizatsii i provedeniiu)*. Perm: ZAO Lukoil-Perm i Fond Evraziia.

Shanin, Teodor. 1986. "Soviet Theories of Ethnicity: The Case of a Missing Term." *New Left Review* 158: 113–22.

Sharafutdinova, Gulnaz, and Arbakhan Magomedov. 2003. "The Volga Federal Okrug." In *The Dynamics of Russian Politics*, vol. 1, edited by Reddaway and Orttung, 153–86.

Shearer, David R. 1996. *Industry, State, and Society in Stalin's Russia, 1926–1934*. Ithaca, NY: Cornell University Press.

Shekhmetov, G. P., E. A. Vshivkova, S. V. Nesterova, A. M. Put′ko, and S. M. Barkov, eds. 2004–2006. *Usol′e: Mozaika vremen*. 2 vols. Perm′: OOO Raritat-Perm.

Shevchenko, Olga. 2008. *Crisis and the Everyday in Postsocialist Moscow*. Bloomington: Indiana University Press.

Shever, Elana. 2010. "Engendering the Company: Corporate Personhood and the 'Face' of an Oil Company in Metropolitan Buenos Aires." *PoLAR* 33 (1): 26–46.

——. 2012. *Resources for Reform: Oil and Neoliberalism in Argentina*. Palo Alto, CA: Stanford University Press.

Shostina, Natalia, ed. 2008. *Kamwa: Al′manakh*. Perm: Master-Flok.

Siegelbaum, Lewis H. 2008. *Cars for Comrades: The Life of the Soviet Automobile*. Ithaca, NY: Cornell University Press.

Sim, Li-Chen. 2008. *The Rise and Fall of Privatization in the Russian Oil Industry*. Hampshire, UK: Palgrave Macmillan.

Sivaramakrishnan, K. 1999. *Modern Forests: Statemaking and Environmental Change in Colonial Eastern India.* Palo Alto, CA: Stanford University Press.

Slavkina, M. V. 2007. *Velikie pobedy i upushchennye vozmozhnosti: Vliianie neftegazovogo kompleksa na sotsial'no-ekonomicheskoe razvitie SSSR v 1945–1991* gg. Moscow: Izdatel'stvo Neft' i Gaz.

Slezkine, Yuri. 1994. "The USSR as a Communal Apartment, or How a Socialist State Promoted Ethnic Particularism." *Slavic Review* 53 (2): 414–52.

——. 1996. *Arctic Mirrors: Russia and the Small Peoples of the North.* Ithaca, NY: Cornell University Press.

Smith, Jessica, and Federico Helfgott. 2010. "Flexibility or Exploitation? Corporate Social Responsibility and the Perils of Universalization." *Anthropology Today* 26 (3): 20–23.

Smith, Julia Cauble. 2010. "Darst Creek Oilfield," *Handbook of Texas Online*, http://www.tshaonline.org/handbook/online/articles/dod05 (accessed February 5, 2015).

Snajdr, Edward. 2008. *Nature Protests: The End of Ecology in Slovakia.* Seattle: University of Washington Press.

Sokolova, Liudmila Evgen'evna. 2003. *Neftianiki Vishery.* Solikamsk: OGUP IPK Solikamsk.

Sokolovskaia, E. 2001. *Pervaia piatiletka.* Perm: ZAO Lukoil Perm.

Spiro, David E. 1999. *The Hidden Hand of American Hegemony: Petrodollar Recycling and International Markets.* Ithaca, NY: Cornell University Press.

Stammler, Florian. 2005. *Reindeer Nomads Meet the Market: Culture, Property and Globalisation at the "End of the Land."* Münster: LIT Verlag.

——. 2011. "Oil without Conflict? The Anthropology of Industrialisation in Northern Russia." In Behrends, Reyna, and Schlee, *Crude Domination*, 243–69.

Stammler, Florian, and Vladislav Peskov. 2008. "Building a 'Culture of Dialogue' Among Stakeholders in North-West Russian Oil Extraction." *Europe-Asia Studies* 60 (5): 831–49.

Stark, David. 1996. "Recombinant Property in East European Capitalism." *American Journal of Sociology* 101 (4): 993–1027.

——. 2001. "Ambiguous Assets for Uncertain Environments: Heterarchy in Postsocialist Firms." In DiMaggiio, *The Twenty-First-Century Firm*, 69–144.

Starodubrovskaya, Irina. 1995. "Financial-Industrial Groups: Illusions and Reality." *Communist Economies and Economic Transformation* 7 (1): 5–19.

Stern, Jonathan P. 1987. *Soviet Oil and Gas Exports to the West: Commercial Transaction or Security Threat?* Hants, UK: Gower.

Stone, Randell. 2002. *Satellites and Commissars: Strategy and Conflict in the Politics of Soviet-Bloc Trade.* Princeton, NJ: Princeton University Press.

Suny, Ronald Grigor. 1972. "A Journeyman for the Revolution: Stalin and the Labour Movement in Baku, June 1907–May 1908." *Soviet Studies* 23 (3): 373–94.

Suny, Ronald Grigor, and Michael D. Kennedy, eds. 1999. *Intellectuals and the Articulation of the Nation.* Ann Arbor: University of Michigan Press.

Sverkal'tseva, O. G. 1998. *Eto nasha s toboi biografiia: Khronika, sobytiia, liudi.* Perm: Ozon.

Tarbell, Ida. 1904. *The History of the Standard Oil Company.* New York: Cosimo, Inc.

Thompson, Niobe. 2009. *Settlers on the Edge: Identity and Modernization on Russia's Arctic Frontier.* Vancouver: University of British Columbia Press.

Thompson, William. 2005. "Putting Yukos in Perspective." *Post-Soviet Affairs* 21 (2): 159–81.

Tinker Salas, Miguel. 2009. *The Enduring Legacy: Oil, Culture, and Society in Venezuela.* Durham, NC: Duke University Press.

Tolf, Robert W. 1976. *The Russian Rockefellers: The Saga of the Nobel Family and the Russian Oil Industry*. Palo Alto, CA: Hoover Institution Press.

Treier, Heie. 2003. "Ethnofuturism: Politics At the Grass Roots Level." Paper presented at International Association of Art Critics "Dakar: Art, Minorities and Majorities" Conference, July. New York: AICA Press.

Trofimuk, A. A. 1957. *Uralo-Povolzh'e: Novaia neftianaia baza SSSR (Istoriia otkrytiia, sostoiianie, perspektivy)*. Moscow: Gosudarstvennoe nauchno-tekhnicheskoe izdatel'stvo neftiianoi i gorno-toplivoi literatury.

Tsepilova, Olga. 2007. "Forging Change in a Contaminated Russian City: A Longitudinal View of Kirishi." In *Cultures of Contamination Legacies of Pollution in Russia and the US*, edited by Michael Edelstein, Maria Tysiachniouk, and Liudmila V. Smirnova, 31–46. Greenwich, CT: JAI Press.

Varese, Federico. 2001. *The Russian Mafia: Private Protection in a New Market Economy*. Oxford: Oxford University Press.

Verdery, Katherine. 1991. *National Ideology under Socialism: Identity and Cultural Politics in Ceaușescu's Romania*. Berkeley: University of California Press.

——. 1996. *What Was Socialism and What Comes Next?* Princeton, NJ: Princeton University Press.

——. 1999. *The Political Lives of Dead Bodies: Reburial and Postsocialist Change*. New York: Columbia University Press.

——. 2003. *The Vanishing Hectare: Property and Value in Postsocialist Romania*. Ithaca, NY: Cornell University Press.

Verdery, Katherine, and Gail Kligman. 2011. *Peasants under Siege: The Collectivization of Romanian Agriculture, 1949–1962*. Princeton, NJ: Princeton University Press.

Vikkel', L., S. Fedotova, and O. Iuzifovich. 2009. *Permskii period: Vagit Alekperov i ego komanda*. Perm: Aster.

Vitalis, Robert. 2006. *America's Kingdom: Mythmaking on the Saudi Oil Frontier*. Palo Alto, CA: Stanford University Press.

Volkov, Vadim. 2002. *Violent Entrepreneurs: The Use of Force in the Making of Russian Capitalism*. Ithaca, NY: Cornell University Press.

——. 2008. "Standard Oil and Yukos in the Context of Early Capitalism in the United States and Russia." *Demokratizatsiya* 16 (3): 240–64.

Vorob'eva, A.I. 2000. *Byl' chusovskikh gorodkov*. Ekaterinburg: Izdatel'stvo Ural'skogo universiteta.

Votinova, A. S., and P. V. Papov. 2006. "Dinamika regional'noi politiki po voprosam mezhbiudzhetnykh otnoshenii Permskoi oblasti v nachale 2000-ykh gg." *Politicheskii al'manakh Prikam'ia* 6: 287–310.

Watts, Michael. 1994. "Oil as Money: The Devil's Excrement and the Spectacle of Black Gold." In *Money, Power, and Space*, edited by Stuart Corbridge, 406–45. Oxford: Blackwell.

——. 2001. "Petro-Violence: Community, Extraction, and Political Ecology of a Mythic Commodity." In *Violent Environments*, edited by Nancy Lee Peluso and Michael Watts, 189–212. Ithaca, NY: Cornell University Press.

——. 2004. "Resource Curse? Governmentality, Oil and Power in the Niger Delta, Nigeria." In *the Geopolitics of Resource Wars*, edited by Philippe Le Billon, 50–80. London: Routledge.

——. 2005. "Righteous Oil? Human Rights, the Oil Complex, and Corporate Social Responsibility." *Annual Review of Environment and Resources* 30: 373–407.

——. 2008. "Blood Oil: The Anatomy of a Petro-Insurgency in the Niger Delta." *Focaal* 52: 18–38.

——. 2012. "A Tale of Two Gulfs: Life, Death, and Dispossession along Two Oil Frontiers." *American Quarterly* 64 (3): 437–67.

Watts, Michael, Hannah Appel, and Arthur Mason, eds. 2015. *Subterranean Estates: Lifeworlds of Oil and Gas.* Ithaca, NY: Cornell University Press.

Weber, Max. [1894] 2000. "Commerce on the Stock and Commodity Exchanges." *Theory and Society* 29 (3): 339–71.

——. [1904–5] 1958. *The Protestant Ethic and the Spirit of Capitalism.* Translated by Talcott Parsons. New York: Charles Scribner's.

Wegren, Stephen K. 1994. "Building Market Institutions." *Communist and Post-Communist Studies* 27 (3): 195–224.

Weiner, Douglas R. 1999. *A Little Corner of Freedom: Russian Nature Protection from Stalin to Gorbachev.* Berkeley: University of California Press.

Welker, Marina. 2009. "'Corporate Security Begins in the Community': Mining, the Corporate Social Responsibility Industry, and Environmental Advocacy in Indonesia." *Cultural Anthropology* 24 (1): 142–79.

——. 2014. *Enacting the Corporation: An American Mining Firm in Post-Authoritarian Indonesia.* Berkeley: University of California Press.

Welker, Marina, and David Wood. 2011. "Shareholder Activism and Alienation." *Current Anthropology* 52 (S3): S57–S69.

Welker, Marina, Damani J. Partridge, and Rebecca Hardin. 2011. "Corporate Lives: New Perspectives on the Social Life of the Corporate Form: An Introduction to Supplement 3." *Current Anthropology* 52 (S3): S3–S16.

Wengle, Susanne A. 2015. *Post-Soviet Power: State-Led Development and Russia's Marketization.* Cambridge: Cambridge University Press.

Werth, Paul. 2002. *At the Margins of Orthodoxy: Mission, Governance, and Confessional Politics in Russia's Volga-Kama Region, 1827–1905.* Ithaca, NY: Cornell University Press.

Weszkalnys, Gisa. 2011. "Cursed Resources, or Articulations of Economic Theory in the Gulf of Guinea." *Economy and Society* 40 (3): 345–72.

——. 2014. "Anticipating Oil: The Temporal Politics of a Disaster Yet to Come." *Sociological Review* 62 (1): 211–35.

Wilson, Thomas. 2012. "The Europe of Regions and Borderlands." In *A Companion to the Anthropology of Europe*, edited by Ullrich Kockel, Máiread Nic Craith, and Jonas Frykman, 163–80. Malden, MA: Wiley-Blackwell.

Winegar, Jessica. 2006. *Creative Reckonings: The Politics of Art and Culture in Contemporary Egypt.* Palo Alto, CA: Stanford University Press.

Woodruff, David. 1999. *Money Unmade: Barter and the Fate of Russian Capitalism.* Ithaca, NY: Cornell University Press.

World Bank. 2004. *Russian Economic Report: February 2004.*

Wu, Chin-Tau. 2002. *Privatising Culture: Corporate Art Intervention since the 1980s.* New York: Verso.

Yaeger, Patricia, Ken Hiltner, Saree Makdisi, Vin Nardizzi, Laurie Shannon, Imre Szeman, and Michael Ziser. 2011. "Editor's Column: Literature in the Ages of Wood, Tallow, Coal, Whale Oil, Gasoline, Atomic Power, and Other Energy Sources." *PMLA* 126 (2): 305–26.

Yakovlev, A. A., ed. 1991. *Birzhi v SSSR: Pervyi god raboty.* Moscow: Institute issledovanie organizovannykh rynkov.

Yanagisako, Sylvia. 2002. *Producing Culture and Capital: Family Firms in Italy.* Princeton: Princeton University Press.

Yergin, Daniel. 1991. *The Prize: The Epic Quest for Oil, Money, and Power.* New York: Touchstone.

Yurchak, Alexei. 2001. "Entreprenurial Governmentality in Postsocialist Russia: A Cultural Investigation of Business Practices." In *The New Entrepreneurs of Europe and Asia*, edited by V. E. Bonnell and T.B. Gold, 278–324. New York: M. E. Sharpe.

——. 2005. *Everything Was Forever, Until It Was No More: The Last Soviet Generation*. Princeton, NJ: Princeton University Press.

Zabolotskii, E. M. 1999. "Delo geolkoma." *Repressirovannye geologi* 3: 398–403.

Zaloom, Caitlin. 2006. *Out of the Pits: Traders and Technology from Chicago to London*. Chicago: University of Chicago Press.

ZAO Lukoil-Perm. 2001. *Pervaia piatiletka*. Format: Ekaterinburg.

——. 2002. *Dinamika razvitiia*. Format: Ekaterinburg.

——. 2004. *Sotsial'nye proekty*. Perm: ZAO Lukoil-Perm.

Zelentsova, Elena. 2010. *Permskii proekt: Kontseptsiia kul'turnoi politiki Permskogo kraia*. Moscow: Klassika.

Zigon, Jarett. 2011. *HIV is God's Blessing: Rehabilitating Morality in Neoliberal Russia*. Berkeley: University of California Press.

Index